# Developments in Strategic Materials and Computational Design IV

# Developments in Strategic Materials and Computational Design IV

*A Collection of Papers Presented at the 37th International Conference on Advanced Ceramics and Composites January 27–February 1, 2013 Daytona Beach, Florida*

Edited by
Waltraud M. Kriven
Jingyang Wang
Yanchun Zhou
Andrew L. Gyekenyesi

Volume Editors
Soshu Kirihara
Sujanto Widjaja

The American Ceramic Society

WILEY

*Library of Congress Cataloging-in-Publication Data is available.*

ISBN: 978-1-118-80727-9
ISSN: 0196-6219

Printed in the United States of America.

10 9 8 7 6 5 4 3 2 1

# Contents

## MATERIALS FOR EXTREME ENVIRONMENTS: ULTRAHIGH TEMPERATURE CERAMICS AND NANOLAMINATED TERNARY CARBIDES AND NITRIDES

## SECOND ANNUAL GLOBAL YOUNG INVESTIGATOR FORUM

# Preface

Contributions from two Symposia, two Focused Sessions, and the Annual Global Young Investigator Forum that were part of the 36th International Conference on Advanced Ceramics and Composites (ICACC), in Daytona Beach, FL, January 27–February 1, 2013 are presented in this volume. The broad range of topics is captured by the Symposia and Focused Session titles, which are listed as follows: Focused Session 1—Geopolymers and Chemically Bonded Ceramics; Focused Session 2—Thermal Management Materials and Technologies; Symposium 10—Virtual Materials (Computational) Design and Ceramic Genome; and, Symposium 12—Materials for Extreme Environments: Ultrahigh Temperature Ceramics and Nanolaminated Ternary Carbides and Nitrides.

This was the 11th consecutive year for the topic covered by Focused Session 1 concerning Geopolymers and Chemically Bonded Ceramics. As in years past, it continued to attract attention from international researchers as well as new application domains. Twelve papers are included in this year's proceedings. The studies focus on processing as well as the associated microstructural and mechanical properties in relevant environments. Such studies are critical in the pursuit of sustainable and environmentally friendly ceramic composites. Focus Session 2 emphasizes new materials and the associated technologies related to thermal management. The topic includes innovations in ceramic or carbon based materials tailored for either high conductivity applications (e.g., graphite foams) or insulation (e.g., ceramic aerogels); heat transfer nanofluids; thermal energy storage devices; phase change materials; and a slew of technologies that are required for system integration. One paper is included here addressing the relatively new application of high conductivity graphite foams for thermal energy storage.

Symposium 10 is dedicated to the modeling of ceramics and composites. This includes property prediction, innovative simulation methods, modeling defects and diffusion in ceramics as well as the study of virtual materials with the aim of further optimizing the behavior to facilitate the design of new ceramics and composites with tailored properties. Nine papers are available within this volume discussing subjects such as stochastic crystal growth, crack modeling, numerical assessment of self-healing composites, multi-scale modeling of CMCs, and laminate property predictions using 3D unit cells. Symposium 12 addresses the many facets related to materials for extreme environments. This includes the relationship between material structures and properties, structural stability under extreme environments, novel characterization methods, and life predictions. Four papers from this symposium are included within this collection. Lastly, a single study is included from the Second Annual Global Young Investigators Forum. The paper is focused on the dielectric and piezoelectric properties of a novel PZT ceramic.

The editors wish to thank the symposium organizers for their time and labor, the authors and presenters for their contributions; and the reviewers for their valuable comments and suggestions. In addition, acknowledgments are due to the officers of the Engineering Ceramics Division of The American Ceramic Society and the 2013 ICACC program chair, Dr. Sujanto Widjaja, for their support. It is the hope that this volume becomes a useful resource for academic, governmental, and industrial efforts.

WALTRAUD M. KRIVEN, *University of Illinois at Urbana-Champaign, USA*
JINGYANG WANG, *Institute of Metal Research, Chinese Academy of Sciences, China*
YANCHUN ZHOU, *Aerospace Research Institute of Materials & Processing Technology, China*
ANDREW L. GYEKENYESI, *NASA Glenn Research Center, USA*

# Introduction

This issue of the Ceramic Engineering and Science Proceedings (CESP) is one of nine issues that has been published based on manuscripts submitted and approved for the proceedings of the 37th International Conference on Advanced Ceramics and Composites (ICACC), held January 27–February 1, 2013 in Daytona Beach, Florida. ICACC is the most prominent international meeting in the area of advanced structural, functional, and nanoscopic ceramics, composites, and other emerging ceramic materials and technologies. This prestigious conference has been organized by The American Ceramic Society's (ACerS) Engineering Ceramics Division (ECD) since 1977.

The 37th ICACC hosted more than 1,000 attendees from 40 countries and approximately 800 presentations. The topics ranged from ceramic nanomaterials to structural reliability of ceramic components which demonstrated the linkage between materials science developments at the atomic level and macro level structural applications. Papers addressed material, model, and component development and investigated the interrelations between the processing, properties, and microstructure of ceramic materials.

The conference was organized into the following 19 symposia and sessions:

| | |
|---|---|
| Symposium 1 | Mechanical Behavior and Performance of Ceramics and Composites |
| Symposium 2 | Advanced Ceramic Coatings for Structural, Environmental, and Functional Applications |
| Symposium 3 | 10th International Symposium on Solid Oxide Fuel Cells (SOFC): Materials, Science, and Technology |
| Symposium 4 | Armor Ceramics |
| Symposium 5 | Next Generation Bioceramics |
| Symposium 6 | International Symposium on Ceramics for Electric Energy Generation, Storage, and Distribution |
| Symposium 7 | 7th International Symposium on Nanostructured Materials and Nanocomposites: Development and Applications |

| | |
|---|---|
| Symposium 8 | 7th International Symposium on Advanced Processing & Manufacturing Technologies for Structural & Multifunctional Materials and Systems (APMT) |
| Symposium 9 | Porous Ceramics: Novel Developments and Applications |
| Symposium 10 | Virtual Materials (Computational) Design and Ceramic Genome |
| Symposium 11 | Next Generation Technologies for Innovative Surface Coatings |
| Symposium 12 | Materials for Extreme Environments: Ultrahigh Temperature Ceramics (UHTCs) and Nanolaminated Ternary Carbides and Nitrides (MAX Phases) |
| Symposium 13 | Advanced Ceramics and Composites for Sustainable Nuclear Energy and Fusion Energy |
| Focused Session 1 | Geopolymers and Chemically Bonded Ceramics |
| Focused Session 2 | Thermal Management Materials and Technologies |
| Focused Session 3 | Nanomaterials for Sensing Applications |
| Focused Session 4 | Advanced Ceramic Materials and Processing for Photonics and Energy |
| Special Session | Engineering Ceramics Summit of the Americas |
| Special Session | 2nd Global Young Investigators Forum |

The proceedings papers from this conference are published in the below nine issues of the 2013 CESP; Volume 34, Issues 2–10:

- Mechanical Properties and Performance of Engineering Ceramics and Composites VIII, CESP Volume 34, Issue 2 (includes papers from Symposium 1)
- Advanced Ceramic Coatings and Materials for Extreme Environments III, Volume 34, Issue 3 (includes papers from Symposia 2 and 11)
- Advances in Solid Oxide Fuel Cells IX, CESP Volume 34, Issue 4 (includes papers from Symposium 3)
- Advances in Ceramic Armor IX, CESP Volume 34, Issue 5 (includes papers from Symposium 4)
- Advances in Bioceramics and Porous Ceramics VI, CESP Volume 34, Issue 6 (includes papers from Symposia 5 and 9)
- Nanostructured Materials and Nanotechnology VII, CESP Volume 34, Issue 7 (includes papers from Symposium 7 and FS3)
- Advanced Processing and Manufacturing Technologies for Structural and Multi functional Materials VII, CESP Volume 34, Issue 8 (includes papers from Symposium 8)
- Ceramic Materials for Energy Applications III, CESP Volume 34, Issue 9 (includes papers from Symposia 6, 13, and FS4)
- Developments in Strategic Materials and Computational Design IV, CESP Volume 34, Issue 10 (includes papers from Symposium 10 and 12 and from Focused Sessions 1 and 2)

The organization of the Daytona Beach meeting and the publication of these proceedings were possible thanks to the professional staff of ACerS and the tireless dedication of many ECD members. We would especially like to express our sincere thanks to the symposia organizers, session chairs, presenters and conference attendees, for their efforts and enthusiastic participation in the vibrant and cutting-edge conference.

ACerS and the ECD invite you to attend the 38th International Conference on Advanced Ceramics and Composites (http://www.ceramics.org/daytona2014) January 26-31, 2014 in Daytona Beach, Florida.

To purchase additional CESP issues as well as other ceramic publications, visit the ACerS-Wiley Publications home page at www.wiley.com/go/ceramics.


SOSHU KIRIHARA, *Osaka University, Japan*
SUJANTO WIDJAJA, *Corning Incorporated, USA*

Volume Editors
August 2013

# Geopolymers and Chemically Bonded Ceramics

# IMPORTANCE OF METAKAOLIN IMPURITIES FOR GEOPOLYMER BASED SYNTHESIS

A. Autef[1], E. Joussein[2], G. Gasgnier[3] and S. Rossignol[1]

[1] Groupe d'Etude des Matériaux Hétérogènes (GEMH-ENSCI) Ecole Nationale Supérieure de Céramique Industrielle, 12 rue Atlantis, 87068 Limoges Cedex, France.
[2] Université de Limoges, GRESE EA 4330, 123 avenue Albert Thomas, 87060 Limoges, France.
[3] Imerys Ceramic Centre, 8 rue de Soyouz, 87000 Limoges, France.
corresponding author: sylvie.rossignol@unilim.fr

ABSTRACT
    Geopolymers are the object of numerous studies because of their low environmental impact. The synthesis of these geomaterials is achieved by the alkaline activation of aluminosilicates. Alkaline activation is typically accomplished by the activation of potassium or sodium silicate. Since these alkaline silicate solutions are relatively expensive. It is imperative to understand all of the phenomena and reactions involved during geopolymer synthesis. We thus attempted to study the role played by siliceous species in the alkaline silicate solutions.
    During the setting of the materials, the reactive mixture forms at least two phases: (i) a solid phase and (ii) a gelified liquid which recovered it. The quantity of gel varies with the Si/Al, Si/K and Si/$H_2O$ molar ratios. Several exchanges take place at the gel-solid interface and involve composition and pH variations. Moreover, the nature and the number of networks depend on the alkaline solution used.

## I.    INTRODUCTION

    The alkaline silicate solutions (waterglass) necessary for the synthesis of geopolymer materials are solutions containing a dissolved glass with an aspect similar to water. Alkaline silicate solutions are widely used in the industry as binders, emulsifying agents, deflocculants or in the paper industry. These sodium or potassium-based solutions, present complex structures, composed of diverse monomeric and polymeric species [1,2,3]. Their composition evolves according to various variables, such as the value of pH [4] or the $SiO_2/M_2O$ molar ratio (where M=Na or K). These parameters allow control of the various species in the mixture which confer variable properties of the solutions, in particular in terms of reactivity. Important differences are also noted between potassium and sodium elements; these differences can be at the origin of variations, both in terms of structure and stability, within geopolymer materials.

    Several studies were recently realized in these alkaline silicate solutions [5] and on their role during geopolymer formation [6,7,8]. These various studies allowed highlighting the existence of two phases within the consolidated material [7]: a geopolymer phase and a gel phase, present in more or less high quantity according to the source of silica used during the manufacture of the alkaline silicate solution. Indeed, the use of sand as a substitute for the amorphous silica leads to a decrease of the Si / Al ratio and of the quantity of geopolymeric phase [7].

    The consolidation of the material is then possible thanks to the important presence of gel but leads to a decrease in the mechanical properties. These materials, synthesized from a commercial metakaolin, also contain impurities initially present in the raw material (e.g. anatase, muscovite, quartz).

The objective of this study is to understand the role played by the siliceous species from the activation solution during the formation of geopolymers. Hence, the role of the alkaline silicate solution was studied by comparing a commercial solution with a laboratory prepared solution with the same Si / K molar ratio. To eliminate the effect of impurities within the consolidated materials, a high purity metakaolin (99 %) was used for both activating solutions.

## II.   EXPERIMENTAL PART

### 1.  Sample preparation

Geopolymer materials were synthesized according to two ways as described by Figure 1.

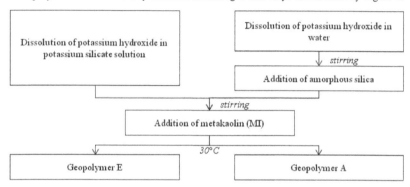

**Figure 1** : Synthesis protocol of geopolymer compounds.

In the case of geopolymers E [9], potassium hydroxide pellets (85.7 % of purity) were dissolved in some potassium silicate solution supplied by the Chemical Labs company (Si / K=1.7; density of 1.20 g / cm$^3$). Dehydroxylated kaolinite MI (98.8 % of purity) supplied by IMERYS France is added to this preparation under stirring. This metakaolin has been calcined at 750°C during 5 hours ($S_{BET}$ = 8 m$^2$.g$^{-1}$). The reactive mixture is then placed in a polystyrene cylindrical mould at room temperature. The characteristics of the raw materials are presented in Table 1.

In the case of geopolymers A [6], the KOH pellets were first dissolved in water at room temperature to form an alkaline solution (pH=14). An amorphous silica, being very fine and highly reactive (denoted S; purity of 99.9%) and supplied by SIGMA ALDRICH was dissolved in the alkaline solution. The continuation of the protocol is similar to what was previously described. Nomenclatures and molar ratios are presented in the Table 2.

**Table 1** Characteristics and nomenclature of raw materials used.

| Nature | Amorphous silica | Metakaolin MI |
|---|---|---|
| Nomenclacture | S | MI |
| $d_{50}$ (μm) | 0.14 | 7.54 |
| BET value (m$^2$/g) | 202 | ~ 7 |
| Chemical analysis (wt. %) | 99.9 SiO$_2$ | 50 SiO$_2$ |
|  |  | 50Al$_2$O$_3$ |

**Table 2** Nomenclature and compositions of compounds.

| Sample | Used metakaolin | Si/Al | Si/K | Si/H$_2$O |
|--------|-----------------|-------|------|-----------|
| A | MI | 1.40 | 2.10 | 0.30 |
| E | MI | 1.40 | 2.10 | 0.23 |

## 2. Characterization

X-ray patterns were performed from powder samples after crushing at 63 μm, and obtained using a Brucker-AXS D 5005 apparatus from 5 to 70° (2 theta). The device is equipped with a cobalt anode ($\lambda$ = 1.79026 Å).The acquisition time is 1 second, and the step is 0.02° (2 theta). XRD patterns were analyzed using EVA software.

The FTIR spectra were obtained using a Thermo Fisher Scientific 380 infrared spectrometer (Nicolet). The IR spectra were gathered over a wave number range of 400 to 4000 cm$^{-1}$ with a resolution of 4 cm$^{-1}$. The atmospheric $CO_2$ contribution was removed with a straight line between 2400 and 2280 cm$^{-1}$. To follow the evolution of the involved bonds within the sample in time, a computer algorithm was used to acquire a spectrum every 10 minutes for 13 hours, producing 64 scans. To allow comparisons of the various spectra, the spectra were corrected with the baseline and then normalized. The characterization of the powders and gels was also conducted by FTIR.

Differential thermal analysis (DTA) and thermo gravimetric analysis (TGA) were performed on a SDT Q600 apparatus from TA Instruments in an atmosphere of flowing dry air (100 mL/min) in platinum crucibles. Signals were measured with Pt/Pt-10%Rh thermocouples. Some milligrams of material are placed in a platinum crucible and the analysis is made from 30 °C to 800 °C, at 10 °C / min.

The chemical analyses were obtained by XRF investigations using a XMET 5500 X from Oxford. Samples are analyzed from pressed pellet.

## III. RESULTS AND DISCUSSION

### 1. Synthesis of materials

In the way to study the influence of the alkaline solution only one sort of metakaolin highly pure (MI metakaolin) was used. Therefore, the influence of the impurities was eliminated. It was chosen to maintain constant the Si / K and Si / H$_2$O ratios, what leads to a decrease in the Si / Al ratio from 1.62 to 1.40 compared to the previous study from an other type of metakaolin raw material [10]. Whatever the composition, (i) a consolidated geopolymer-like material was synthesized, and (ii) a demixing brought in the reactive mixture from the first hour after the synthesis: a fine coat of transparent liquid appeared slowly at the surface. The polycondensation phase is effective in 6 at 10 hours. The viscosity of the supernatant liquid increased until to form a gel, 5 to 8 days after the synthesis. Afterward, the supernatant phase will be named "gel" and the solid phase "solid". As an example, the reactive mixture E gave a sample consisting of the "solid E" recovered from the "gel E". The same results were observed for A samples.

According to previous results the variations of the various molar ratios during the substitution of the other metakaolin by the MI metakaolin led to the formation of a more important quantity of gel on the solids surface. This increase seemed to be inversely proportional to the Si / H$_2$O ratio leading to an extension of the gelation time. The quantity of gel formed also increased with the Si/Al ratio. This observation highlighted the role of the aluminum as a networking agent. The variation of these molar ratios influenced the formed gel quantity and the gelation speed without affecting the local structure of the gel. There was thus no modification of the formation mechanisms but only an influence on their kinetics.

## 2. Comparison of A and E compounds (open mold)

The difference between both A and E solutions was in the nature of the $Q^n$ species in solution [6]. Indeed, studies by $^{29}Si$ NMR revealed a quantity of $Q^0$ species superior to $Q^1$ in the laboratory prepared solution while the opposite was observed in the commercial solution [8]. The silicon availability compared with the aluminum thus differs, what led to variations of reactivity and thus to the formation of the various networks. However, whatever was the alkaline solution used all the prepared samples within the framework of the present study led to the formation of a solid phase and a gel. To characterize the obtained materials, all of them were analyzed by infrared spectroscopy and thermal analysis. The mass variations were different between these two compounds (Figure 2 (a)). The 24 % weight loss noticed in the case of the solid is characteristic of this type of material [11]. Whatever the composition (A or E) the behavior in temperature was similar. The Figure 2 (b) presents the heat flow profiles (endothermic peak) for each of the materials characteristic of the elimination of the water.

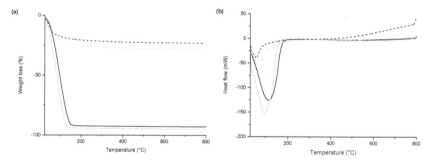

**Figure 2 :** Curves of thermal analyses of gels E (—) and A (—) and solids E (---) and A (---), 20 days of synthesis (a) weight loss and (b) heat flow as a function of temperature.

For geopolymer compounds the loss of water intervened for nearby temperatures of 100°C [10]. The gel appeared as a compound very rich in water with a 93 % weight loss, characteristic of a silica gel [12]. In that case, the water elimination intervened in a temperature domain vaster than for the solid and went on until neighboring temperatures of 200°C. The very similar weight loss behavior for temperatures lower than 100°C showed that a part of the gel behaved as a "solid". There would be in that case a common interface which would make the interpretations difficult.

The results obtained by infrared spectroscopy on drops of reactive mixture at various times are grouped on the Figure 3 (a and b). The absorption band shift due to the siliceous species is similar in the case of mixtures E and A which consolidates the idea that a polycondensation reaction intervened at the beginning.

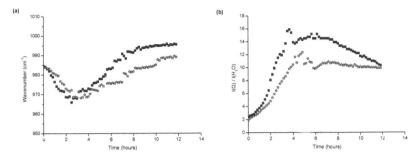

**Figure 3 :** Variation in time of the Si-O-M band (a) and of the I (Q) / I(H₂O) (b) ratio during the formation of E (■) and A (☆) solids in open mold.

### 3. Focus on the composition E
#### a. Characterization of the supernatant liquid

To determine the gel and solid formation mechanisms, FTIR spectroscopy studies and chemical analyses were realized on materials E synthesized with the commercial potassium silicate solution. This supernatant liquid found on the solid surface was expelled during the consolidation of the latter. As previously observed, the quantity of liquid formed above the solid compound was stable with time until the gelation moment which intervened at the end of 7 days. To follow this transformation in the gel, the supernatant phase was punctually analyzed by infrared spectroscopy (Figure 4).

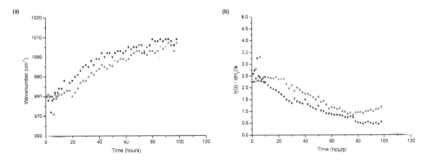

**Figure 4 :** Variation in time of the (a) Si-O-M band and (b) I(Q²) / I(H₂O) ratio in the supernatant liquid 1 (■) and 2 (Δ) produced by sample E, in open mold during the gelation.

The observed absorption bands were attributed to the changes of vibrational and rotational states of the various bonds from the literature [13,14, 15]. The drift of the absorption peak position caused by the siliceous species (denoted $Q^2$) was raised (Figure 4 (a)) because it was defined as one of the characteristics of the polycondensation phenomena [16].Indeed, a polycondensation reaction was characterized by an evolution of the position of the vibration band relative to the bond $Q^2$. An increase of the intensity of this vibration band was also noted.

The $I(Q^2)$ / $I(H_2O)$ ratio of the absorbed intensities by the change of vibrational states of the siliceous and aqueous species was also determined (Figure 4 (b)) by considering the maxima of absorption situated respectively between 1660-1630 cm$^{-1}$ and 1100-970 cm$^{-1}$. Whatever was the considered sample (gel 1 or 2) issue from two preparations, the same evolutions were noticed, evidencing the reproducibility of the synthesis. The increase of the wavenumber corresponding to the Si-O-M peak, due to a replacement of the Si-O-Al bound by the Si-O-Si bound, suggests that the reactions were essentially made between siliceous species [10].The decrease of the $I(Q)$ / $I(H_2O)$ ratio (Figure 4 (b)) resulted mainly in the increase of the $H_2O$ vibration band compared with that attributed to Si-O-M which decreased very slowly. This phenomenon is in agreement with the hypothesis of local reorganizations between the siliceous species which polycondensate to form a gel [5]. From approximately 70 hours of reaction, stabilization was observed; it could be explained by a ripening of the gel [5].

**b. Influence of the reactivity (comparison between open and closed molds)**

In the case of the solid, acquisitions were not any more made in a punctual way but by automated follow-up during polycondensation. The acquisitions were realized on a drop of reactive mixture in an open mold. The variations of the Si-O-M vibration band position as well as the $I(Q)$ / $I(H_2O)$ are represented in the Figure 5A.The Si-O-M band shift with time towards lower wavenumbers (Figure 5A (a)) was due to the polycondensation reactions which began for the formation of the geopolymer, as shown in previous works [17]. The sudden evolution of its position towards the high wavenumber which arose beyond the first 3 hours following the synthesis could result from the formation of Si-O-Si recombination between the various $Q^n$ species within the network to form a second network or a crystallized compound. Indeed, during such a formation, there was gradually consumption of the species $Q^2$ for the benefit of the $Q^3$ and $Q^4$ formation [18].This evolution would thus be characteristic of the formation of a compound for which the nature was not identified. This compound was formed from the available species in solution; it could also be a potassium aluminosilicate or a potassium silicate. It would seem that this phenomenon is comparable to that observed for the supernatant liquid but according to different reaction kinetics. According to this hypothesis, the formation of this compound would contribute to enrich the supernatant liquid, until led to saturation in the middle causing the gel formation. The increase in the $I(Q)$ / $I(H_2O)$ ratio (Figure 5A (b)) at the beginning of the reaction resulted from the decrease in the water contribution. This revealed that the solid network formation led to an elimination of water which was necessary for the polycondensation reaction. The small decrease which then appeared translates a weak increase of the vibration band attributed to $H_2O$ which suggests that the formation of the compound revealed by the shift of the Si-O-M vibration band is accompanied by a weak water discharge. These first results being obtained in a drop of mixture in an open mold not allowing observation of the demixing phenomenon.

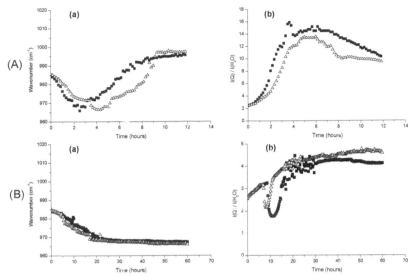

**Figure 5 :** (A) Variation in time of the (a) Si-O-M band and (b) I(Q²) / I(H₂O) ratio during the formation of solids E1(■) and E2 (Δ) in open mold. (B) Variation in time of the (a) Si-O-M band and (b) I(Q²) / I(H₂O) the formation of solids E1 (■) and E2 (Δ) in closed mold.

To reproduce the conditions of formation of a geopolymer material, an *in situ* analysis in a closed mold was realized by infrared spectroscopy (cf. experimental part). The obtained results are given on the Figure 5B. The observed differences between both samples can be understood by temperature variations, around 5°C, during the acquisitions which modified the reaction kinetics. The observed wavenumber diminution (Figure 5B (a)) is characteristic of a polymerization reaction during geopolymer formation [10]. Nevertheless, for the solid E the evolution of the wavenumber seemed to be disturbed near 10 hours of reaction. Indeed, the band position increased during a few hours before starting its decrease again. This means that another phenomenon disrupted the polycondensation reactions introduced during the first 10 hours following the synthesis. This disturbance could correspond to the formation of a crystallized phase or to a silica gel [5]. A silicate network could form very quickly, and then siliceous species would react to form a siliceous compound. Once this compound formed, the charged species in the middle can again govern and lead to the formation of a solid geopolymer. On the other hand, the slowing down and/or the inversion of the progressive variation of the I(Q) / I(H₂O) ratio can again be characteristic of the formation of a compound (Figure 5B (b)). Indeed, the I(Q) / I(H₂O) ratio presents a non linear behavior: a rough fall of the I(Q)/I(H₂O) ratio is observed approximately ten hours after the synthesis. The siliceous species contribution being constant in time, it is the water contribution that imposes the observed evolution. There is first water consumption which corresponds to the formation of the solid network. Then the water production (decrease of the I(Q) / I(H₂O) ratio) translates the formation of a compound. This phenomenon corresponds to the formation of a gel. Finally, when the gel is totally formed the polycondensation of the solid network continues and consumes some water during its nucleation.

## IV.    EVIDENCE FOR THE PRESENCE OF CRYSTALLIZED PHASE

To determine the presence of a possible crystallized compound, XRD studies were realized on the gels (dried at 110°C) and on solids. Figure 6 presents the patterns obtained.

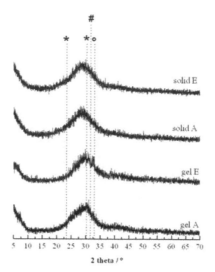

**Figure 6 :** XRD patterns obtained on E and A gels and solids (PDF files: * $K_2Si_2O_5$ (00-026-1331); # $K_2SiO_3$ (04-013-3814); ° $K_2Si_4O_9$ (00-039-0212).

The gels and solids being amorphous, it is difficult to determine exactly their nature; however the appearance of certain shoulder can translates the implementation of particular local orders within these materials. XRD patterns obtained on the various solids were similar and corresponded to amorphous materials. Their diffraction dome was centered on 28°($2\theta$). However, for a 24°($2\theta$) value a shoulder was observed for all the compounds. This shoulder could be attributed to the presence of the $K_2Si_2O_5$ compound. Indeed, this compound has already been identified in geopolymer materials prepared from a commercial alkaline silicate [19]. Besides, in the case of dried gels, the maximum diffracted intensity corresponded to superior values of $2\theta$. Furthermore, there was an appearance of a diffraction peak badly defined at 32°($2\theta$). The latter, particularly visible in the gel E, could be attributed to a network with $K_2Si_4O_9$. In the gels, there would thus be existence of one or several alkaline silicates and a siliceous amorphous network.

These observations were confirmed by chemical analyses of solid phases. Figure 7 groups the molar percentages calculated from these data for the compositions E and A. The superior and lower part of samples, that are the gel and the solid bottom, as well as the interface between the gel and the solid were analyzed. The results show that the molar proportions (close to some reactive mixture) were very similar for gels and solids and whatever the activating solution that was used. This suggests that the distribution in elements was homogeneous within the solid and within the gel, although the distribution of phases and their characteristics were different. On the other hand, the interfaces were mainly rich in silicon and potassium. This disparity could mean that the migration of the species between the solid phase and the gel was partially assured by a potassium silicate solution, which the molar composition could be Si / K = 0.5.

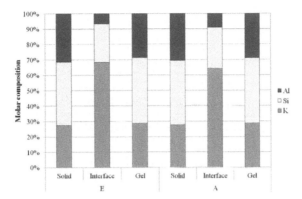

**Figure 7** : Recalculated molar compositions from XRF analyses performed on E and A compounds.

V.   CONCLUSION
Geopolymers present a growing interest because of their low environmental impact. These materials are obtained by alkaline activation of alumino-silicate. The high cost of the necessary alkaline solutions constitutes a break in the development of these innovative binders. It is thus indispensable today to understand all the phenomena and the reactions occurring during their synthesis to promote the development of these materials. The present research work ensues from this imperative: its objective was to study the role played by the siliceous species within the activating solution in the presence of pure metakaolin.
The results highlighted the formation of a gel phase during the synthesis of geopolymers from pure metakaolin.
From these studies, it is possible to provide that (i) the variations of Si / Al, Si / K and Si / $H_2O$ molar ratios influence the quantity of gel formed and the gelation kinetics in the presence of a very pure metakaolin and (ii) the change in the initial alkaline solution does not lead to modification of the phenomena but the nature and the number of existing networks differ.

ACKNOWLEDGEMENTS
The authors thank the IMERYS Ceramic Centre for the metakaolin samples.

REFERENCES

[1] J. G. Vail, Soluble silicate: Their properties and uses. Reinhold, New York, 1952.
[2] G. Lagerström, Equilibrium studies of polyanions. III. Silicate ions in NaClO4 medium, *Acta Chemica Scandinavica*, **13**, 722–736 (1959).
[3] G. Engelhardt, D. Zeigan, H. Jancke, D. Hoebbel, W. Wieker, $^{29}$Si NMR spectroscopy of silicate solutions. II. On the dependence of structure of silicate anions in water solutions from the Na/Si ratio, *Zeitschrift für anorganische Chemie*, **418**, 17–28 (1975).
[4] M. Dietzel, Dissolution of silicates and the stability of polysilicic acid, *Geochim. Cosmochim. Acta*, **64**, 19, 3275 (2000).
[5] M. Tohoue Tognonvi, S. Rossignol, J. P. Bonnet, Effect of alkali cation on irreversible gel formation in basic medium, *Journal of Non-Crystalline Solids*, **357**, 43–49 (2011).
[6] A. Autef, E. Joussein, G. Gasgnier, S. Rossignol, Parameters that influence silica dissolution in alkaline. *35th International Conference on Advanced Ceramics and Composites*, accepted.
[7] A. Autef, E. Joussein, G. Gasgnier, S. Rossignol, Role of the silica source on the geopolymerization rate, *Journal of Non-crystalline solids*, **358**, 2886–2893 (2012).
[8] E. Prud'homme, A. Autef, N. Essaidi, P. Michaud, B. Samet, E. Joussein, S. Rossignol, Defining existence domains in geopolymers through their physicochemical properties, Applied Clay Science, http://dx.doi.org/10.1016/j.clay.2012.10.013 (2012).
[9] E. Prud'homme, P. Michaud, E. Joussein, C. Peyratout, A. Smith, S. Arrii-Clacens, J. M. Clacens, S. Rossignol, Silica fume as porogent agent in geo-materials at low temperature, *Journal of the European Ceramic Society*, **30**, 1641–1648 (2010).
[10] E. Prud'homme, Rôles du cation alcalin et des renforts minéraux et végétaux sur les mécanismes de formation de géopolymères poreux ou denses, thèse de doctorat. Université de Limoges, France, 2011.
[11] P. Duxson, G. C. Lukey, J. S. J. van Deventer, Physical evolution of Na-geopolymer derived from metakaolin up to 1000 °C, *Journal of Materials Science*, **42**, 3044–3054 (2007).
[12] M. Tohoue Tognonvi, J. Soro, S. Rossignol, Physical-chemistry of silica/alkaline silicate interactions during consolidation. Part1: Effect of cation size, *Journal of Non-Crystalline Solids*, **358** (1) 81–87 (2012).
[13] J. R. Marinez, Infrared spectroscopy analysis of the local atomic structure in silica prepared by sol-gel, *Journal of Chemical Physics*, **109** (17), 7511–7517 (1998).

[14] W. K. W. Lee, J. S. J. Van Deventer, Use of Infrared Spectroscopy to Study Geopolymerization of Heterogeneous Amorphous Aluminosilicate, *Langmuir*, **19** (21), 8726–8734 (2003).
[15] P. Innocenzi, Infrared spectroscopy of sol-gel derived silica-based films: a spectra microstructure overview, *Journal of Non-Crystalline Solids*, **316**, 309–319 (2003).
[16] C. A. Rees, J. L. Provis, G. C. Luckey, J. S. J. Van Deventer, Attenuated Total Reflectance Fourier Transform Infrared Analysis of Fly Ash Geopolymeer Gel Aging, *Langmuir*, **23** (15), 8170–8179 (2007).
[17] E. Prud'homme, P. Michaud, E. Joussein, C. Peyratout, A. Smith, S. Rossignol, Consolidated geo-materials from sand or industrial waste, *Ceramic Engineering and Science Proceedings*, **30** (2),313–324 (2010).
[18] S. Puyam, P. S. Singh, M. Trigg I. Burgar, T. Bastow, Geopolymer formation processes at room temperature studied by $^{29}$Si and $^{27}$Al MAS-NMR, *Materials Science Engineering A*, **396**, 392–402 (2005).
[19] S. Delair, E. Prud'homme, C. Peyratout, A. Smith, P. Michaud, L. Eloy, E. Joussein, S. Rossignol, Durability of inorganic foam in solution: The role of alkali elements in the geopolymer network, *Corrosion Science*, **59**, 213–221 (2012).

MECHANICAL STRENGTH DEVELOPMENT OF GEOPOLYMER BINDER AND THE EFFECT OF QUARTZ
CONTENT

C. H. Rüscher[1], A. Schulz[1], M. H. Gougazeh[2], A. Ritzmann[3]
[1]Institut für Mineralogie, Leibniz Universität Hannover
[2]Natural Resources and Chemical Engineering Department, Tafila Technical University
[3]GNF Berlin Adlershof e.V.

ABSTRACT
        The development of compressive strength of alkali activated metakaolin in dependence of
waterglass to metakaolin ratios and quartz additions were investigated during ageing at room
temperature. The compressive strength could be optimized for maximal strength for nominal
Si/Al ratios of about 1.8-2 of the binder for ageing between 7 and 10 days. However this
composition invariably leads to weakening for longer ageing time (weakening effect). Moreover,
the optimum in strength for Si/Al $\approx$ 1.9 could be seen as a shoulder or weaker peak in an
extended field of Si/Al and alkali/Al ratios. Additions of quartz, either as "sand" for mortars or
as fine grained quartz or as significant virgin contents in the kaolin source could avoid the
weakening effect and improve, therefore, the geopolymer property.

INTRODUCTION
        Many investigations concern the relationship between geopolymer composition and the
development of the microstructure and nanostructure of the binder with respect to their
mechanical properties development. Just concentrating on geopolymer binders produced by the
reaction of metakaolin and waterglass a few systematical studies may be described in some more
detail. Rowles and O'Connor [1] showed that highest compressive strength of about 64 MPa could
be achieved at nominal compositions of Si/Al $\approx$ 2.5 and Na/Al $\approx$ 1.25 in the field of varying
ratios Na/Al between 0.5 and 2 and Si/Al between 1 and 3. The tests were performed using
samples heated for 24 h at 75°C in closed moulds and holding them another 7 days closed before
measurement. In a following study Rowles and O'Connor [2] showed that all the binders could be
separated into two phases by variations in Si/Al and Na/Al ratios. Duxson et al. [3, 4] obtained
highest strength of about 80 MPa for Si/Al ratios of about 1.9 and Na/Al = 1 in the field of Si/Al
ratios between 1.15 and 2.15. This main picture becomes not much modified using K or Na or
both in the waterglass solutions. These authors cured the sample at 40°C for 20 h under sealed
conditions in Teflon cups. Tsitouras et al. [5] used conditions comparable to those in [1] (70°C for
48 h and testing after 7 days) obtaining an increase in compressive strength for Na/Al = 1 from
Si/Al = 1.25 to 2. This small survey of literature values could show that similar trends for
optimizations of maximal mechanical strength could be met concerning the compositions.
Moreover, Duxson et al. [4] pointed out that structural changes occur within the first month of
ageing and that the increase in strength as well as the reduction in compressive strength are also
subject to alteration in the alkali composition. The striking feature of optimal strength of a
"geopolymer binder" of composition $Na_2O \cdot Al_2O_3 \cdot 4SiO_2 \cdot 13H_2O$ could be the strong increase in
strength up to about 80 MPa within the first days of ageing at 23°C, followed by a more or less
smooth weakening during further ageing [6]. As further shown by Lloyd [6] the weakening could
also be accelerated by ageing the 28 days at 23°C cured sample, i.e. at the highest strength
reached, and ageing it further at elevated temperature (e.g. 95°C). This treatment reveals
obviously a strong decrease in strength down to 30 MPa within 5 days, where it remains stable
during further ageing. Possibly, there are also somewhat other observations described in the

literature using different geopolymer compositons and additions, either in question of this problem or just by chance. In fact just "the weakening effect" of alkali activated metakaolin were found to be a direct consequence of permanently running reactions in the geopolymer body due to shortening of polysiloxo chains (-Si-O-Si-O-Si-...) [7-10].

"Binder formation" of alkali-waterglass activated metakaolin may follow always the same mechanism for waterglass/metakaolin compositions [7-10]. Hardening occurs generally completely inhomogeneous due to condensation of polysiloxo chains from the waterglass solution from that moment when hydroxide is consumed by the solution of metakaolin. The amount of available hydroxide solution may govern how much $Al^{3+}$ could be transferred into the waterglass and how the initial waterglass condensate becomes modified. Accordingly the more or less rapid increase in strength is just a consequence of the reaction kinetics of crosslinking the long chains via sialate bondings (-Si-O-Al-O-). Unfortunatelly by the same time long polysiloxo chains become more and more shortened due to the strong increase of sialate bonds during polycondensation. It has been shown that a protection of the long polysiloxo chains, could become available in the presence of high amounts of unreacted metakaolin or also by the presence of Ca-ions [10]. Here the compressive strength development is discussed in some details considering some additional results in the field of Si/Al and alkali/Al ratios. Additionally the development of compressive strength in the presence of quartz, either as addition or virgin in the kaolinite source, will be considered.

EXPERIMENTAL

Series of cements M2-M7 were prepared using commercially available potassium waterglass (KWG$_{sil}$: Cognis, Germany, Silerit M60, 19.1 % $SiO_2$, 24.8% $K_2O$, 56.1% $H_2O$, all in wt%) and metakaolin (Metastar 501 denoted MKM, composition is discussed below in 3.1) with the following mass ratios: KWG/MKM = 0.8 (M2), 1.0 (M3), 1.3 (M4), 1.66 (M5), 2.5 (M6), and 5.0 (M7). Samples M1 were prepared using KOH solution (50%) in mass ratio 1:1 for KOH-solution/MKM. Cements of composition M5 were used for preparation of mortars using "norm sand" (CEN, DIN EN196 part 1) with mass ratio 1:3 (M5/sand) with the addition of some arbitrary amount of $H_2O$. Some ordinary portland cement (OPC) with (OPC/sand = 1:3) and without the addition of sand were prepared for comparison. All samples were given into cube forms of 1 cm edge length and aged under open conditions in the laboratory between 1 and 90 days. Compressive strength measurements of the 1*1*1 cm$^3$ cubes were carried out using an universal testing machine (MEGA 2-3000-100 D Seidner and Co. GmbH, Riedlingen, Germany).

Further experiments were carried out using metakaolin of different sources with and without the addition of quartz powders (Fluka 83340). One type of metakaolin (Mephisto K05, Prague, denoted MKMph) were used together with self prepared (KWG$_{sp}$) waterglass using KOH pellets (Fluka 60370), silica fume (Merck 00657) and deionized water, obtaining at room temperature slightly yellowish clear solution of nominal composition: 19.3 % $SiO_2$, 29.2 % $K_2O$, 51.5 $H_2O$, i.e. with a slightly higher content in $K_2O$ compared to the composition for the KWG$_{sil}$ above (Silerit M60). Cements were prepared in series A in mass ratio 1.66 = KWG$_{sp}$/MKMph. Mortars were prepared using cements as described for A with the addition of quartz powder in mass ratio 1:1, series B. For samples of series C1, C2 and D1, D2 metakaolin has been prepared from Fluka kaolinite (No. 83340) by calcinations for 3 h at 650°C in atmospheric conditions (MKF). Series C1 and C2 were prepared using mass ratio KWG$_{sil}$/MKF of 1.75 (C1) and 1.22 (C2). For D1 and D2 30 mass % of manually mortared quartz powder were mixed into the Fluka kaolinite before calcinations. The obtained metakaolin were mixed in mass ratios KWGsil/MKFqz of 1.22 (D1) and 1 (D2).

All these series of cements (A-D) were prepared by hand mixing the different metakaolin powders in the alkaline solution of potassium waterglass for about 10 min, forming homogeneous slurry. The slurry was then filled to cylindrical PE-containers (18 mm in diameter and 12 mm in height), which were closed. These containers were held under laboratory conditions (22+/-1°C) during ageing. The containers were opened just before the compressive force measurements. These samples were then used for other characterizations (XRD, IR). For the compressive strength measurements the size of the sample could be used as given by the size of the boxes in a manually driven testing machine (ENERPAC P392, USA). The compressive force was monitored digitally (HBM, Scout 55 and U3 force transducer).

Thermogravimetric analysis were carried out using a Setaram equipment (Setsys Evolution). The measurements were done by heating to 1000°C, holding for 30 minutes and cooling with a flow of 20 ml/min of technical air. For heating and cooling a rate of 5°C/min were used. XRD pattern were recorded on a Bruker AXS D4 ENDEAVOR diffractometer (Ni filtered Cu-$K\alpha$ radiation). The measurements were carried out with a step width of 0.03° (2θ) and 1 second per step. The powder data were analysed with the Stoe WinXPOW software package. FTIR analysis was performed on a Bruker IFS66v FTIR spectrometer by using KBr pellet techniques in the 370–4000 cm$^{-1}$ range, with a resolution 2 cm$^{-1}$ (1 mg sample diluted in 200 mg of KBr). SEM/EDX investigations were carried out on a JEOL machine (JSM-6390A) equipped with Bruker XFlash 410-11-200.

RESULTS AND DISCUSSION

Nominal compositions of metakaolins and geopolymers

Bulk XRF and large area EDX analysis generally reveal the average composition of samples. The use of these values could lead to some systematic faults in "nominal" Si/Al and alkali/Al ratios of geopolymer compositions. For example bulk chemical analysis for MKM revealed about (by weight): 55.0% $SiO_2$, 40 % $Al_2O_3$, 3.4 % $K_2O$, and others below 0.6% (MgO, $Fe_2O_3$, CaO, $Na_2O$). Own EDX analysis reveal. 50.60 $SiO_2$, 45.99 $Al_2O_3$, 2.4 $K_2O$. Tsitouras et al. [5] also using MK metastar 501 noted. 54.56 % $SiO_2$, 43.56 $Al_2O_3$, 1.47% $Fe_2O_3$. Since we are dealing with more or less natural products a variation in composition can be expected. Impurity phases like feldspar, muscovite and quartz can easily be found in XRD pattern (Fig. 1). The mentioned impurity phases remain unaltered during alkali activation and further ageing. Generally the amount of impurity phases – as it is usually in the percentage range - should be subtracted for further calculations of nominal compositions of geopolymer. As an approximation metakaolin may considered with ideal molar ratio $SiO_2/AlO_{1.5} = 1$ or mass ratio 1.18. The consequence from this can be seen in Tab. 1 where the Si/Al, K/Al, K/Si and $H_2O$/Si molar ratios are given on i. the basis of bulk chemical analysis with 55% $SiO_2$, 40% $Al_2O_3$ and ii. assuming just 90% of the total mass used to be as metakaolin of ideal Si/Al = 1 molar ratio. As could be seen below the later values would systematically shift the data points given for M1-M7 in Fig. 2a and 3 towards lower Si/Al and alkali/Al contents. However, it could be that data given by Duxson et al. [3,4], Tsitouris et al. [5], and Rowles and O'Connor [1,2] require similar corrections. Therefore, this effect will be ignored in the following discussion of compressive strength data for series of samples M1-M7.

Using as usual about 1 mg sample in 200 mg KBr IR spectra of MKF and MKM do not show any significant quartz content. However the amount detected in XRD indicates about up to

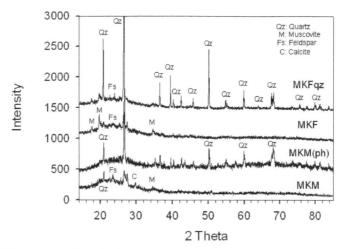

Figure 1. XRD pattern of metakaolin samples as used in this study showing some impurity phases as denoted. MKFqz is with the addition of 30 wt% Qz, whereas MKF is nearly free of Qz.

Table 1 (left). Geopolymer-cements of series M as described in the text. KWG/MK: used mass ratios waterglass/metakaolin, MK: used nominal compositions of MK (by wt%) as given by i, ii for calculations of nominal molar ratios of the geopolymer: Si/Al, K/Al, K/Si, $H_2O/Si$. (*Note: For M1 1.0 denotes only the ratio of 50%-KOH solution/MK)

Table 2 (right). Geopolymer-cements of series A, C and D, notations as described in Tab. 1.

| Geo-Cem | MassRatio KWG/MK | MK | Molar Ratio Geop. | | | |
|---|---|---|---|---|---|---|
| | | | Si/Al | K/Al | K/Si | $H_2O/Si$ |
| M2 | 0.8 | i | 1.49 | 0.54 | 0.36 | 2.13 |
| | | ii | 1.31 | 0.52 | 0.39 | 2.34 |
| M3 | 1 | i | 1.57 | 0.67 | 0.42 | 2.53 |
| | | ii | 1.39 | 0.65 | 0.47 | 2.76 |
| M4 | 1.3 | i | 1.69 | 0.87 | 0.52 | 3.05 |
| | | ii | 1.51 | 0.84 | 0.56 | 3.31 |
| M5 | 1.7 | i | 1.85 | 1.12 | 0.61 | 3.60 |
| | | ii | 1.65 | 1.08 | 0.65 | 3.87 |
| M6 | 2.5 | i | 2.18 | 1.68 | 0.77 | 4.56 |
| | | ii | 1.98 | 1.62 | 0.82 | 4.80 |
| M7 | 5.0 | i | 3.19 | 3.35 | 1.05 | 6.22 |
| | | ii | 2.96 | 3.25 | 1.10 | 6.49 |
| M1 | 1.0* | i | 1.17 | 1.14 | 0.97 | 3.03 |
| | | ii | 1.00 | 1.10 | 1.10 | 3.43 |
| i: MKM (55% $SiO_2$, 40% $Al_2O_3$) | | | | | | |
| ii: 0.9 * MK$_{ideal}$ (54.1% $SiO_2$, 45.9% $Al_2O_3$) | | | | | | |

| Geo-Cem | MassRatio KWG/MK | MK | Molar Ratio Geop. | | | |
|---|---|---|---|---|---|---|
| | | | Si/Al | K/Al | K/Si | $H_2O/Si$ |
| A | 1.7 | i | 1.69 | 1.23 | 0.73 | 3.36 |
| | | ii | 1.75 | 1.45 | 0.83 | 3.81 |
| i: 54.06% $SiO_2$, 43.64% $Al_2O_3$ | | | | | | |
| ii: 0.81 [54.1% $SiO_2$, 45.9% $Al_2O_3$=MK$_{ideal}$] | | | | | | |
| C1 | 1.75 | i | 1.55 | 1.02 | 0.66 | 3.91 |
| | | ii | 1.68 | 1.13 | 0.67 | 3.97 |
| C2 | 1.22 | i | 1.36 | 0.71 | 0.52 | 3.10 |
| | | ii | 1.47 | 0.78 | 0.53 | 3.15 |
| i: FKM (50.38% $SiO_2$, 45.85% $Al_2O_3$) | | | | | | |
| ii: 0.91 [MK$_{ideal}$] | | | | | | |
| D1 | 1.22 | i | 1.55 | 1.02 | 0.66 | 3.90 |
| | | ii | 1.66 | 1.10 | 0.66 | 3.97 |
| D2 | 1.00 | i | 1.44 | 0.84 | 0.58 | 3.44 |
| | | ii | 1.54 | 0.90 | 0.58 | 3.45 |
| i: 0.7 * FKM (50.38% $SiO_2$, 45.85% $Al_2O_3$) | | | | | | |
| ii: 0.65 * [MK$_{ideal}$] | | | | | | |

4% in both cases. For MKMph a quartz content of about 10-15% could be estimated from IR spectra. According to Rietveld phase analysis the crystalline contribution could be about 19% which could be separated as about 13.5% and 5.5% to be due to quartz and alumina, respectively. This implies that not more than about 81% of the MKMph sample could be activated. Corresponding Si/Al, K/Al, K/Si, $H_2O/Si$ ratios are given in Tab. 2 in comparison to values obtained using results of large area bulk EDX analysis (54.06 $SiO_2$, 43.64 $Al_2O_3$, 0.57 $K_2O$, 0.81 $Fe_2O_3$).

For determination of the nominal molar ratios of Si/Al, K/Al, K/Si, $H_2O$/Si ratios of the MKF and MKFqz related geopolymer (Tab. 2) results of thermogravimetric analysis of Fluka kaolinite could be used. A mass loss of 12.71% were obtained to be compared to an expected mass loss due to dehydration of ideal kaolinite $Al_2Si_2O_5(OH)_4$ of 13.9%. This implies that only 91% of the sample could be activated for geopolymerisation ignoring a possible weight loss related to muscovite and other phases which could be dehydrated. A cross check of the quality of TG data could be given considering the mass loss of about 8.98% obtained for the Fluka kaolinite with the addition of quartz, i.e. 29.3% loss in good agreement with the amount of quartz added.

Compressive strength data of alkali activated metakaolin, series M1-M7.

The compressive strength of a series of alkali activation of metakaolin of nominal compositions as given in Tab. 1 (M1,..,M7) aged for 7, 14, and 28 days show highest strength at Si/Al ratio of 1.85 (Fig. 2 a). These data were obtained during ageing shown in Fig. 2 b. There is

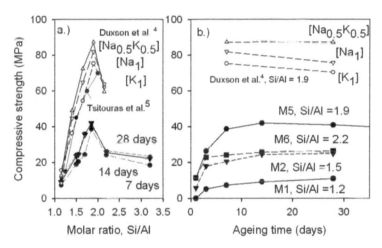

Fig. 2 Compressive strength development of geopolymer cements of series M as a function of a.) nominal molar ration Si/Al at 7, 14 and 28 days of ageing and b.) ageing time. Data given by Duxson et al. [4] and Tsitouras et al [5] are included.

a relative steep increase in strength within the first 7 days of ageing and than a rather flat behavior for further ageing. For the cement M5 with the highest strength a gradual weakening is observed, which becomes more pronounced with further ageing as will be discussed further below (compare Fig. 4). It is interesting to note that a compressive strength of about 8 MPA is also obtained for metakaolin activated just with KOH solution without the addition of waterglass. As mentioned in the introduction in the presence of waterglass quasi instantaneous formation of polysiloxo chains occur during the solution process of metakaolin. This effect could be seen using the "molybdate method". This method is able to distinct between the content of polymeric silicate units, and shorter ones and were used for series of samples where all the metakaolin could be consumed during binder formation [7-9]. Since a maximum in the concentration of polymeric silicate units occurs always at much shorter time compared to the time required for reaching maximal strength this condensation effect can be distinct from further network formation which just encloses the preformed chains. Notably during network formation the concentration of polymeric silicate units decreases leading finally to an even complete loss of strength initially gained. In the absence of waterglass this "waterglass effect" is absent. Conclusively it may be argued that using mixtures of alkali-hydroxide solution and metakaolin just the basic strength of an aluminosilicate network is attained.

Duxson et al. [4] observed a significant strength of alkali activated metakaolin without the addition of waterglass, too. For better comparison some data are also shown in Fig. 2 a for 7 days aged samples for Si/Al = 1.2. Values for higher Si/Al ratios are also given which are obtained using appropriate alkali-waterglass solutions. Related on the Si/Al ratio the compressive strength values are by a factor of 2 higher in all cases, showing highest values at about the same nominal Si/Al ratio (1.9) as obtained within the series of composition M1-M7. This trend is also not altered using KOH/NaOH mixed alkali activators, as shown for compositions $K_{1-x}Na_xOH$, x = 0, 0.5, 1. As shown in Fig. 2 b Duxson et al. [4] also observed slight weakening of compressive strength for samples with nominal Si/Al = 1.9 prepared with NaOH and KOH containing activators. For activators including NaOH/KOH mixtures a different behavior were related to significant influences related to the alkaline ionic mobilities for geopolymerisation, too.

The by a factor of two higher specific values obtained by Duxson et al. [4] could first of all be related to further differences in compositions, alterations during synthesis and ageing, and testing conditions, e.g. 40°C curing for 20 h. Duxson et al. [4] prepared the alkali activator so as to reveal a nominal Na/Al ratio = 1 in the cement, whereas our compositions M2-M7 were just varied by different mass ratios of the waterglass and metakaolin, which systematically increases the Si/Al and Na/Al ratios. The effect of different chemical compositions of Si/Al and alkali/Al ratios could be visualized replotting our data in direct comparison to a contour-plot developed by Rowles and O'Connor [1]. In their systematic study on the influence of nominal Na/Al ratios between 0.5 and 2.0 and Si/Al ratios between 1.00 and 3.00 in the starting mixtures on the compressive strength these authors obtained contours for 15, 30, 45 and 60 MPa as shown in Fig. 3. A maximum in strength of about 64 MPa is observed at Na/Al of 1.25 and Si/Al of 2.5. Following however the line of constant Na/Al = 1 shows that a maximum in compressive strength would appear at Si/Al = 2. This trend could well be supported by the data obtained by Duxson et al. [4] and also by Tsitouras et al. [5] as mentioned in the introduction, too. The new data for M2 to M6 are also not in contradiction to the contours given by Rowles and O'Connor [1]. Here the alkali/Al ratio steeply increases related on Si/Al ratio. It can be seen that the highest value for M5 of about 43 MPA plots close to the contour for 45 MPa, M6 well to the contour for 30 MPa and M4 to M3 to M2 also closely follows the decrease in strength extracted by Rowles and O'Connor [1]. It could be interesting to note that the specific values seems to be closely related. This turns out to be surprising since Rowles and O'Connor [1] cured their samples at 75°C for 24 hours, whereas at the lower curing temperature of 40°C by a factor of about 2 higher values are obtained in the work of Duxson et al. (compare Fig. 2 a, 3). This could imply that some intermediate temperature between room temperature and 70°C could improve the compressive strength. This suggestion is not supported by results of Tsitouras et al. [5] who cured their cements at 70°C obtaining strength values as given by Duxson et al. (Fig. 2). A conclusive explanation about the difference in absolute values remains open. However, it can be seen that irrespectively of differences in absolute values and using either Na or K in the activator the trends in compressive strength related on Si/Al and alkali/Al ratios well support the contour plot given by Rowles and O'Connor [1]. We may presently also not rule out that depending on the alkali/Al and Si/Al ratios two or more maxima in compressive strength could exist. Further systematic study seems to be required to explore the whole field of compositions in closer steps. This concerns also the effect of variations in alkali compositions as investigated by Duxson et al. [4].

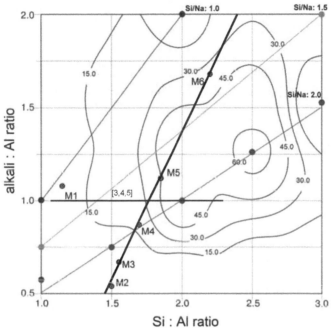

Fig. 3 Contour plot of compressive strength data in the field of Si/Al and Na/Al ratio taken from Rowles and O'Connor 1. Included are points of composition M1-M6 from this work. Thick solid line marks the systematic variation in composition obtained by KWG/MKM ratios (Tab. 1, for i). Thinner solid line denoted [3, 4, 5] corresponds to series of nominal constant Na/Al ratio worked on by Duxson et al. [2, 3] and Tsitouras et al. [4].

Using scanning electron microscopy and energy dispersive analysis (SEM/EDX) Rowles and O'Connor [2] obtained a separation into two phases concerning the Si/Al and Na/Al ratios in all cases of their cements. In particular regions of higher and lower Si/Al ratios with respect to the nominal Si/Al content was observed on a micrometer scale. Those regions with lower Si/Al were identified as "grains" relating it to metakaolin in its origin, being still strongly altered in compositions. The matrix possess always higher Si/Al ratios. These observations could well support the mechanism described above concerning the initial condensation of waterglass, followed by a transport of $Al^{3+}$ ions from the metakaolin to the waterglass condensates. In MAS NMR investigations Rowles et al. [11] also observed residual unreacted metakaolin for series of composition with Si:Al/Na:Al ratios 1.1/0.6, 1.5/0.8, 2.0/1.0, 2.5/1.3 and 3.0/1.5 as 42%, 28 %, 18%, 0%, 25%, respectively. A similar effect of decreasing contents of unreacted metakaolin is also indicated along the series of samples M2 to M5 considering earlier XRD and IR investigations [8, 9]. This could also imply that the amount of unreacted metakaolin could not be influenced significantly by temperature, i.e. 75°C compared to 25°C. Instead the constitution of the initial waterglass (degree of condensation, water and hydroxide content) and the waterglass to metakaolin ratio sensitively govern the reaction volume whereas the temperature strongly influences the reaction kinetics [8].

The effect of quartz in the binder
        Significant differences in the development of compressive strength of the geo-cement M5, the geo-mortar (M5 plus sand) and ordinary Portland cement (OPC) and its mortar counterpart are observed during ageing (Fig. 4). OPC shows significantly higher strength compared to the OPC-Mortar, i.e. for the binder diluted by sand. Contrary to this the geo-mortar

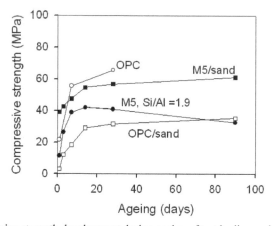

Figure 4. Compressive strength development during ageing of test bodies as denoted (see text).

shows higher strength compared to the geo-cement alone. Additionally the compressive strength of the geo-mortar increases in strength during ageing whereas the geo-cement becomes weaker. Thus it seems likely that an interaction between the geo-cement and the sand improves the compressive strength development of the composite. This implication could be supported in some more detail following the compressive strength development of geopolymer cements series A and B (Fig. 5) and C1, C2 and D1, D2 (Fig. 6). The compressive strength values of all these samples range between 50 and 100 MPa, i.e. on a significant higher level as obtained for series M1 - M7 (Fig. 2). This could be due to the much smaller aspect ratio for series A-D (0.5) compared to samples M1-M7 (1) and to the manually conducted measurement technique. However, this should not influence the ageing behavior. Series A and B almost show the same absolute values and ageing behavior. A steep increase is observed up to 8 days of ageing followed by no change anymore up to 40 days. This implies that it makes no difference either to have an amount of around 13% or 63 % of quartz in the binder. However, a sound of cracking could be noted before the breakdown strength was reached for the series of sample B (denoted as

B* in Fig. 5). The difference between absolute breaking and the sound of cracking becomes smaller with ageing time. A different development in compressive strength could be observed for series of samples C1, C2 and D1, D2. For C1 and C2 a flat increase in strength is seen for ageing between 3 and 7 days followed by a more or less smooth weakening. Contrary to this a steep increase in strength up to 7 days of ageing is observed followed by a further significant increase in strength for series D1 and D2.

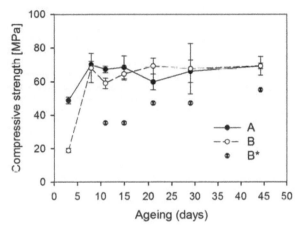

Figure 5: Compressive strength development during ageing of test series A, B. B* denotes sound of cracking noted for series B samples. (see text).

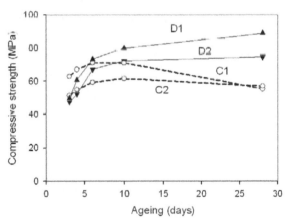

Figure 6: Compressive strength development during ageing of test series D (FKMqz, with quartz addition) and C (FKM, free of quartz).

SUMMARY AND CONCLUSIONS

It could be shown that the determination and quantification of impurity phases, if present in the percentage range in the metakaolin, is required to correctly calculate nominal composition ratio of the geo-cement, which cannot be obtained using bulk chemical analysis (XRF, EDX, wet chemical analysis). However, ignoring this the trend in development of maximal compressive strength obtained along variations of alkali-waterglass/metakaolin mass ratios could well support the contour plot in the field of Si/Al versus Na/Al ratio obtained by Rowles and O'Connor [1]. Support for this could also be given including the compressive strength development related on Si/Al ratio obtained by Duxson et al. [4] and Tsitouras et al. [5]. The agreement in optimal strength at Si/Al at about 1.8-2 within each series of investigation in [4,5] can, however, only be seen considering indeed the whole field of nominal compositions as explored by Rowles and O'Connor [1] (Fig. 3). The maximum at Si/Al = 1.9 indicates only shoulder or weaker peak towards the global maximum. It may be pointed out here that "a basis strength of geo-cement" could be obtained in a network consisting of sialate bonds as formed without the addition of waterglass in the activator. The addition of waterglass invariably leads to formation of polysiloxo chains during the solution of metakaolin. The presence and support of these chains is required to obtain significant increase in mechanical strength. Conclusively any shortening – fragmentation - of these chains must lead to mechanical weakening of the geo-binder.

Following this line of arguments the addition of quartz may well support the existence of polysiloxo chains in the geo-cement. The results for series C and D and also for A and B could well confirm the finding in Fig. 4. In addition with the high strength contribution of quartz itself, the improved strength development of the geo-mortar with respect to the geo-cement can be related to a significant improvement in the condensation of the sialate network and a protection of polysiloxo chains, too.

ACKNOWLEDGMENTS

MG gratefully acknowledges the German Science Foundation "Deutsche Forschungsgemeinschaft" (DFG) for financially support of his research visits in Hannover. AS thanks the "Graduierten Akademie, LUH" for financial support.

**References**

[1] M. Rowles and B. O'Connor, Chemical optimization of the compressive strength of aluminosilicate geopolymers synthesized by sodium silicate activation of metakaolinite, J. Mater. Chem., 13, 1161-1165 (2003).

[2] M. Rowles and B. O'Connor, Chemical and Structural Microanalysis of Aluminosilicate Geopolymers Synthesized by Sodium Silicate Activation of Metakaolinite, J. Am. Ceram. Soc., 92, 2354-2361 (2009).

[3] P. Duxson, J.L. Provis, G.C. Lukey, S.W. Mallicoat, W.M. Kriven, J.S.J. van Deventer, Understanding the relationship between geopolymer composition, microstructure and mechanical properties, Colloids and Surfaces A: Physicochem. Eng. Aspects 269, 47-58 (2005).

[4] P. Duxson, S.W. Mallicoat, G.C. Lukey, W.M. Kriven, J.S.J. van Deventer, The effect of alkali and Si/Al ratio on the development of mechanical properties of metakaolin-based geopolymers, Colloids and Surfaces A: Physicochem. Eng. Aspects 292, 8-20 (2007).

[5] A. Tsitouras, T. Perraki, M. Perraki, S. Tsivilis, G. Kakali, The Effect of Synthesis Parameters on the Structure and Properties of Metakaolin Based Geopolymers, Mat. Sc. F. Vols. 636-637 (2010) pp 149-154.

[6] Lloyd, R.R. (2009): Accelerated ageing of geopolymers. In "Geopolymers-Structure, processing, properties and industrial applications", Provis, J. L., & van Deventer, J.S.J., eds. Woodhead Publishing Limited, Oxford, 139-166.

[7] Rüscher, C.H., Mielcarek, E., Lutz, W. Ritzmann, A., Kriven, W.M. (2010a): Weakening of alkali activated metakaolin during ageing investigated by molybdate method and infrared absorption spectroscopy. J. Am. Ceram. Soc., 93, 2585-2590.

[8] Rüscher, C.H., Mielcarek, E., Lutz, W. Ritzmann, A., Kriven, W.M. (2010b): The ageing process of alkali activated metakaolin. Ceram. Trans., 215, 1-10.

[9] Rüscher, C.H., Mielcarek, E., Lutz, W., Jirasit, F., Wongpa, J. (2010c): New insights on geopolymerisation using molybdate, Raman, and infrared spectroscopy. 34th international conference on advanced ceramics and composites, Daytona Beach, Florida, 24th-29th, January 2010, invited lecture, Ceramic Engineering and Science Proceedings, 31,19-35.

[10] Claus H. Rüscher, Elzbieta M. Mielcarek, Jakrapan Wongpa, Chai Jaturapitakkul, Fongjan Jirasit and Ludger Lohaus, Silicate-,aluminosilicate and calciumsilicate gels for building materials: chemical and mechanical properties during ageing, Eur. J. Mineral. 23, 111-124 (2011).

[11] M. R. Rowles, J. V. Hanna, K.J. Pike, M.E. Smith, B.H. O'Connor, $^{29}$Si, $^{27}$Al, $^{1}$H and $^{23}$Na MAS NMR Study oft he Bonding Character in Aluminosilicate Inorganic Polymers, Appl. Magn. Reson., 32, 663-689 (2007).

# THE ROLE OF $SiO_2$ & $Al_2O_3$ ON THE PROPERTIES OF GEOPOLYMERS WITH AND WITHOUT CALCIUM

P. De Silva[1], S. Hanjitsuwan[2], P. Chindaprasirt[2]
[1]School of Arts and Sciences, Australian Catholic University, North Sydney, Australia
[2]Department of Civil Engineering, Khon Kaen University, Thailand

ABSTRACT
    The rapid setting properties of high calcium fly ash based geopolymers (HCFG) have restricted potential application of these binders compared to those derived from bituminous coal or metakaolin. In calcium-free systems, initial $SiO_2$ and $Al_2O_3$ contents are known to play a key role in controlling setting and final properties of these binders. This study investigates the effect of $SiO_2$ and $Al_2O_3$ contents on setting and hardening of HCFG and the outcomes are compared with calcium free metakaolin-based systems (MG).
    A series of HCFG mixes were formulated with varying amounts of $SiO_2$ and $Al_2O_3$ to achieve $SiO_2/Al_2O_3$ in the range 2.7 –4.8. It has been shown that the effect of varying $SiO_2$ and $Al_2O_3$ in HCFG on setting and hardening is markedly different from that observed for MG systems. Increases in either $SiO_2$ or $Al_2O_3$ content appear to shorten the setting time of calcium-based systems unlike MG where increasing $Al_2O_3$ accelerates setting. When calcium is present, the setting process was associated primarily with CSH or CASH formation and there appears to be a prevailing $SiO_2/Al_2O_3$ ratio that prolongs setting. Furthermore, in HCFG, $SiO_2/Al_2O_3$ ratios in the range of 3.20–3.70 resulted in products with highest strengths and longest setting times. The results suggest that the initial predominance of $Ca^{2+}$ and its reactions effectively help maintaining a $SiO_2/Al_2O_3$ ratio at which amorphous geopolymer phase is stable.

INTRODUCTION
    Geopolymers are a product of alkali activated alumina-silica based materials, hence a system of $Na_2O$ (or $K_2O$)-$SiO_2$-$Al_2O_3$-$H_2O$ (ordinary geopolymers). A great deal of information, including chemistry of formation and chemical & physical properties of these systems have been published elsewhere[1-4]. The formation of geopolymers occurs via a series of dissolution and precipitation reactions. Condensation between dissolved silicate and aluminate species result in the 3-D, strong polymeric network of geopolymers. The $SiO_2/Al_2O_3$ of ratio is an important parameter in geopolymer synthesis as the dissolution rates of alumina and silica of raw materials and rate of condensation between aluminate and silicate species in essence control the setting, hardening and strength development of geopolymer systems[5]. In general, increases in $Al_2O_3$ tend to accelerate the setting of ordinary geopolymers while addition of $SiO_2$ inhibits the setting[6]. Nevertheless, increase in $SiO_2$ content tends to produce microstructures with low porosity and hence enhance the strength of these systems[6,7].
    The most common aluminosilicate source materials investigated in geopolymer synthesis are metakaloin (calcined kaolinite) and fly ash (a by-product of coal combustion process)[3,8-10]. While the major components of metakaolin and fly ash are $Al_2O_3$ and $SiO_2$, fly ash compositions can be varied depending on the composition of original coal. In addition to $SiO_2$ and $Al_2O_3$, Class C fly ash derived from lignite coal sources contains considerable amounts CaO (>20 wt%). Calcium oxide gives fly ash pozzolanic properties and calcium containing cement systems generally have relatively shorter setting times. Therefore, when calcium is present in a $Na_2O$ (or $K_2O$)-$SiO_2$-$Al_2O_3$-$H_2O$ system, some changes in the chemistry, especially with respect to setting and hardening processes, are to be expected.

The main problem when Class C fly ash is used in geopolymer synthesis is its rapid setting property and hence low workability. In general, at room temperature, final setting can be achieved within 1-2 hrs. The setting is believed to be due to the rapid formation of CSH phase[11-13,] similar to that obtained during hydration of Portland cement. Some studies also reported the presence of zeolitic phases such as gismondine (calcium alumino silicate zeolite)[14]. Ettringite was also identified as a hydration product when a mixture of Class C fly ash and Portland cement was subjected to alkaline activation[15]. Furthermore, the addition of extra water has been also reported to improve workability of Class C based geopolymer systems (high calcium fly ash based geopolymers – HCFG) more than conventional superplasticizers[15].

It has been shown that calcium sources can be used to accelerate the setting of low calcium fly ash-based geopolymers[16] and metakaolin based geopolymer systems (MG)[17,18]. The coexistence of CSH phase with the geopolymeric gel has been shown to improve the mechanical properties of the final product[16]. The role of calcium sources such as Ground granulated blast furnace slag (GGBFS), cement, wollastonite (CaSiO$_3$) and Ca(OH)$_2$ on mechanical properties of MG is reported elsewhere[17]. Specifically, the effects of different calcium silicate sources on geopolymerisation seems to be highly dependent on the crystallinity and thermal history of the calcium silicate sources, as well as the alkalinity of the alkaline activator.

This study examines the effects of changes in SiO$_2$/Al$_2$O$_3$ ratio on the setting, microstructure, phase development and physical properties of high calcium fly ash based geopolymer systems and results are compared with those of metakaolin based systems.

MATERIALS AND METHODS

The class C fly ash used in this study had the chemical composition of SiO$_2$- 35.21% , Al$_2$O$_3$ – 16.57%, CaO – 25.52%, Fe$_2$O$_3$ – 13.66%, MgO – 3.28%, Na$_2$O – 2.73% and K$_2$O – 1.99%. The nano-silica, rice husk ash (RHA) and nano-alumina were used as additional SiO$_2$ and Al$_2$O$_3$ source, respectively. Sodium silicate with 28.70% SiO$_2$, 8.90% Na$_2$O, and 62.50% H$_2$O by weight and NaOH were used as the alkaline activators.

The mix formulation having SiO$_2$/Al$_2$O$_3$ = 4, the recommended molar ratio[7] for ordinary geopolymers, Na$_2$O.1.17Al$_2$O$_3$.4.80SiO$_2$.11.45H$_2$O was used as the control mix. The control mix was prepared using fly ash, sodium silicate and NaOH in required quantities. The other mixtures were prepared by keeping the fly ash and water content constant but changing the silica and alumina contents, by adding nano-silica, RHA, or nano-alumina as shown in Table 1,. The molar quantities of silica and alumina in the formulations were in the range of 4.80-5.60 and 1.17-1.87, respectively. The final mix formulations had the SiO$_2$/Al$_2$O$_3$, CaO/SiO$_2$ and CaO/Al$_2$O$_3$ ratios in the range 2.57 – 4.79, 0.59-0.69 and 1.76-2.81 respectively. The details of the mixing procedure are given elsewhere[19].

Alkali activated fly ash pastes were cast into 2.5×2.5×2.5 cm$^3$ cube moulds and cured at 60°C, 95% RH for 24 hrs. The X-ray diffraction (XRD), scanning electron microscopy (SEM) with energy dispersive X-ray analysis (EDXA) and compressive strength analyses were studied on the hardened geopolymer paste. A separate set of mixtures of the same mix formulation were prepared for setting time measurements. Setting time trials were performed at room temperature (22°C), using standard Vicat needle apparatus. XRD and SEM analyses were also conducted on the samples at the final setting time.

Table 1 - Mixt formulations of HCFG pastes

| Added SiO$_2$ or Al$_2$O$_3$ (moles) | Molar composition | SiO$_2$/ Al$_2$O$_3$ | Na$_2$O/ SiO$_2$ | Na$_2$O/ Al$_2$O$_3$ |
|---|---|---|---|---|
| Control mix | Na$_2$O.1.17Al$_2$O$_3$.4.80 SiO$_2$.11.45H$_2$O | 4.00 | 0.21 | 0.85 |
| *SiO$_2$ additions:* | | | | |
| 0.14SiO$_2$ (nano-SiO$_2$, RHA) | Na$_2$O.1.17Al$_2$O$_3$.4.94 SiO$_2$.11.45H$_2$O | 4.22 | 0.20 | 0.85 |
| 0.18SiO$_2$ (nano-SiO$_2$, RHA)) | Na$_2$O.1.17Al$_2$O$_3$.4.98 SiO$_2$.11.45H$_2$O | 4.26 | 0.20 | 0.85 |
| 0.25SiO$_2$ (nano-SiO$_2$, RHA) | Na$_2$O.1.17Al$_2$O$_3$.5.05 SiO$_2$.11.45H$_2$O | 4.32 | 0.20 | 0.85 |
| 0.50 SiO$_2$ (RHA) | Na$_2$O.1.17Al$_2$O$_3$.5.30 SiO$_2$.11.45H$_2$O | 4.53 | 0.19 | 0.85 |
| 0.80 SiO$_2$ (RHA) | Na$_2$O.1.17Al$_2$O$_3$.5.60 SiO$_2$.11.45H$_2$O | 4.79 | 0.18 | 0.85 |
| *Al$_2$O$_3$ additions:* | | | | |
| 0.20 (nano-Al$_2$O$_3$ ) | Na$_2$O.1.37Al$_2$O$_3$.4.80 SiO$_2$.11.45H$_2$O | 3.50 | 0.21 | 0.78 |
| 0.30 (nano-Al$_2$O$_3$ ) | Na$_2$O.1.47Al$_2$O$_3$.4.80 SiO$_2$.11.45H$_2$O | 3.27 | 0.21 | 0.68 |
| 0.33 (nano-Al$_2$O$_3$ ) | Na$_2$O.1.50Al$_2$O$_3$.4.80 SiO$_2$.11.45H$_2$O | 3.20 | 0.21 | 0.67 |
| 0.50 (nano-Al$_2$O$_3$ ) | Na$_2$O.1.67Al$_2$O$_3$.4.80 SiO$_2$.11.45H$_2$O | 2.87 | 0.21 | 0.60 |
| 0.70 (nano-Al$_2$O$_3$ ) | Na$_2$O.1.87Al$_2$O$_3$.4.80 SiO$_2$.11.45H$_2$O | 2.57 | 0.21 | 0.53 |

RESULTS

Setting Time
The initial and final setting times of HCFG mix formulations are shown in Table 2.

Table 2 - Setting time of mix formulations at 22°C

| Added SiO$_2$ or Al$_2$O$_3$ amount (moles) | Initial composition | Initial set (min) | Final set (min) |
|---|---|---|---|
| Control | Na$_2$O. 1.17Al$_2$O$_3$. 4.80 SiO$_2$. 11.45H$_2$O | 60 | 100 |
| *SiO$_2$ added samples:* | | | |
| 0.14SiO$_2$ (nano-SiO2)) | Na$_2$O. 1.17Al$_2$O$_3$. 4.94 SiO$_2$. 11.45H$_2$O | 38 | 70 |
| 0.14SiO$_2$ (RHA) | Na$_2$O. 1.17Al$_2$O$_3$. 4.94 SiO$_2$. 11.45H$_2$O | 47 | 78 |
| 0.50 SiO$_2$ (RHA) | Na$_2$O. 1.17Al$_2$O$_3$. 5.30 SiO$_2$. 11.45H$_2$O | 16 | 33 |
| 0.80 SiO$_2$ (RHA) | Na$_2$O. 1.17Al$_2$O$_3$. 5.60 SiO$_2$. 11.45H$_2$O | 9 | 21 |
| *Al$_2$O$_3$ added samples:* | | | |
| 0.20 Al$_2$O$_3$ | Na$_2$O. 1.37Al$_2$O$_3$. 4.80 SiO$_2$. 11.45H$_2$O | 55 | 90 |
| 0.30 Al$_2$O$_3$ | Na$_2$O. 1.47Al$_2$O$_3$. 4.80 SiO$_2$. 11.45H$_2$O | 65 | 91 |
| 0.33 Al$_2$O$_3$ | Na$_2$O. 1.50Al$_2$O$_3$. 4.80 SiO$_2$. 11.45H$_2$O | 40 | 65 |
| 0.50 Al$_2$O$_3$ | Na$_2$O. 1.67Al$_2$O$_3$. 4.80 SiO$_2$. 11.45H$_2$O | 35 | 60 |
| 0.70 Al$_2$O$_3$ | Na$_2$O. 1.87Al$_2$O$_3$. 4.80 SiO$_2$. 11.45H$_2$O | 9 | 20 |

The final setting time of the control mix ($SiO_2/Al_2O_3$ = 4.00) was 100 min and increase of $SiO_2$ content of the original mix accelerated both initial and final setting, irrespective of the source of silica. The trend is also very similar for both silica sources. At the same time, increasing $Al_2O_3$ content of the original mix also led to a decrease in both initial and final setting times of mixes. The fastest setting of less than 20 min recorded for the mix with 0.7 $Al_2O_3$ addition ($SiO_2/Al_2O_3$ = 2.57).

Compressive Strength

The compressive strength results of paste after curing at 60°C for 24 hrs are shown in Table 3. The control sample with $SiO_2/Al_2O_3$ = 4.00 had strength of 62.6 MPa. Compared to the control, 0.14 moles of $SiO_2$ addition ($SiO_2/Al_2O_3$ = 4.22) shows a decrease in strength with both types of silica. However, the effect of decreasing strength is also higher with increasing $SiO_2$ content; increase in total $SiO_2$ content from 4.8 to 5.8 ($SiO_2/Al_2O_3$ from 4.00 to 4.79) decreased the strength by about 50% (from 62.6 to 30.9 MPa) and the trend is similar with the both sources of silica. Increasing $Al_2O_3$ content did not show any significant effect on the compressive strength, both the control ($SiO_2/Al_2O_3$ = 4.00) and 0.50 $Al_2O_3$ addition ($SiO_2/Al_2O_3$ = 2.87) had comparable strengths, around 63 MPa.

Table 3. Compressive strength of mix formulations

| Added SiO₂ or Al₂O₃ amount (moles) | Molar composition | Compressive strength (MPa) |
|---|---|---|
| Control mix | Na₂O. 1.17Al₂O₃. 4.80 SiO₂. 11.45H₂O | 62.6 |
| *SiO₂ added samples:* | | |
| 0.14 SiO₂ (nano-SiO₂) | Na₂O. 1.17Al₂O₃. 4.94 SiO₂. 11.45H₂O | 60.4 |
| 0.14 SiO₂ (RHA) | Na₂O. 1.17Al₂O₃. 4.94 SiO₂. 11.45H₂O | 58.3 |
| 0.50 SiO₂ (RHA) | Na₂O. 1.17Al₂O₃. 5.30 SiO₂. 11.45H₂O | 34.6 |
| 0.80 SiO₂ (RHA) | Na₂O. 1 17Al₂O₃. 5.60 SiO₂. 11.45H₂O | 30.9 |
| *Al₂O₃ added samples:* | | |
| 0.20 Al₂O₃ | Na₂O. 1.37Al₂O₃. 4.80 SiO₂. 11.45H₂O | 64.0 |
| 0.30 Al₂O₃ | Na₂O. 1.47Al₂O₃. 4.80 SiO₂. 11.45H₂O | 60.0 |
| 0.50 Al₂O₃ | Na₂O. 1.67Al₂O₃. 4.80 SiO₂. 11.45H₂O | 63.2 |

A comparison of Setting time and Strength development of HCFG and MG systems

Combined final setting times and compressive strength with respect to $SiO_2/Al_2O_3$ molar ratio of metakaolin based geopolymer systems[6] and the HCFG mixes in the present study and are shown in Fig. 1a and 1b respectively. In MG systems (Fig 1a), there is a somewhat linear relationship between $SiO_2/Al_2O_3$ ratio and setting time[6]. Increasing ratio also increases the strength but only up to about $SiO_2/Al_2O_3$ = 4 and strength decreases thereafter. $SiO_2/Al_2O_3$ ratios around 3.50-4.00 are regarded as the most suitable for optimal strength characteristics.

In HCFG systems (Fig 1b) setting time characteristics and strength development does not follow the same trend as MG systems. In HCFG systems increasing both $SiO_2$ and $Al_2O_3$ tend to accelerate setting in HCFG systems and there exists a maximum $SiO_2/Al_2O_3$ ratio, at which the system display longest setting times. The most favourable $SiO_2/Al_2O_3$ ratio for setting falls within the range of 3.20-3.70. Decreasing $SiO_2/Al_2O_3$ ratio (or increasing $Al_2O_3$) favours higher strengths reaching a maximum at around $SiO_2/Al_2O_3$ ratio 3.50 and remains constant thereafter. It was particularly interesting to note that in HCFG systems, unlike MG systems, increasing Al$_2$O$_3$ content did not show any significant effect on the compressive strength, both SiO$_2$/Al$_2$O$_3$ = 4.00 and SiO$_2$/Al$_2$O$_3$ = 2.87 had comparable strengths. These results also show that in HCFG systems an optimum of SiO$_2$/Al$_2$O$_3$ ratio (3.20-3.70) leads to products with longest setting time with reasonably high strengths.

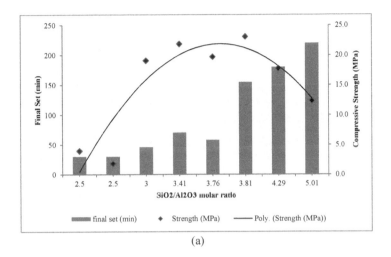

(a)

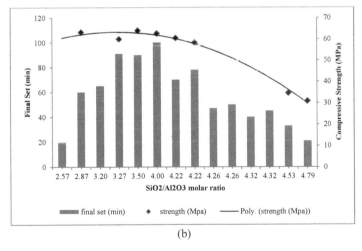

(b)

Figure 1. Final setting times and compressive strength with respect to SiO$_2$/Al$_2$O$_3$ ratio (a) MG system (b) HCFG system

Solid phase development of HCFG systems by XRD

The XRD patterns of HCFG systems of control, high SiO₂ addition ($SiO_2/Al_2O_3$ = 4.79) and high Al₂O₃ addition ($SiO_2/Al_2O_3$ = 2.87) at the final setting time (22°C) and after 24 hrs curing at 60°C are shown in Fig. 2 (a-d) and Fig. 3 (a-d) respectively[19]. The XRD pattern of raw fly ash (Fig 2a) comprises mainly of amorphous phase with traces of crystalline anhydrite ($CaSO_4$), quartz ($SiO_2$), magnesioferrite ($MgFe_2O_4$), calcium oxide (CaO) and hematite ($Fe_2O_3$). In control sample (Fig. 2b), the main changes occurring during alkali activation are the disappearance of $CaSO_4$ and CaO (originally present in fly ash) suggesting initial involvement of $Ca^{+2}$ ions in the setting process. Trace amounts of $Na_2SO_4$ and $Ca(OH)_2$ are also detectable at this stage. The XRD patterns of SiO₂ rich sample (Fig. 2c) showed the presence of $Na_2SO_4$ only, with no sign of $Ca(OH)_2$. This suggests greater consumption of of $Ca^{+2}$ ions in the setting process when extra silica is present in the system. As with the control sample, the XRD pattern of alumina-added sample also showed the formation of new phases $Na_2SO_4$ and $Ca(OH)_2$. There is no unreacted Al₂O₃ present in the alumina-added sample (Fig 2d) indicating involvement of the added alumina in the reactions during the setting period.

It has been reported[20] that in calcium containing geopolymer systems, the possible phases expected to form at the setting time are calcium silicate hydrate (CSH), calcium aluminate silicate hydrate (CASH) and sodium aluminate silicate hydrate (NASH, geopolymer gel phase). The presence of CSH (and or CASH), and NASH gel phases in the XRD patterns are usually associated with a broad peak around 28-30° (2θ) for the geopolymer phase and main peak at 30° (2θ) for CSH, CASH phases[18,20-22]. There is a weak indication of the presence of humps around 28-30°(2θ) in the XRD patterns of all 3 samples studied though this evidence is not firm and would require further investigation. The amorphous nature of the gel phases and the interference between peaks of other phases makes it difficult to fully identify trace reaction products. However, the involvement of $Ca^{+2}$ ions in the setting time process can be corroborated by the disappearance of calcium containing phases in the fly ash and conversion into hydration products.

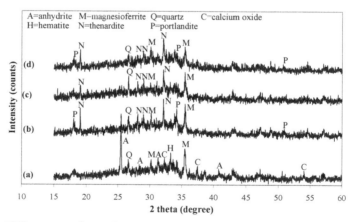

Figure 2. XRD patterns of geopolymer pastes with different additives at final setting at 22°C[19]
(a) fly ash (b) control (c) $SiO_2/Al_2O_3$ = 4.79 (d) $SiO_2/Al_2O_3$ = 2.87

The XRD patterns of HCFG systems of control, high SiO₂ addition ($SiO_2/Al_2O_3$ = 4.79) and high Al₂O₃ addition ($SiO_2/Al_2O_3$ = 2.87) at 60°C are shown in Fig. 3 (a-d). In all 3 samples examined the formation of NASH geopolymer phase is strongly evident by the presence of the broad peak around 28° (2θ) and presence of CSH phase can be confirmed by the presence of peaks at 29.35° and 32.05° (2θ). In addition, the control and Al$_2$O$_3$ added samples (Fig. 3b & 3d) also contain traces of unreacted Ca(OH)$_2$. However, there is no indication of presence of Ca(OH)$_2$ in SiO$_2$ added sample (Fig. 3c). This further supports the argument that more Ca$^{+2}$ ions are consumed during setting and hardening process of high calcium based systems when extra silica is present in the system, possibly via reacting with silica. There is no presence of Na$_2$SO$_4$ which had been previously detected at the setting time analysis of these samples (Fig. 2 b-d) which indicates the involvement of SO$_4^{-2}$ ions in the reaction process during curing stage. As there was no presence of ettringite-like phases detected in these samples, it is conceivable that SO$_4^{-2}$ ions could also be involved in the geopolymer reaction. Possible incorporation of SO$_4^{-2}$ ions in the geopolymer reaction and hence in the polymeric network structure has also been suggested in previous studies[23].

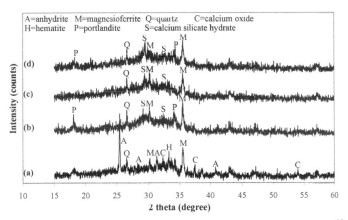

Figure 3. XRD patterns of HCFG pastes with different additives after curing at 60°C[19] (a) fly ash (b) control (c) SiO$_2$/Al$_2$O$_3$ = 4.79 (d) SiO$_2$/Al$_2$O$_3$ – 2.87

It is important to note that usually in calcium free MG systems, when SiO$_2$/Al$_2$O$_3$ < 3 the presence of zeolitic phases is a common feature in XRD patterns[7]. But this was not observed with the present HCFG systems.

SEM/EDAX analysis of HCFG pastes cured at 60°C

Figures 4a, 4b and 4c respectively show the SEM images of the control sample, and samples high in SiO$_2$ (SiO$_2$/Al$_2$O$_3$ = 4.79) and high in Al$_2$O$_3$ (SiO$_2$/Al$_2$O$_3$ = 2.87). Many unreacted fly ash particles embedded in a continuous matrix is a dominant feature in the control mix (Fig. 4a). It also contains white globules of CSH or CASH (as analyzed by EDXA) in the matrix. CSH appears to be more abundant in the silica rich sample (Fig. 4b) and this sample also consists of more developed cracks. This perhaps is responsible for the low strengths observed with this sample. Fig. 4d shows a higher magnification of a fly ash particle surface with CSH formations. The microstructure of samples containing excess alumina (Fig. 4c) appears to be somewhat different from that of the excess silica sample. It appears to contain white globules of CSH or CASH concentrated more in the matrix rather than on fly ash surfaces. The matrix also looks somewhat homogeneous and continuous with embedded fly ash particles and has minimal cracks. This sample produced higher strengths.

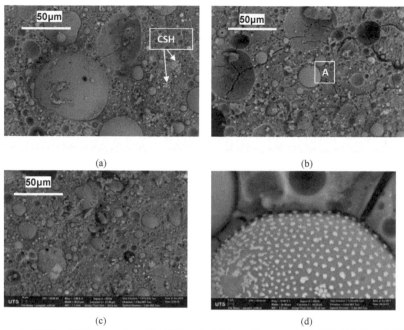

(a)                                                    (b)

(c)                                                    (d)

Figure 4. SEM of HCFG pastes with different additives (a) control (b) SiO$_2$/Al$_2$O$_3$ = 4.79 (c) SiO$_2$/Al$_2$O$_3$ = 2.87 (d) CSH in enlarged area A

It is understood that EDXA provides semi-qualitative analysis of compositions and not a reliable indicator of exact compositions but instead represents elemental distribution of the matrix. However, the EDXA spot analyses on the dense matrix area of HCFG samples have been carried out and it indicated the presence of considerable amount of (6-12%) SO$_4^{-2}$ in the matrix in conjunction with mainly Na, Al and Si. This further confirms that CaSO$_4$, originally present in the fly ash, dissolves during reaction and SO$_4^{-2}$ becomes a part of the matrix and is present in conjunction with other elements in the matrix. The %CaO was also fairly high and consistent (around 20%) in the matrix in all mix combinations confirming the presence of Ca$^{+2}$ as other phases, such as CSH and CASH.

DISCUSSION

For metakaolin based geopolymer systems it has been shown that the setting and hardening is mainly associated with the sodium aluminate silicate hydrate gel (NASH) formation[5,6]. The dissolution and condensation reactions occur in a metakaolin based geopolymer system can be shown as follows:

i. Dissolution of SiO$_2$ and Al$_2$O$_3$ sources:

$$Al_2O_3 + 3H_2O + 2OH^{-1} \rightarrow 2Al(OH)_4^{-1}$$
$$SiO_2 + H_2O + OH^{-1} \rightarrow SiO(OH)_3^{-1}$$
$$SiO_2 + 2OH^{-1} \rightarrow SiO_2(OH)_2^{2-}$$

ii.     Condensation of silicate and aluminate species:

$Na^+ + SiO_2(OH)_2^{2-}$ or $SiO(OH)_3^{-1} + Al(OH)_4^{-1}$     $\rightarrow$     NASH gel

When the system is high in SiO$_2$, more silicate species are available for condensation and reaction between silicate species, resulting oligomeric silicates, becomes dominant. And further condensation between oligomeric silicates and aluminates result in a rigid 3D net works[5]. The rate of condensation between silicate species is slow and, this normally leads to longer setting times. Correspondingly, when the system has excess Al$_2$O$_3$, increased aluminate (Al(OH)$_4^-$) species become available for the reaction leading to a faster rate of condensation between Al(OH)$_4^-$ and silicate species and shorter setting times[5].

This results presented shows that the effect of changing silica and alumina contents of HCFG systems different from that of MG systems in many ways. The key difference in high calcium based systems is the availability of Ca$^{+2}$ ions arising from dissolution of CaSO$_4$ and CaO. Hence, during the setting process, and considering the Na$_2$O-SiO$_2$-Al$_2$O$_3$-CaO-H$_2$O system as a whole, the dissolution-precipitation reactions and hence the stable equilibrium phase assemblages involving these oxides mainly have been shown to depend on the pH of the medium and oxide concentration[20]. This initial reaction pathway is typically governed by the CSH, CASH or NASH gel formation process. As has been shown previously[20], at high pH (>12) and with CaO/SiO$_2$ (0.24-3.20) and SiO$_2$/Al$_2$O$_3$ (2.00-6.00), the presence of calcium degrades NASH in favour of CASH or (NC)ASH formation; CaO/SiO$_2$ = 1.10 with SiO$_2$/Al$_2$O$_3$ = 6.00 favouring CASH phase and CaO/SiO$_2$ = 0.24 with SiO$_2$/Al$_2$O$_3$ = 2.00 resulting (NC)ASH phase.

With a CaO/SiO$_2$ ratio of 0.59-0.69, SiO$_2$/Al$_2$O$_3$ ratio of 2.57-4.79 and with high pH (>13) during early hydration for the HCFG systems investigated in the present study, it is apparent that the setting process is governed by the formation of CSH or CASH-like phases rather than pure NASH. Fast dissolution of highly active Al$_2$O$_3$ and SiO$_2$ sources in high pH medium provides high initial concentrations of silicate (SiO$_2$(OH)$_2^{2-}$ or SiO(OH)$_3^{-1}$) and aluminate (Al(OH)$_4^{-1}$) to react with Ca$^{+2}$ forming CASH phase resulting shorter setting times. Therefore, increase in both SiO$_2$ and Al$_2$O$_3$ tend to decrease the setting time giving an optimum SiO$_2$/Al$_2$O$_3$ ratio for the longest setting time. The reaction between Ca$^{+2}$ and aluminate, silicate species continues until all the available Ca$^{+2}$ ions are exhausted and, with time, concentration of Ca$^{+2}$ becomes the limiting factor for this reaction. The pH of the system also decreases with time, due to the consumption of OH$^-$ during hydrolysis to further form silicate and aluminate species. The low pH and limited Ca$^{+2}$ environment facilitate the polymerization reaction between silicate and aluminate species producing NASH gel.

Possible dissolution-precipitation reactions occurring during setting and hardening processes of high calcium based systems can be presented as follows.

i.     Dissolution of SiO$_2$, Al$_2$O$_3$ and calcium sources (CaSO$_4$ & CaO).

$SiO_2 + Al_2O_3$     $\xrightarrow{\text{OH}^-}$     $SiO_2(OH)_2^{-2}$ or $SiO(OH)_3^{-1} + Al(OH)_4^{-1}$

$CaSO_4$ , $CaO$     $\xrightarrow{\text{H}_2\text{O}}$     $Ca^{+2} + SO_4^{-2} + OH^{-1}$

ii.     Precipitation/Condensation reactions

$Ca^{+2} + SiO_2(OH)_2^{2-}$ or $SiO(OH)_3^{-1} + Al(OH)_4^{-1}$     $\longrightarrow$     CASH gel     (1)

$Na^+ + SiO_2(OH)_2^{2-}$ or $SiO(OH)_3^{-1} + Al(OH)_4^{-1}$     $\longrightarrow$     NASH gel     (2)

As CASH gel is stable at high pH (>12) environments and reaction 1 is dominant in the setting process where as reaction 2 becomes feasible at lower pH (9-12) where NASH is stable. Therefore reaction 2 becomes secondary in these systems with mainly responsible for strength development.

As shown by this study, sulfate, coming from dissolution of CaSO$_4$, first tend to form Na$_2$SO$_4$ rather than ettringite-like phase during early hydration and then disappear during higher temperature curing. It was evident from EDXA results that sulfate ions are most likely incorporated in the geopolymer network structure at later stages of the curing process.

Although increase in Al$_2$O$_3$ causes rapid setting of ordinary geopolymer systems (MG) it also has a detrimental effect on strength development[7]. This effect on the strength has been attributed to the formation of crystalline sodium aluminate silicate (zeolitic) phases originated from NASH phases with low SiO$_2$/Al$_2$O$_3$ (< 3.8) ratios. The high calcium fly ash systems investigated in the present study disclose that increasing alumina (SiO$_2$/Al$_2$O$_3$ = 2.87) neither induced any zeolitic phase development (as evident from XRD results) nor showed any significant effect on initial strength development. The observation that increasing Al$_2$O$_3$ had minimal detrimental effect on the strength may also be attributed to the presence of high amounts of Ca$^{+2}$ ions in these systems. Accordingly, the excess Al$_2$O$_3$ is rapidly consumed by Ca$^{+2}$ during the setting process resulting in SiO$_2$/Al$_2$O$_3$ ratio at a higher level at which amorphous sodium aluminate silicate phase (geopolymer gel) is stable. In other words, the presence of Ca$^{+2}$ may hinder the formation of sodium-based zeolitic phases possibly reducing potential detrimental effect on the strength. However, further investigations are needed to confirm this hypothesis. As with conventional geopolymer systems, increase in silica (SiO$_2$/Al$_2$O$_3$ > 4) decreased the strength of high calcium based geopolymers irrespective of silica source with no significant differences in strength trends between the two silica sources investigated.

CONCLUSION
1. Unlike in MG systems, in HCFG systems, increase in either alumina or silica accelerate the setting with an optimal SiO$_2$/Al$_2$O$_3$ ratio in the range 3.20-3.70 controlling setting rather than Ca$^{+2}$ ions itself.
2. In HCFG systems, the setting process is conceivably regulated by initial CASH formation and formation of NASH at later stages of curing is mainly responsible for strength development. NASH is responsible both setting and strength development of MG systems.
3. In HCFG systems, increasing alumina (SiO$_2$/Al$_2$O$_3$ up to 2.87) neither induced any zeolitic phase development nor showed significant influence on strength development contrary to MG based systems. Furthermore, the presence of Ca$^{+2}$ appears to hinder the formation of sodium based zeolitic phases thereby reducing its detrimental effect on the strength.
4. As with MG systems, increase in silica (SiO$_2$/Al$_2$O$_3$ > 4.0) delivers a corresponding strength reduction for high calcium based geopolymers. This trend is independent on silica source for the two silica sources investigated.

REFERENCES
[1]J. Davidovits, Mineral polymers and methods of making them *US Patent* 4,349,386 (1982)
[2]A. Palomo, M.W. Grutzeck and M.T.Blanco, Alkali-activated Fly ashes. A cement for the future, *Cement and Concrete Research* 29, 1323-1329 (1999).

[3]J.G.S. Jaarsveld, J.S.J. Deventer and G.C. Lukey, The effect of composition and temperature on the properties of Fly ash and Kaolinite- based geopolymers *Chemical Engineering Journal* 89, 63-73 (2002).

[4] J. Davidovits, Alkaline Cements and Concretes, *KIEV Ukraine*, 131- 149 (1994).

[5] M. Steveson, K. Sagoe-Crenstil, Relationships between composition, structure and strength of inorganic polymers Part 1 metakolin-derived inorganic polymers. *J Mater Sci* 40:2023-2036 (2005)

[6] P. De Silva, K. Sagoe-Crenstil, V. Sirivivatnanon Kinetics of geopolymerization: Role of Al$_2$O$_3$ and SiO$_2$. *Cem Concr Res* 37:512-218 (2007)

[7] P. De Silva, K. Sagoe-Crenstil Medium-term phase stability of Na$_2$O-Al$_2$O$_3$-SiO$_2$-H$_2$O geopolymer systems. *Cem Concr Res* 38:870-876 (2008)

[8] A. Palomo, M.W.Grutzeck, and M.T. Blanco, Alkali-activated Fly ashes. A cement for the future *Cement and Concrete Research* 29, 1323-1329 (1999).

[9] JGS Van Jaarsveld, JSJ Van Deventer, G.C. Lukey, The characterisation of source materials in fly ash-based geopolymers. *Mater Lett* 57:1272-1280 (2003)

[10] P. Chindaprasirt, T. Chareerat, S. Hatanaka, T. Cao High-strength geopolymer using fine high-calcium fly ash. *J Mater Civ Eng (ASCE)* 23:264-270 (2011)

[11] K. Somna, C. Jaturapitakkul, P. Kajitvichyanukul, P. Chindaprasirt NaOH-activated ground fly ash cured at ambient. *Fuel* 90:2118-2124 (2011)

[12] U. Rattanasak, K. Pankhet, P. Chindaprasirt Effect of chemical admixtures on properties of high-calcium fly ash geopolymer. *Int J Miner Metall Mater* 18:364-369 (2011)

[13] P. Chindaprasirt, U. Rattanasak, C. Jaturapitakkul, Utilization of fly ash blends from pulverized coal and fluidized bed combustions in geopolymerics materials. *Cem Concr Compos* 33:55-60 (2011)

[14] X. Guo, H. Shi, W.A. Dick, Compressive strength and microstructural characteristics of class C fly ash geopolymer. *Cem Concr Compos* 32:142-147 (2010)

[15] X . Guo, H. Shi, L. Chen, W.A. Dick Alkali-activated complex binders from class C fly ash and Ca-containing admixtures. *J Hazard Mater* 173:480-486 (2010)

[16] J .Temuujin, A.V. Riessen, R. Williams, Influence of calcium compounds on the mechanical properties of fly ash geopolymer pastes. *J Hazard Mater* 167:82-88 (2009)

[17] C.K. Yip, G.C. Lukey, J.L. Provis, J.S.J. Van Deventer, Effect of calcium silicate sources on geopolymerisation. *Cem Concr Res* 38:554–564 (2008)

[18] C.K Yip, G.C. Lukey, J.S.J. Van Deventer, The coexistence of geopolymeric gel and calcium silicate hydrate at the early stage of alkaline activation. *Cem Concr Res* 35:1688-1697 (2005)

[19] P.Chindaprasirt, P. De Silva, K. Sagoe-Crentsil and S.Hanjitsuwan, Effect of SiO$_2$ and Al$_2$O$_3$ on the setting and hardening of high calcium fly ash-based geopolymer systems. *Journal of Materials Science,* Volume 47:4876-4883(2012)

[20] I. Garcia-Lodeiro, A. Palomo, A. Ferna'ndez-Jime'nez, D.E. Macphee, Compatibility studies between N-A-S-H and C-A-S-H gels. Study in the ternary diagram Na$_2$O-CaO-Al$_2$O$_3$-SiO$_2$-H$_2$O. *Cem Concr Res* 41:923-931 (2011)

[21] I. Lecomte, C. Henrist, M. Li'egeois, F. Maseri, A. Rulmont, R. Cloots, (Micro)-structural comparison between geopolymers, alkali-activated slag cement and Portland cement. *J Eur Ceram Soc* 26:3789-3797 (2006)

[22] X . Pardal, I. Pochard, A. Nonat. Experimental study of Si–Al substitution in calcium-silicate-hydrate (C-S-H) prepared under equilibrium conditions. *Cem Concr Res* 39:637–664 (2009)

[23] C. Desbats-Le Chequer, F. Frizon, Impact of sulfate and nitrate incorporation on potassium- and sodium- based geopolymers: geopolymerization and materials properties. *J Mater Sci* 46:5657-5664 (2011).

# SYNTHESIS OF THERMOSTABLE GEOPOLYMER-TYPE MATERIAL FROM WASTE GLASS

Qin Li, Zengqing Sun, Dejing Tao, Hao Cui, and Jianping Zhai*
State Key Laboratory of Pollution Control and Resource Reuse, and School of the Environment,
Nanjing University, Nanjing 210023, P. R. China

ABSTRACT

Waste glass was activated by a series of alkali hydroxide and/or sodium/potassium silicate solutions. The synthesized products were characterized by mechanical testing, scanning electron microscopy, X-ray diffraction, and Fourier transform infrared spectroscopy, and their thermal stability was determined in terms of compressive strength evolution at up to 800 °C. Compressive strength measurements showed a maximum strength of 119 MPa after 28 days. Thermal treatment at 100, 200, 400, 600 and 800 °C is necessary but sufficient to obtain strength of more than 79 MPa. The results indicate that waste glass could serve as a suitable source material for thermostable geopolymers, with promiseing in applications requiring mechanical strength and in areas where fire hazards are a concern.

INTRODUCTION

The process design and synthesis of aluminosilicate refractory ceramics, termed geopolymers, at room temperature has attracted great attention among the industrial and scientific communities for their superior mechanical strength, high-temperature performance, and low production cost[1-3]. Geopolymers are a class of inorganic polymers synthesized by the alkali activation of aluminosilicate materials[2] firstly introduced by Joseph Davidovits in the late 1970s[3]. According to Davidovits, geopolymers are three-dimensional polymeric silicon-oxygen-aluminum materials containing various amorphous to semi-crystalline phases[4]. Deventer deduced that geopolymers have a monolithic structure that is bound by hydrated aluminosilicate gel formed after the alkali breaks O linkages of the tetrahedral Si and Al monomers in the aluminosilicate starting material[5]. The empirical formula of geopolymer gel material is therefore $Mn(-(SiO_2)z-AlO_2)n \cdot wH_2O$, where M is a cation such as $Na^+$, $K^+$ or $Ca^{2+}$, z is 1, 2, or 3, and n is the degree of poly-condensation[5-7].

Large numbers of scholars have developed a variety of raw materials, such as metakaolin[8-9], fly ash[10-12], or slag[13], to synthesis geopolymer matrixes of excellent performance. All of these materials are able to supply the polymerization reaction with sufficient silica and aluminum[3, 14]. Since our daily life produces large amounts of waste glass. Secondary uses for recycled waste glass have been developed, most of the time, they are collected, sorted, and crushed to be used as a raw material for re-smelting[15-17], or used to synthesize water filtration media[18], abrasives[19] etc. However, all of these applications can only reuse specific waste glasses, and lead to high energy consumption and secondary pollution[15, 16, 19]. In contrast, using waste glass as an aluminosilicate source material for geopolymer manufacture would be a creative, efficient, and environmentally friendly alternative.

In this study, geopolymers are creatively synthesized using waste glass as raw material. The characteristics of the optimal geopolymer product obtained are discussed in detail. No previous study has systematically characterized the synthesis and thermal properties of geopolymeric matrices made from waste glass source material. The high temperature performance of the optimum geopolymer was also studied in terms of compressive strength gains or losses after exposure to high temperatures (100, 200, 400, 600, and 800 °C). The synthesized products were characterized by mechanical testing, scanning electron microscopy (SEM), X-ray diffraction (XRD), and Fourier transform infrared spectroscopy (FT-IR).

EXPERIMENTAL

Materials and Sample Preparation

The glass used in this study was sourced from municipal waste collection and was a mix of different colors. The glasses were first ultrasonically washed to remove contaminants such as paper scraps, metal, plastic, and organic matter to avoid unwanted chemical or physical interactions during the geopolymerization. The dried glasses were crushed, pulverized in a ball mill for 75 min and then screened using a 200 mesh standard sieve. The particle size distribution of the screened waste glass was measured on a Mastersizer 2000 laser analyzer(Malvern, UK), showing an average particle size (d50) of 26 μm, as shown in Fig. 1. Chemical composition of the waste glass was determined using an ARL 9800XP$^+$ X-ray fluorescence spectrometer. The major and trace element contents are listed in Table I.

Industrial grade sodium silicate solution (8.5 wt% $Na_2O$ and 26.5 wt% $SiO_2$, with a $SiO_2/Na_2O$ molar ratio of, 3.2), sodium hydroxide and potassium hydroxide (analytical grade), and deionized water were used throughout this investigation.

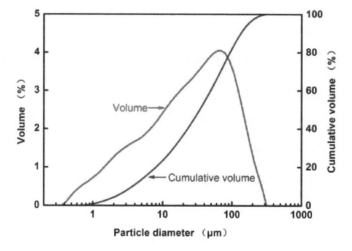

Fig.1. Particle size distribution of ground waste glass

Table I. Major and trace element contents of waste glass

| Major elements as oxides (wt%) | | | | | | | | |
| --- | --- | --- | --- | --- | --- | --- | --- | --- |
| $SiO_2$ | $Al_2O_3$ | CaO | $Na_2O$ | $K_2O$ | MgO | $Fe_2O_3$ | $SO_3$ | LOI [a] |
| 68.41 | 7.18 | 7.75 | 11.63 | 1.71 | 1.05 | 0.47 | 0.09 | 1.22 |

| Trace element contents ($\mu g/g$) | | | | | | | | | | |
| --- | --- | --- | --- | --- | --- | --- | --- | --- | --- | --- |
| Ba | Cr | Cu | Sr | As | Pb | Zr | Zn | Rb | V | Ni |
| 161.73 | 44.72 | 17.38 | 17.25 | 14.53 | 15.71 | 11.14 | 10.62 | 8.85 | 2.48 | 3.41 |

[a] LOI, loss on ignition at 960 °C.

Table II. Composition of activating solutions (M, Na and K; R, molar ratio of $SiO_2/Na_2O$)

| Activator ID | Activator type | Content (wt%) | | | |
| --- | --- | --- | --- | --- | --- |
| | | NaOH | KOH | $SiO_2$ | $H_2O$ |
| A1 | 10N NaOH | 30.10 | - | - | 69.90 |
| A2 | 10N KOH | - | 40.38 | - | 59.62 |
| A3 | 10N MOH ($N_K/N_{Na}=1$) | 16.88 | 23.15 | - | 59.97 |
| B1 | $Na_2SiO_3$ (R=1.0) | 28.41 | - | 21.37 | 50.22 |
| B2 | $Na_2SiO_3$ (R=1.5) | 20.74 | - | 23.66 | 55.60 |
| B3 | $Na_2SiO_3$ (R=2.0) | 16.59 | - | 24.91 | 58.51 |
| C1 | $M_2SiO_3$ (R=1.0) | 13.45 | 18.80 | 20.22 | 47.53 |
| C2 | $M_2SiO_3$ (R=1.5) | 9.61 | 14.38 | 22.69 | 53.32 |
| C3 | $M_2SiO_3$ (R=2.0) | 10.31 | 8.22 | 24.32 | 57.15 |

The activating solutions were divided into three groups: A, B, and C. Each group consisted of three solutions, making a total of nine activators. Activators A1–A3 were sodium and/or potassium hydroxide solutions, B1–B3 were mixtures of sodium silicate and sodium hydroxide, and C1–C3 were mixtures of sodium silicate and potassium hydroxide. The compositions of all activating solutions are listed in Table II.

Geopolymer Synthesis

Geopolymers were synthesized by mixing the ground waste glass with each activating solution with a liquid/solid ratio of 0.4, as summarized in Table III. Each mixing process lasted 5 min and was followed by casting the slurries into 20 mm cube triplet molds and 5 min vibration to remove entrapped air bubbles. The molds were   sealed with polyethylene film and set in a standard curing box at 60 °C with 99 ±1% humidity under ambient pressure for 24 h. The samples were then demolded and subjected to curing under the same conditions in sealed polypropylene boxes.

Table III. Mix design and calculated molar ratios for waste glass-based geopolymers (M, Na and/or K)

| Sample ID | Liquid/solid (mass ratio) | Si/Al | Al/M | Si/M | Si/Ca | Al/Ca | $H_2O$/M |
|---|---|---|---|---|---|---|---|
| G1 | 0.4 | 8.14 | 0.19 | 1.55 | 5.77 | 0.71 | 2.18 |
| G2 | 0.4 | 8.14 | 0.20 | 1.63 | 5.77 | 0.71 | 1.89 |
| G3 | 0.4 | 8.14 | 0.18 | 1.47 | 5.77 | 0.71 | 1.79 |
| G4 | 0.4 | 9.14 | 0.20 | 1.83 | 6.49 | 0.71 | 1.60 |
| G5 | 0.4 | 9.28 | 0.23 | 2.13 | 6.56 | 0.71 | 1.99 |
| G6 | 0.4 | 9.33 | 0.24 | 2.24 | 6.61 | 0.71 | 2.25 |
| G7 | 0.4 | 9.10 | 0.21 | 1.91 | 6.45 | 0.71 | 1.55 |
| G8 | 0.4 | 9.21 | 0.23 | 2.12 | 6.53 | 0.71 | 1.94 |
| G9 | 0.4 | 9.30 | 0.25 | 2.33 | 6.59 | 0.71 | 2.21 |

Analysis Methods

The compressive strength values of synthesized geopolymers of all mix designs cured for 7, 14, and 28 days were measured using a NYL-300 compressive strength testing apparatus (Wuxi Jianyi, China) with force applied at a rate of 1.0 KN/s. The results are reported as the average of three replicates.

Geopolymer of the highest compressive strength was subjected to high-temperature performance tests. Specimens of the selected geopolymer were heated in a muffle furnace from ambient temperature to 100, 200, 400, 600, and 800 °C at a heating rate of 5 °C/min. After heat treatment at the selected temperature for 2 h, the specimens were left to cool naturally in the furnace with the door of the furnace closed. The compressive strength gains or losses of the specimens were then measured.

The ground waste glass, geopolymers of optimal group after 28 d standard curing and those subjected to thermal exposure were characterized via XRD, SEM, and FT-IR. XRD was conducted using an ARL X'TRA high-performance powder X-ray diffractometer with Cu Kα radiation generated at 40 mA and 40 kV with a scanning step of 0.02° and scanning rate of 10° min⁻¹ from 5 to 65° 2θ. Microstructural images were observed using a Hitachi S-340N scanning electron microscope. FT-IR spectra were recorded using a Nicolet 6700 FTIR Spectrometer. The KBr pellet method was used to prepare the samples, which was scanned at a range of from 4000 to 400 cm⁻¹.

RESULTS AND DISCUSSION

Compressive Strength

The compressive strengths of the obtained geopolymers ranged from 23.85 MPa to 119.42 MPa, as illustrated in Table IV. From these data, it can be seen that the compressive strength of all mix design increased at different degrees with curing time, and that the best mechanical compressive strength was achieved by the geopolymer matrix formed from waste glass activated with group C solutions. In particular, that activated by solution C2 exhibited a 28 d compressive strength of 119.42 MPa. The strengths of the group B activated mortar samples were only lower than those of the corresponding group C samples. The results indicate that geopolymer matrices activated by alkali silicate solutions containing $K^+$ have a higher compressive strength, owing to the stronger basicity of $K^+$ which allows higher rates of silicate dissolution[20, 21], consistent with the reports of Phair and van Deventer[5, 21]. In contrast, geopolymers activated by group A did not show good compressive strength. The optimum compressive strength of group GA was the 28 d compressive strength value of GA2, i.e., 38.15 MPa, which was still 8.95 MPa lower than the minimum 7 d compressive strength of group GB. While in GA, geopolymers manufactured by $K^+$ containing activator had a lower compressive strength than those activated by NaOH, which is different with the results obtained for GC, but consistent with Davidovits's study[22]. Geopolymers activated by silicate solutions exhibited higher compressive strength than those activated by hydroxide solutions; this could be because of the presence of a proper amount of soluble Si in the activation solution benefits the development of the compressive strength, just as Xu[11] suggested. As a result, the C2 activated geopolymer was chosen for further study of high-temperature performance.

Table IV. Compressive strength of geopolymer matrixes (M, Na and K)

| ID | Activator | Compressive Strength | | |
|----|-----------|------|------|------|
| | | 7 d | 14 d | 28 d |
| GA1 | 10N NaOH | 34.43 | 35.58 | 38.15 |
| GA2 | 10N KOH | 23.85 | 25.50 | 28.73 |
| GA3 | 10N MOH ($N_K/N_{Na}=1$) | 24.65 | 28.62 | 33.67 |
| GB1 | $Na_2SiO_3$ (R=1.0) | 49.63 | 57.46 | 58.95 |
| GB2 | $Na_2SiO_3$ (R=1.5) | 58.13 | 70.79 | 87.20 |
| GB3 | $Na_2SiO_3$ (R=2.0) | 47.10 | 56.73 | 63.15 |
| GC1 | $M_2SiO_3$ (R=1.0) | 65.46 | 67.48 | 85.65 |
| GC2 | $M_2SiO_3$ (R=1.5) | 75.46 | 94.25 | 119.42 |
| GC3 | $M_2SiO_3$ (R=2.0) | 48.43 | 57.64 | 65.56 |

High Temperature Performance Test

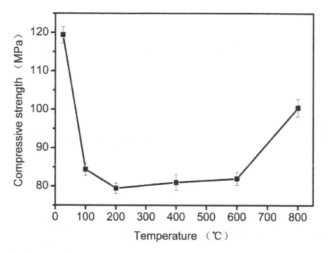

Fig. 2. Compressive strength of selected geopolymer after thermal exposure at different temperatures.

The effect of heat on the waste glass-based geopolymer was investigated in terms of compressive strength evolution after exposure to five different temperatures. Fig.2 shows the compressive strength gains or losses after heating procedures. The compressive strength reduced drastically in the initial stage (below 200 °C), from 119.42 MPa for the as-cured sample to 79.47 MPa at 200 °C. This reduced strength was maintained until 600 °C. Above this temperature, the strength increased to about 100 MPa at 800 °C. It is believed that[22] the strength evolution of geopolymers after exposure to elevated temperatures depends on the dominant process of (1) damages caused by thermal incompatibility due to non-uniform temperature distribution; and (2) further geopolymerization and/or sintering leading to strength increase.

The reduction of strength at the initial heating stage was probably caused by the loss of structural water, and weakening of the structure by possible concomitant development of micro cracks due to the accumulated effects of volume expansion derived from dehydration reactions[23]. The increase of compressive strength at 800 °C can mainly be attributed to further geopolymerization and/or sintering[22-23], which densified the internal structure of geopolymer matrix as discussed below (Scanning Electron Microscopy Section). After the heating process, the selected geopolymer specimens generally evolved their compressive strengths to a certain extent up to 79.47 MPa, demonstrating that the present glass-based geopolymer has good high-temperature performance.

Scanning Electron Microscopy Analysis

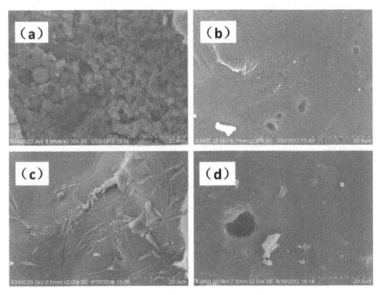

Fig. 3. Scanning electron microscope images of (a) waste glass powder and geopolymer GC2 after (b) 28 d of standard curing; (c) exposure to 100 °C; and (d) exposure to 800 °C.

SEM images of ground waste glass and the selected geopolymer before and after thermal exposures are shown in Fig.3. As shown in Fig.3(a), waste glass consists exclusively of irregular, coarse and angular particles, which is far different from conventional pulverized coal combustion fly ashes[24]. While typical pulverized coal combustion fly ash largely consist of small glassy spheres. Fig. 3(b) displays a well-formed geopolymer matrix of specimen GC2 after 28 d standard curing. The compact structure suggests a well geopolymerization behavior and demonstrates that waste glass could be a suitable alternative source material for geopolymer production. Fig. 3(c) shows the SEM image of the selected geopolymer after calcination at 100 °C for 2 h. It can be seen that many small and narrow cracks appeared, which indicates the breakage of an integrated structure after heating owing to volume expansion caused by water vapor diffusion as well as thermal incompatibilities arising from non-uniform temperature distribution[22]. The presence of these cracks is valid evidence to explain the compressive strength decrease of geopolymer after heating at 100 °C, as discussed in Section High Temperature Performance Test. In Fig. 3(d) the morphology of the selected geopolymer after calcination at 800 °C is relatively smooth with a luster typical of high temperature sintering. The much smoother and more compact microstructure of geopolymer heated at 800 °C than that heated at 100 °C corresponds to the remarkable increase of mechanical strength as discussed above. Fig. 3(d) also shows traces of the formation of some new crystalline phases, which is further supported by the corresponding XRD analysis discussed below.

X-ray Diffraction Analysis

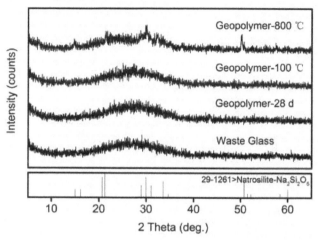

Fig. 4. X-ray diffraction data for waste glass and selected geopolymers before and after thermal exposure

The structure changes that occurred during the geopolymerisation and high temperature process were evaluated by XRD analysis. As shown in Fig. 4, the XRD patterns of raw waste glass and the selected geopolymer sample before and after heat exposures are simple and similar, with typical amorphous curves of glass, indicating that the geopolymerization of waste glass does not lead to substantial formation of new crystalline phases. It is also seen in Fig. 4 that the XRD patterns of the samples unheated and heated at 100 °C are almost the same. However, there are traces of crystalline Natrosilite ($Na_2Si_2O_5$) after treatment at 800 °C, as supported by the SEM image shown in Fig. 3(d) and in agreement with the reports of Bakharev[25], who studied the thermal behavior of geopolymers prepared using fly ash. The analysis of both XRD and SEM supported the compressive strength variations from a microscopic way.

Infrared Spectroscopy Analysis

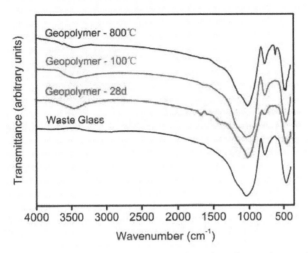

Fig. 5. Infrared spectra of waste glass and the selected geopolymers before and after thermal exposures.

The FT-IR spectra of the initial glass powder and resulting geopolymers, both unheated and heated to different temperatures, are shown in Fig. 5. All samples show two broad absorbance bands, at 1250–820 cm$^{-1}$ and 730–450 cm$^{-1}$. Verdolotti et al.[26] reported that the broadness of the absorbance band at 1250–820 cm$^{-1}$ reflects the variability of the bond angles and bond lengths of the tetrahedral structures around the silicon atoms, and that the absorbance at 730–450 cm$^{-1}$ arises from the stretching vibration of Al-O-Si bonds. The bands at 3500 cm$^{-1}$ and 1700 cm$^{-1}$ correspond to O-H stretching and O-H bending, respectively, and arose from the presence of structural water[11 27]. The absorption peak with a maximum at about 1630 cm$^{-1}$ is assigned to $H_2O$ bending and the broader distribution of absorption intensities in the range between 3000 and 3700 cm$^{-1}$ is due to OH stretching[11 27]. These peaks may be related to water enclosed in the pores of the geopolymer-type binder, which became rehydrated even after heat treatment of the samples. As a result, these peaks were observed for both the unheated and heated geopolymer samples, but were absent for the waste glass raw material, because the glass powder was fully dried prior to any analysis. The formation of a new phase could be spotted in the FTIR spectrum of the 800 °C treated sample. The characteristic absorption peaks of residual waste glass and the newly formed geopolymer-type binder became superimposed and could not be further separated here[27-28]. Compared with that of the raw glass powder the selected geopolymer undergoes a very small shift of its Si-O-Si position to lower frequency with alternating Si-O and Al-O bonds, which is owing to polycondensation process[29-30].

CONCLUSION

This work has proven that waste glass is able to serve as an alternative aluminosilicate source material for the preparation of geopolymers with high compressive strength and good high-temperature performance.

The mechanical properties of waste glass-based geopolymers depend strongly upon the chemical composition of the initial reacting system. In this study, the highest compressive strength of all mix designs was 119.42 MPa. The optimal geopolymer exhibited superior high-temperature performance in terms of compressive strength after thermal exposure (79.47 MPa after heating at 800 °C for 2 h).

SEM micrographs indicated that the microstructure of geopolymer remained almost the same after exposure to 800 °C, with the formation of only traces of a crystalline Natrosilite phase in the geopolymer matrix, as confirmed by XRD analyses.

The results of this study suggest that waste glass geopolymers may be suitable for a broad range of applications, especially where excellent high-temperature performance is essential.

ACKNOWLEDGEMENT

The authors gracefully acknowledge financial support from the Foundation of State Key Laboratory of Pollution Control and Resource Reuse of China, the Natural Science Foundation of China (No. 51008154), the Research Fund for the Doctoral Program of Higher Education of China (No.20090091120007), the Fundamental Research Funds for the Central University (No.1112021101), and the Research Projects on Environmental Protection of Jiangsu Province (2012030).

REFERENCES

[1] J.Davidovits, Geopolymers Inorganic Polymeric Materials, *J. Therm. Anal.* ,**37**, 1633-6 (1991) .

[2] K.Ikeda, K.Onikura, Y.Nakamura, and S.Vedanand, Optical Spectra of Nickel-Bearing Silicate Gels Prepared by the Geopolymer Technique, with Special Reference to the Low-Temperature Formation of Liebenbergite (Ni2SiO4), *J. Amer. Ceram. Soc.*,**84**, 1717-20 (2001).

[3] K. Komnitsas, and D. Zaharaki, Geopolymerisation: A Review and Prospects for the Minerals Industry, *Miner. Eng.*. **20**, 1261–77 (2007) .

[4] J.Davisovits, Geopolymer Chemistry and Applications : A Practical and Scientific Approach to Sustainable Development (3$^{rd}$ Ed), *Lavoisier Librairie., 07-2011.*

[5] J.W.Phair, and J.S.J.van Deventer, Characterization of Fly-Ash-Based Geopolymeric Binders Activated with Sodium Aluminate, *Ind.Eng.Chem.Res.*,**41**, 4242-51 (2002).

[6] J. Davidovits, Geopolymers and Geopolymeric Materials, *J .Therm. Anal.*, **35**, 429-41 (1989).

[7] B. D. Oswaldo, I. E. G. Jose, A. A. Raul, and A. G. Bustos, Statistical Analysis of Strength Development as a Function of Various Parameters on Activated Metakaolin/Slag Cements, *J. Amer. Ceram. Soc.*, **196**, 86-92 (2011).

[8] P. Duxson, S.W. Mallicoat, and G. C. Lukey, The Effect of Alkali and Si/Al Ratio on the Development of Mechanical Properties of Metakaolin-Based Geopolymers, *Colloids Surf. A.*, **292**, 8-20 (2007).

[9] Z. Aly, E. R. Vance, D. S. Perera, J. V. Hanna, and D. Durce, Aqueous Leachability of Metakaolin-Based Geopolymers with Molar Ratios of Si/Al = 1.5–4, *J Nucl. Mater.*, **378**, 172-9 (2008).

[10] D. L. Kong, Y, Sanjayan , and S. C. Kwesi, Comparative Performance of Geopolymers Made with Metakaolin and Fly Ash After Exposure to Elevated Temperatures, *Cem. Concr. Res.*, **37**, 1583-9

(2007).

[11]H. Xu, Q. Li, L. Shen, W. Wang, J. Zhai, Synthesis of Thermostable Geopolymer from Circulating Fluidized Bed Combustion (CFBC) Bottom Ashes, *J. Hazard. Mater.*, **175**, 198-204 (2010).

[12]D. Panias, P. Giannopoulou, and T. Perraki, Effect of Synthesis Parameters on the Mechanical Properties of Fly Ash-Based Geopolymers, *Colloids Surf. A.*, **301**, 246-54 (2007).

[13]I. Maragkos, P. Giannopoulou, D. Panias, Synthesis of Ferronickel Slag-Based Geopolymers, *Miner. Eng.*, **22**, 196-203 (2009).

[14]H. Xu, and J.S.J. van Deventer, The Geopolymerisation of Alumino-Silicate Minerals, *J. Miner. Proce.*, **59**, 247-66 (2000).

[15]M.I.Yaziz, F.L.Chin, S.N.Tang, and N.N.Bich, Heavy Metal Leaching of Solified Sludge from a Glass Components Industry, *J.Environ.Sci. Health A.*, **34**, 853-61(1999).

[16]T. Nishida, M. Seto, S. Kubuki,. O. Miyaji, T. Ariga, and Y. Matsumoto, Solidification of Hazardous Heavy Metal Ions With Soda-Lime Glass-Characterization of Iron and Zinc in the Waste Glass, *J. Ceram. Soc. Jap.*, **108**, 245-8 (2000).

[17]I.W.Donald, B.L.Metcalfe, and R.N.J.Taylor, The Immobilization of High Level Radioactive Wastes Using Ceramics and Glasses, *J. Mater. Sci.*, **32**, 5851-87 (1997).

[18]N. J. Coleman, A Tobermorite Ion Exchanger from Recycled Container Glass, *J. Envir. Waste. Manag.*, **8**, 366-82 (2011).

[19]N. Chand, A. Naik, and S. Neogi, Three-Body Abrasive Wear of Short Glass Fibre Polyester Composite, *Wear*, **24**, 238-46 (2000).

[20]J.W.Phair, and J.S.J.Van Deventer, Effect of the Silicate Activator pH on the Microstructural Characteristics of Waste-Based Geopolymers, *J. Miner. Proce.*, **66**, 121-43 (2002) .

[21]J.W.Phair, and J.S.J.Van Deventer, Effect of Silicate Activator pH on the Leaching and Material Characteristics of Waste-Based Inorganic Polymers, *Miner. Engin.*, **14**, 289-304 (2001).

[22]J.Davidovits, Geopolymer Chemistry and Applications 3rd edition, France, 2011.

[23]R.E.Lyon, P.N.Balaguru, A.Foden, U.Sorarhia, and J.Davidovits, Fire-Resistant Aluminosilicate Composites, *Fire and Mater.*, **21**, 67-73(1997).

[24]P. Duxson, A. Fernandez-Jimenez, J. L. Provis, G. C. Lukey, A. Palomo, and J.S.J.Van Deventer, Geopolymer Technology: the Current State of the Art, *J. Mater. Sci.*, **42**, 2917-33 (2007).

[25]T. Bakharev, Thermal Behavior of Geopolymers Prepared Using Class F Fly Ash and Elevated Temperature Curing, *Cem. Conr. Res.*, **36**, 1134-47 (2006).

[26]L. Verdolotti, S. Iannace, M. Lavorgna, and R. Lamanna, Geopolymerization Reaction to Consolidate Incoherent Pozzolanic Soil, *J. Mater. Sci.*, **43**, 865-73 (2008).

[27]C. H. Ruscher, E. M. Mielcarek, J. Wongpa, C. Jaturapitakkul, F. Jirasit and L. Lohaus, Silicate-, Aluminosilicate and Calciumsilicate Gels for Building Materials: Chemical and Mechanical Properties During Ageing, *Eur. J. Mineral.*, 23, 111-24 (2011).

[28]C. H. Ruscher, E. Mielcarek, W. Lutz, A. Ritzmann, and W. M. Kriven, Weakening of Alkali-Activated Metakaolin During Aging Investigated by the Molybdate Method and Infrared Absorption Spectroscopy, *J. Am. Ceram. Soc.*, 93, 2585-90 (2010)

[29]P. Chindaprasirt, C. Jaturapitakkul, W. Chalee, and U. Rattanasak, Comparative Study on the Characteristics of Fly Ash and Bottom Aash Geopolymers, *Waste Manag.*, **29**, 539–43(2009) .

[30]S. Andini, R. Cioffi, F. Colangelo, T. Grieco, F. Montagnaro, and L. Santoro, Coal Fly Ash as Raw Material for the Manufacture of Geopolymer-Based Products, *Waste Manag.*, **28**, 416–23 (2008).

# THE EFFECT OF CURING CONDITIONS ON COMPRESSION STRENGTH AND POROSITY OF METAKAOLIN-BASED GEOPOLYMERS

Bing Cai[a], Torbjörn Mellgren[a], Susanne Bredenberg[a,b] ,Håkan Engqvist[a]*

[a]Division for Applied Materials Science, Department of Engineering Sciences, The Ångström Laboratory, Uppsala University, Box 534, SE-751 21 Uppsala, Sweden

[b]Orexo AB, P.O. Box 303, SE-751 05 Uppsala, Sweden

ABSTRACT

Geopolymers have been suggested to use as construction, waste treatment and fire proof materials and even drug delivery material due to its excellent mechanical strength, chemical stability and flame resistance. The aim of this study was to investigate the influence of temperature, time and humidity during curing on mechanical strength and porosity of geopolymers.

The geopolymer precursor paste was obtained by mixing metakaolin, waterglass and de-ionized water. The paste was molded into cylindrical rubber moulds (6 ⬜ 12 mm) and cured under different conditions: i.e. temperatures (ambient temperature, 37°C and 90°C), humidity and time (24, 48 and 96 hours). The compressive strength was determined using a universal testing machine. Helium pycnometer was used to measure the porosity. Via x-ray diffraction the phase composition of the cured samples was determined.

Elongated curing slightly decreased the total porosity of the tested geopolymers. Higher curing temperature increased the compressive strength after 24 hour but did not affect strength for longer curing times. In general, the samples cured in moisture had higher mechanical strength than those cured in air. But low compression strength of samples cured under high temperature and long time showed that some water content in the geopolymer was essential to retaining its microstructure.

INTRODUCTION

Geopolymer is an alkali activated inorganic polymer, which is widely used in various fields, due to its excellent properties of mechanical strength, fire resistance, acid resistance, heavy metal immobilization and biological compatibility[1,2]. Geopolymer is composed of the three-dimensional polysialate framework containing $SiO_4$ and $AlO_4$ linked alternately[1,3]. The alkali cations balance the charge of the tetrahedral Al in the structure. The process of geopolymerization was described by Davidovits: the reaction is initiated by liberation of aluminates and silicates on the surface of a metakaolin particle by the penetration of alkali solution; the process continues on polymerization of silalate and condensation of the gel phase[1,2]. Reaction conditions, i.e. temperature, curing time and humidity, are influential factors on the mechanical strength and porosity of geopolymer [3-8]. But various conclusions have been drawn using different compositions of geopolymer, the range of variants and testing methods. The aim of this study is to investigate the influence of temperature, curing time and humidity on mechanical strength and porosity of one metakaolin-based geopolymer composition.

EXPERIMENTAL

Waterglass was prepared by mixing sodium hydroxide (Sigma-Aldrich, Sweden) with sodium silicate solution (Sigma-Aldrich, Sweden) until a clear solution was formed (Na2O/SiO2 wt ratio: 0.541). Metakaolin was obtained by dehydrating kaolin (Sigma-Aldrich, Sweden) under 800 °C for 2 h. Metakaolin, waterglass and de-ionized water were mixed vigorously into a homogenous paste by in a glass mortar. The paste was subsequently cured in cylindrical rubber moulds (diameter, height: 6, 12 mm) under different temperatures (22°C, 37°C and 90°C), periods of time (24, 48 and 96 hours) and relative humidity (air and 100% humidity). Samples were formulated with the following molar ratio: $Al_2O_3/SiO_2$=0.353, $Na_2O/SiO_2$=0.202, $H_2O/SiO_2$=2.977.

Compression strength test

The compression strength was measured using an Autograph AGS-H universal testing machine (Shimadzu Corp., Japan) and determined as the maximum pressure on the rods before breakage. Six cylindrical rods of each sample with the dimensions 6⊡12mm(diameter×height) were tested and the average calculated.

Helium pycnometry

All samples were first drenched in ethanol (99.5% Kemetyl AB) for 10 minutes and air-dried to remove all the moisture. A helium pycnometer (AccuPyc 1340, Micromeritics Corp., USA) was used to measure the true volume of the sample. Three repeats were tested for each group.

Scanning electron microscopy (SEM)

The pore structures in the geopolymer pellets were observed by Leo 1550 FEG microscope (Zeiss, UK) with an inlens detector. The samples were grinded into powder and coated with a thin layer of gold/palladium prior to analysis.

X-ray diffraction (XRD)

The samples were grinded into powder before analysis. The diffraction patterns of the geopolymer samples were determined by D5000 diffractometer (Siemens/Bruker, Germany).

RESULTS AND DISCUSSION

Curing time is an important factors of geopolymerization process[6]. Previous studies also showed that samples cured for a longer period of time have less unreacted raw material and denser structure in the final hardened body[7,9]. XRD results did, however, show little difference between samples in the amount of remaining raw material, see Fig 2. In all of the geopolymer samples, the characteristic hump of metakaolin was reduced and the amorphous peak shifted to the position. Some quartz inherited from metakaolin were presented in all geopolymer products, showing peaks at 21.1, 27.1

and 34.5 degree[10]. Longer curing time also influences the strength and microstructures of geopolymer. As shown in Fig 1, the compression strength of geopolymers was improved with hardening time but the increase was not significant after 48 hours. It indicated that the geopolymer requires at least 24 hours to acquire full compression strength. Observed under SEM, the geopolymer has coarser microstructures after longer curing time, see Fig 3 A-D. However, the samples cured for 96 hours had slightly lower porosity than the samples cured for shorter time at the same condition, see Fig. 4. It implies the extended curing provided the liquid silicate more time to fill pores and resulted in a more dense material[7].

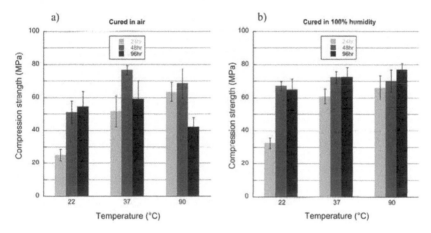

Figure 1 Mechanical strength of the samples cured in air (a) and 100% humidity (b)

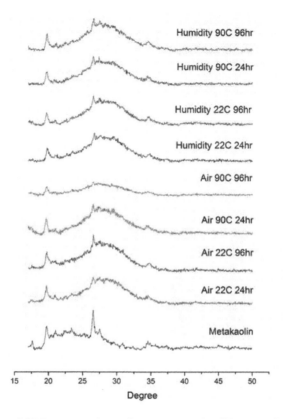

Figure 2 XRD patterns of geopolymers cured under different conditions.

Figure 3 SEM pictures of A) Cured in air 22°C 24 hours; B) Cured in air 22°C 96 hours; C) Cured in air 90°C 24 hours; D) Cured in air 90°C 96 hours E) Cured in 100% humidity 22°C 24 hours F) Cured in 100% humidity 22°C 96 hours.

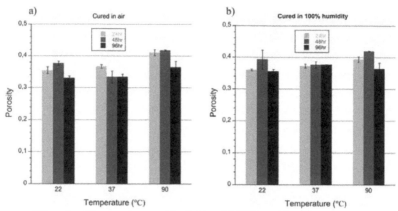

Figure 4 Porosity of the samples cured in air (a) and 100% humidity (b)

Compared to the curing time, the effect of the temperature on the strength and the microstructure was more obvious. Heat improved the mechanical strength of the geopolymers, Fig 1. Elevated temperature increased the dissolution rate of aluminates and silicates from the metakaolin, raised the reaction rate and caused the fast development of strength[5-9]. It is particularly for samples cured for 24 hours since the geopolymers haven't fully developed yet. SEM pictures showed that the samples cured at 22°C formed denser structure than when cured at higher temperatures, see Fig 3 A-D. Similar conclusion can be drawn from the results of the porosity measurements: the samples cured in higher temperature had slightly higher porosity, Fig 4. The high temperature curing in low humidity leads to rapid water evaporation and increases the microcavity in structures. This rapid drying can also lead to increased crack formation in the geopolymer[11]. Therefore, the geopolymers cured at higher temperature have slightly higher porosity. In this study, the curing temperature did not influence the reaction yield, as the geopolymer after curing under different temperatures did not show much variance in XRD results.

Samples cured in air at 37°C and 90°C for 48 and 96 hours did not follow the trend mentioned above. They had lower mechanical strength even though they were cured for a long time at a higher temperature. It could due to the water evaporation during the synthesis. There has been evidence showing that a small amount of water content is essential in retaining the microstructure[4,12]. Dehydration could cause cracking and shrinkage during geopolymerization, forming weaker products[3,12]. Thus, evaporation needs to be limited when curing for a long time period at a high temperature. Another possible reason is that the rapid reaction rate at high temperatures leads to a less dense and compact structure[7]. On contrary, by comparing SEM pictures, Fig 3 C and D, and porosity measurements, Fig 4, these samples did not show much difference to the samples cured at low temperature for a long time. Hence, dehydration should be the major reason for the decrease in mechanical strength of these samples.

There are two opinions in the literature on how humidity influences the geopolymerization. The humidity can promote the hydration of $Al^{3+}$ and $Si^{4+}$ ions from the metakaolin and increase the reaction yield[4]. Humidity also prevents the loss of the essential structural water content and reduces shrinkage and cracks. Another opinion is that humidity suppress the water evaporation necessary to polycondensation producing siloxane polymers with limited length and terminated with -O-Na and -O-H[13]. In this study, the samples cured in moisture had higher mechanical strength but also slightly higher porosity than those cured in air (Fig 1). The results in this study showed that the first explanation of the effect might take a major role over the second one. Since the samples cured in moisture showed no obvious decrease of strength after curing at high temperature and long time (96 hours), it implies the curing in high humidity represents the best condition to produce mechanically strong geopolymers.

CONCLUSIONS

The influences of curing time, temperature and humidity during geopoymerization were studied on a metakaolin-based geopolymer. Properties, i.e. compression strength, porosity, microstructure morphology and x-ray diffraction pattern, were measured for different samples. It showed that geopolymers needed at least 24 hours to gain full compression strength under all temperatures. The porosity of the samples slightly decreased as curing time increased. Higher temperature could increase strength in early curing stage but might result in a porous structure. However, geopolymer cured in higher temperature and for a long period of time had lower compression strength. Curing under humidity improved the strength but meanwhile gave slightly higher porosity.

REFERENCES

1      J, Davidovits, *Geopolymer Chemistry and Applications*. (Institut Géopolymère, 2011).

2      P. Duxson, A. Fernández-Jiménez, J. L. Provis, G. C. Lukey, A. Palomo, and J. S. J. Deventer, "Geopolymer technology: the current state of the art," J Mater Sci **42** (9), 2917-2933 (2007).

3      Divya Khale and Rubina Chaudhary, "Mechanism of geopolymerization and factors influencing its development: a review," J Mater Sci **42** (3), 729-746 (2007).

4      J. G. S. van Jaarsveld, J. S. J. van Deventer, and G. C. Lukey, "The effect of composition and temperature on the properties of fly ash- and kaolinite-based geopolymers," Chemical Engineering Journal **89** (1–3), 63-73 (2002).

5      M. S. Muñiz-Villarreal, A. Manzano-Ramírez, S. Sampieri-Bulbarela, J. Ramón Gasca-Tirado, J. L. Reyes-Araiza, J. C. Rubio-Ávalos, J. J. Pérez-Bueno, L. M. Apatiga, A. Zaldivar-Cadena, and V. Amigó-Borrás, "The effect of temperature on the

geopolymerization process of a metakaolin-based geopolymer," Materials Letters **65** (6), 995-998 (2011).

6      Xiao Yao, Zuhua Zhang, Huajun Zhu, and Yue Chen, "Geopolymerization process of alkali–metakaolinite characterized by isothermal calorimetry," Thermochimica Acta **493** (1–2), 49-54 (2009).

7      Pavel Rovnaník, "Effect of curing temperature on the development of hard structure of metakaolin-based geopolymer," Construction and building materials **24** (7), 1176-1183 (2010).

8      M. Y. Khalil and E. Merz, "Immobilization of intermediate-level wastes in geopolymers," journal of nuclear materials **211** (2), 141-148 (1994).

9      Ángel Palomo, Santiago Alonso, Ana Fernandez-Jiménez, Isabel Sobrados, and Jesús Sanz, "Alkaline Activation of Fly Ashes: NMR Study of the Reaction Products," Journal of the American Ceramic Society **87** (6), 1141-1145 (2004).

10      Valeria F. F. Barbosa, Kenneth J. D. MacKenzie, and Clelio Thaumaturgo, "Synthesis and characterisation of materials based on inorganic polymers of alumina and silica: sodium polysialate polymers," International Journal of Inorganic Materials **2** (4), 309-317 (2000).

11      Kostas Komnitsas and Dimitra Zaharaki, "Geopolymerisation: A review and prospects for the minerals industry," Minerals Engineering **20** (14), 1261-1277 (2007).

12      T. Ramlochan, P. Zacarias, M. D. A. Thomas, and R. D. Hooton, "The effect of pozzolans and slag on the expansion of mortars cured at elevated temperature: Part I: Expansive behaviour," Cement and Concrete Research **33** (6), 807-814 (2003).

13      FENG, #160, Dang, MIKUNI, Akira, HIRANO, Yoshinobu, KOMATSU, Ryuichi, IKEDA, and Ko, *Preparation of geopolymeric materials from fly ash filler by steam curing with special reference to binder products*. (Nippon seramikkusu kyokai, Tokyo, JAPON, 2005).

# CHEMICALLY BONDED PHOSPHATE CERAMICS SUBJECT TO TEMPERATURES UP TO 1000° C

H. A. Colorado[a,b,1], C. Hiel[c], J. M. Yang[a]

[a]Materials Science and Engineering Department, University of California, Los Angeles, CA 90095, USA
[b]Universidad de Antioquia, Mechanical Engineering. Medellín-Colombia
[c]Composite Support and Solutions Inc. San Pedro, California.

ABSTRACT

In this paper wollastonite-based chemically bonded phosphate ceramic (CBPC) composites were exposed to temperatures up to 1000°C in an oxidation environment. Glass, graphite and basalt fibers have been used as reinforcements. CBPCs were fully synthesized at room temperature. Samples were exposed to different temperatures (200, 400, 600, 800 and 1000°C) in air atmosphere in order to determine their effect on their bending strength and microstructure. The characterization was conducted by scanning electron microscopy, by x-ray diffraction, and three point bending tests. Weibull distribution statistics showed a high variability for the bending strength, which was associated with the deterioration produced by the heat exposure.

INTRODUCTION

Wollastonite-based Chemically Bonded Phosphate Ceramics (Wo-CBPCs) can reach a compressive strength of over 100 MPa in minutes and their density is usually below 2.2 $g/cm^3$. These materials bridge the gap between the attributes of sintered ceramics and traditional hydraulic cements: CBPCs have mechanical properties approaching sintered ceramics and a high stability in acidic and high temperature environments; its processing is inexpensive, castable, and environmentally friendly. The bonding in CBPCs is a mixture of ionic, covalent, and van der Waals bonding, with the ionic and covalent dominating. In traditional cement hydration products, van der Waals and hydrogen bonding dominate [1-3].

There is an increasing interest in building materials under the worst conditions such as fire and high strain rates. Further improvements in the concrete technology make it possible to increase building heights such as skyscraper: the Ingalls Building of 64m tall (built in 1903 in Cincinnati, Ohio, US) to the Petronas Towers from 452 m tall (built in 1994 Kuala Lumpur, Malaysia). A new generation of skyscrapers and public buildings demand designs that consider extreme conditions. The current designs do not take these design criteria into the account, which is an opportunity to develop new materials that work well under fire and high pressure conditions.

In this study, Wollastonite-based Chemically Bonded Phosphate Ceramic (Wo-CBPCs) composites were fabricated by pultrusion [4-5]. The resin was obtained by mixing calcium silicate (wollastonite, $CaSiO_3$) and a phosphoric acid ($H_3PO_4$) formulation. The ceramics obtained are multiphase materials containing calcium phosphates (i. e. brushite or monetite), silica, quartz, and residual wollastonite powder. The pot life of the resin, the reaction rate, and the reaction heat are a function of the chemical composition of the raw materials (acidic liquid and powder), and the wollastonite powder size with its aging grade ([1], [6] and [3]). The objective of this project is to produce high performance pultruded fiber reinforced chemically bonded phosphate ceramics, by using glass, basalt, and carbon fibers.

EXPERIMENTAL PROCEDURE

The Wo-CBPC samples were fabricated by mixing 120g of a phosphoric acid formulation (from Composites Support and Solutions-CS&S) and 100g of mineral wollastonite powder (from NYCO Minerals). The mixing was conducted in a Thinky mixer apparatus for 2 min. Next, the impregnation of the fibers in a resin bath and setting of the ceramic by heating in a die completed the pultrusion process. The die temperature was 110°C and the pull speed was 30.5 cm/min.

Ceramic matrix composites of glass, carbon and basalt fibers were fabricated. In all cases, 15 % volume of fibers were used. Glass fibers Flexstrand 225 from Fiber Glass Industries Inc. [8], Carbon fibers HTS40 from Toho Tenax [9], and Basalt fibers BCF13-1200KV12 from Kammemy Vek [10], were used in this research. Table 1 summarizes the main data provided by the manufacturers.

Table 1 Properties of the glass, carbon, and basalt fibers used in this research.

| Fiber type | E-glass | Basalt | Carbon |
|---|---|---|---|
| Tensile strength of single filaments [MPa] | 3100-3800 [11] | 4000-4300 | 4205-4620 |
| Tensile modulus of single filaments [CBP-Ca] | 72-76 | 87-92 | 236-239 |
| Fiber diameter (μm) | 23.5 | 10-22 | 7.1 |

In order to stabilize the weight loss of the ceramic due to the water present in the acidic formulation, all samples were kept in a furnace at 50°C for 24 hours, then at 100°C for 24 hours. This procedure was conducted in order to minimize the damage caused by the shrinkage (due to the release of water), which leads in micro-cracking which decreases the strength of the material.

Ceramic composites were exposed for 1h to temperatures of 200°C, 400°C, 600°C, 800°C and 1000°C then slowly cooled in the oven. For each composite and temperature, twenty samples were thermally exposed. After each thermal process, three point bending tests were conducted and results were analyzed by Weibull statistics. Scanning electron microscopy was conducted over samples polished and mounted on an aluminum stub (after sputtered in a Hummer 6.2 system at conditions of 15mA AC for 30 sec to obtain a thin film of Au) in a SEM (JEOL JSM 6700R). X-Ray Diffraction (XRD) experiments were conducted for the CBPC without fibers at different temperatures with X'Pert PRO (Cu Kα radiation, λ=1.5406 Å) equipment, at 45KV and scanned between 10° and 80°.

Three point bending tests were performed on an Instron 4411 apparatus at a crosshead speed of 2.5 mm/min. The span length was 100 mm and sample dimensions were 7.0 mm x 13 mm x 200 mm.

ANALYSIS AND RESULTS

Figure 1 shows fiber reinforced CBPCs fabricated by pultrusion and by hand. All samples were fabricated with 15 % volume of fibers and the fibers used were glass, carbon and basalt. All samples were dried 1 day at 50°C, 1 day at 100°C and 1 day at 200°C in order to remove the water completely. Glass fibers Textrand 225 from Fiber Glass Industries, Graphite fibers Tenax(R)-A 511, and Basalt fibers BCF13-1200KV12 Int from Kammemy Vek were used in this research. In order to plot the Weibull distributions, twenty samples were tested at each temperature and composition and then tested in three point bending tests.

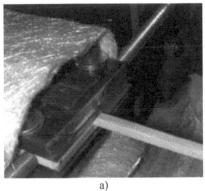 

a)                                                          b)

Figure 1 a) pultrusion die during manufacturing of CBPCs reinforced with glass fibers,
b) pultruded bars of CBPCs reinforced with glass fibers (white) and with carbon fibers
(black).

Figure 2 shows XRD results for the Wo-CBPC samples after the exposure to different temperatures. All reactions are presented in equations 1 to 4. The mixing of wollastonite with the phosphoric acid produces silica ($SiO_2$) and brushite ($CaHPO_4 \cdot 2H_2O$), as shown in equation 1. XRD shows that the Wo-CBPC without processing (as made) and Wo-CBPC after a drying process at 200°C, only water is removed and no further changes in the material were found. After the exposure at 400°C, all brushite transformed to monetite ($CaHPO_4$) by loosing the bonded water to the structure as described in equation 2. For temperature exposures over 600°C, the monetite transforms to calcium pyrophosphate ($Ca_2P_2O_7$), which is a calcium phosphate stable at high temperature, see equation 3. At higher temperatures tricalcium phosphate ($Ca_3(PO_4)_2$) was formed, as a byproduct of CaO and $Ca_2P_2O_7$. Finally, it was found that after the exposure at 1000°C parawollastonite appears. Since it was found several hundred of degrees below the prediction in the phase diagram, it is proposed that this is due to the interdiffusion phenomena of metals ions between the phosphates phases and the residual wollastonite grains.

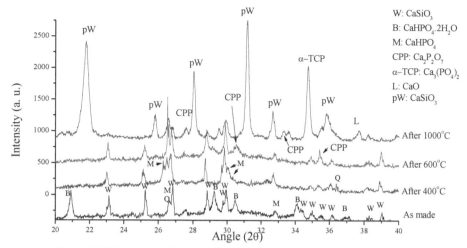

Figure 2 XRD od the wollastonite based CBPC under different temperatures.

$$CaSiO_3 + H_3PO_4 + H_2O = SiO_2 + CaHPO_4. 2H_2O \qquad (1)$$

$$2CaHPO_4.2H_2O \rightarrow 2CaHPO_4 + 4H_2O \qquad (2)$$

$$CaHPO_4 \rightarrow Ca_2P_2O_7 + 2H_2O \qquad (3)$$

$$Ca_2P_2O_7 + CaO \rightarrow Ca_3(PO_4)_2 \qquad (4)$$

Glass fibers

The Weibull distribution of pultruded glass fiber reinforced Wo-CBPC composites at different exposure temperatures is shown in Figure 3. For all samples, exposure time was 1h. As temperature is increased, the curves move toward low bending strength values. The variability is reduced as temperature is reduced as well.

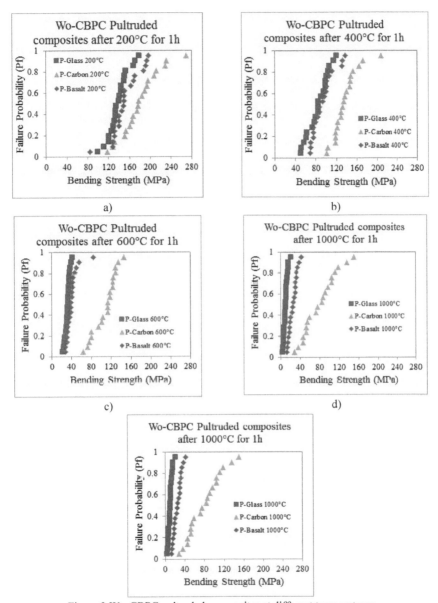

Figure 3 Wo-CBPC pultruded composites at different temperatures

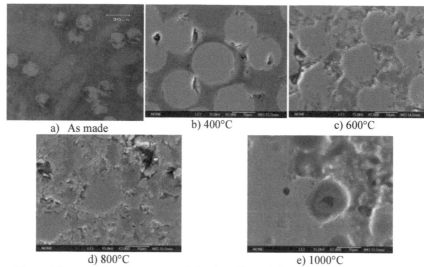

a) As made      b) 400°C      c) 600°C

d) 800°C      e) 1000°C

Figure 4 Cross section view images of the glass fiber-reinforced composite at different exposure temperatures.

Figure 4a shows the CBPC with glass fibers as made. It can be observed that residual wollastonite grains certainly limit the fiber impregnation and the amount of fiber added to the composite.

After exposure to 400°C (also at 500°C but not shown since not significant changes were found), Figure 4b, the fiber is still in good condition. However, after the exposure to 600°C, Figure 4c, the fiber borders are deteriorated and partial melting is observed. After exposure to 800°C, Figure 4d, it is hard to recognize the fiber and the porosity of the matrix increased significantly. After exposure to 1000°C, Figure 4e, the fiber is completely melted and in most cases dissapeared, which is reflected in the empty holes formerly occupied by fibers. The melting or partial softening of the fibers and silica phase in the composite can impregnate some micro deffects, like flaws from the matrix, which reduce the flaw number, and decreases the variability in the strength values (and increases in the Weibull modulus).

Carbon fibers

The Weibull distribution of pultruded carbon fiber reinforced Wo-CBPC composites at different exposure temperatures is shown in Figure 3. As occurred for glass fibers, as temperature is increased, the curves move toward low bending strength values and the variability is reduced as well.

Figure 5a, b, c, d, f and h show cross section view images of the composite as made, after 200°C, 400°C, 600°C, 800°C, and 1000°C respectively. It is observed that after exposed to 400°C, the fiber interface started to deteriorate. The damage increases as temperature increases. After the exposurure to 800°C, part of the fiber dissapeared due to the oxidation in byproducts of $CO$ and $CO_2$. After exposed to 1000°C, there is no evidence of carbon fibers.

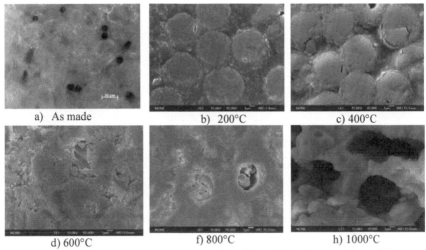

a) As made      b) 200°C      c) 400°C

d) 600°C      f) 800°C      h) 1000°C

Figure 5 Cross section view images of the carbon fiber-reinforced composite at different exposure temperatures.

Basalt fibers

The Weibull distribution of pultruded basalt fiber reinforced Wo-CBPC composites at different exposure temperatures is shown in Figure 3. As occurred for glass and basalt fibers, as temperature is increased, the curves move toward low bending strength values and the variability is reduced as well.

Figure 6 shows SEM images also at different exposure temperatures. It was observed that after exposure to 1000°C some fibers remain without major structural damages while others show significant degradation. This behavior shows a heterogenous oxidation-degradation of the fiber due to the variability in the matrix composition surrounding the fiber. The most significant change in this composite appears after exposure to 1000°C. Partial melting is revealed by the color degradation and its visually spread out into the matrix.

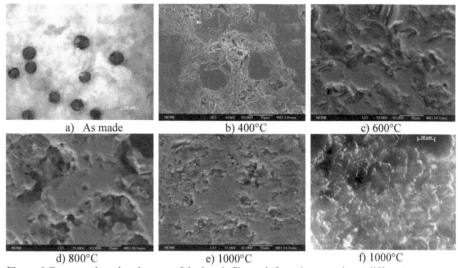

|  |  |  |
|:---:|:---:|:---:|
| a) As made | b) 400°C | c) 600°C |
| d) 800°C | e) 1000°C | f) 1000°C |

Figure 6 Cross section view images of the basalt fiber-reinforced composite at different exposure temperatures.

Figure 7 shows the summary of Weibull modulus for Wo-CBPC pultruded composites at different temperatures. The variability of the Weibull parameter with temperature was discussed before. The most stable Weibull parameter is for basalt fiber composites. The highest modulus for the composite after the thermal exposure treatment, which is best as this indicates less variability, was found to be for the fibers as glass at 200°C, carbon at 400°C, glass at 700°C and 800°C, and basalt at 1000°C.

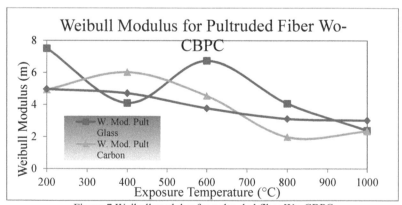

Figure 7 Weibull modulus for pultruded fiber Wo-CBPC

DISCUSSION

Figure 8a shows a representation of the initial raw materials in contact (acidic liquid with wollastonite grains). As soon as the acidic liquid is in contact with the wollastonite powder, calcium ions $Ca^{+2}$ start to dissolve in the liquid. These ions come mostly from the surface of the particles, of about less than $1\mu m$ depth [12] and therefore in the final ceramic, the residual grains have nearly a silica layer, as represented in Figure 8b. As time is going on, the 'sol' gradually evolves towards the formation of a 'gel' that by polymerization forms the main binding phase, crystalline and amorphous calcium phosphate ($CaHPO_4.2H_2O$). This acid-base reaction is exothermic. Since there are residual wollastonite grains, the CBPC described is a composite material in which wollastonite acts as a reinforcement of the matrix.

During the pultrusion process, fibers were impregnated with the resin (far from its gel-point, [12]), Figure 8c. At the very beginning of the pultrusion process the residual wollastonite particles ($CaSiO_3$), the acidic liquid (based on phosphoric acid $H_3PO_4$ and metal oxides, used to control de setting time) impregnate the fibers. The acidic liquid reacts with the calcium ions and form by a polymerization process a continuous calcium phosphate matrix composed of brushite ($CaHPO_4.2H_2O$) and amorphous calcium phosphates. Figure 8d shows a representation of this fiber reinforced Wo-CBPC. A thin layer of amorphous silica ($SiO_2$) around the unreacted part of the wollastonite grain is shown. In general, almost no reaction was found between the fibers and the matrix after the room temperature process. However, as temperature increases, deterioration in the fibers, matrix and interface matrix/fiber increases as well.

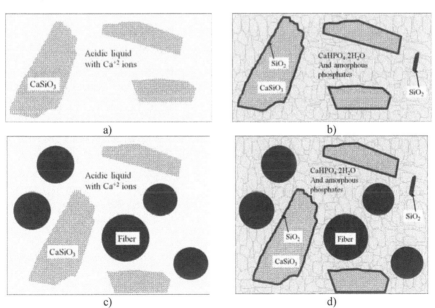

Figure 8 a) representation of the first instant of the wollastonite ($CaSiO_3$), the acidic formulation and the fibers; b) Wo-CBPC; c) resin with fibers; d) fiber reinforced Wo-CBPC.

CONCLUSION
    The wollastonite-based CBPC was studied after the exposure to different temperatures in air environment. The CBPC fabricated in study goes over several phase tranformations as temperature is increased. Due to the presence of many phases and due to the oxidation environment under which the thermal exposure treatments were conducted, parawollaatonite was found after 1000°C for 1h.
    It was oberved that all fibers tested are unstable after 600°C. Glass and basalt fibers are melted after 1000°C. In the case of carbon fibers, they are completely evaporated after 1000°C. Weibull distribution showed high variability, which has been associated with the structural deterioration due to thermal exposure, phases changes, and deterioration of the fibers itself.

ACKNOWLEDGEMENTS
    The authors wish to Colciencias from Colombia for the grant to Henry A. Colorado.

REFERENCES
[1]. Della M. Roy. (1987). New Strong Cement Materials: Chemically Bonded Ceramics, Science, Vol. 235: 651-58.
[2]. Henry Colorado, Clem Hiel, H. Thomas Hahn and Jenn-Ming Yang Chemically Bonded Phosphate Ceramic Composites, Metal, Ceramic and Polymeric Composites for Various Uses, John Cuppoletti (Ed.), ISBN: 978-953-307-353-8, InTech, 265-282 (2011).
[3]. H. A. Colorado, C. Hiel and H. T. Hahn. (2009). Processing-structure-property relations of chemically bonded phosphate ceramics composites, *Bull. Mater. Sc.* Ref.: Ms No. BOMS-D-09-00499R1.
[4]. H.A. Colorado, J. Pleitt, C. Hiel, J.M. Yang, H.T. Hahn, C.H. Castano. Wollastonite based-Chemically Bonded Phosphate Ceramics with lead oxide contents under gamma irradiation. Journal of Nuclear Materials, 425, 197–204 (2012).
[5]. H. A. Colorado, C. Hiel and H. T. Hahn. (2011). Pultruded glass fiber-and pultruded carbon fiber-reinforced chemically bonded phosphate ceramics. Journal of Composite Materials, 45(23), 2391-2399.
[6]. H. A. Colorado, C. Hiel and H. T. Hahn. (2010). Influence of Particle Size Distribution of Wollastonite on the Mechanical Properties of CBPCs (Chemically Bonded Phosphate Ceramics). Processing and Properties of Advanced Ceramics and Composites III: 85-98.
[7]. Meyers, M. A. and Chawla, C. (1999). Mechanical behavior of materials. Prentince-Hall Inc. New Jersey, USA, 680.
[8]. Fiber Glass Industries, Inc., fiber glass manufacturer. New York, USA. Website: http://fiberglassindustries.com.
[9]. Toho Tenax America Inc, carbon fiber manufacturer. Tennessee, USA. Website: http://www.tohotenaxamerica.com/contfil.php
[10]. Kamenny Vek, basalt fiber manufacturer. Moscow, Russia. Website: http://www.basfiber.com/
[11]. Frederick T. Wallenberger, James C. Watson, and Hong Li. Glass Fibers. 2001 ASM International. ASM Handbook, Vol. 21: Composites (#06781G): 27-34.
[12]. H. A. Colorado. PhD thesis. Mechanical behavior and thermal stability of acid-base cements and composites fabricated at ambient temperature (2013).

MECHANICAL PROPERTIES OF GEOPOLYMER COMPOSITE REINFORCED BY
ORGANIC OR INORGANIC ADDITIVES

E. Prud'homme, P. Michaud and S. Rossignol
Groupe d'Etude des Matériaux Hétérogènes (CEC-GEMH-ENSCI)
12 rue Atlantis, 87068 Limoges Cedex, France.

E. Joussein
GRESE, EA 3040
123 avenue Albert Thomas, 87060 Limoges, France.

corresponding author - sylvie.rossignol@unilim.fr – tel.: 33 5 87 50 25 64

ABSTRACT
Metakaolin-based geopolymer materials reinforced with mineral or vegetable additives were developed, and the evolution of the mechanical properties was investigated. Three parameters are investigated in this study, the alkali element used for synthesis (Na or K), role of additives (sands, fine silica or fibers) and the impact of aging. All the results evidence the role played by the various interfaces between geopolymer matrix and the additive. The nature of potassium alkali appears promising in the presence of vegetable additive.

1. INTRODUCTION
    The principle of composite materials is to combine materials with complementary properties (thermal, mechanical, etc.) to optimize their performances and to obtain a material with unique properties[1]. In the case of geopolymers, the goal is to find a solution to the brittle nature of the matrix and thus improve the toughness of the material. These composites can be obtained by impregnating fiber preforms of different types (metal, glass, carbon, ceramic) or by adding aggregates.
    Geopolymers are three dimensional amorphous aluminosilicate binders. They were introduced into the cementitious world by Davidovits in 1978[2]. Their synthesis occurs at low temperature by alkali activation of aluminosilicate obtained from industrial coproducts, calcined clays or raw mineral. In highly alkaline solution, reactive materials are rapidly dissolved into solution to form free $SiO_4$ and $AlO_4^-$ tetrahedral units[3,4]. During the polycondensation reaction, the tetrahedral units are linked in an alternate manner to yield amorphous geopolymers. Mechanical properties of this material depends on various parameters such as alkali element used for synthesis or the ratio between silicon and aluminum, etc.[5] Indeed the geopolymer molecular structure is very sensitive to alkali cation, particularly at low Si/Al ratios[6,7]. For example in terms of compressive strength, potassium-based samples present more interesting properties than do sodium-based geopolymer for a Si/Al ratio from 1.4 to 1.9[5].
    The increase of mechanical properties is an important area of research in geopolymers and an important variety of additives has been tested (carbon, glass or steel fibers or PVA)[8,9,10]. Addition of sand is commonly used for mechanical propertiy modification. Indeed the work of Dove et al.[11] has evidenced the hydrolysis reaction of silica polymorphs in a alkali medium (Eq. 1).

$$SiO_2(s) + 2H_2O \leftrightarrow H_4SiO_4 \qquad \text{Eq. 1}$$

This reaction is generally catalyzed by the alkali action of $K^+$ and $Na^{+12}$. The formation of interface between geopolymer matrix and sands then becomes possible, leading to the modification of mechanical properties.

Addition of fibers has known important development in the 1990's due to health problems related to asbestos for example. These one are indeed much healthier[13]. The incorporation of fiber in a mineral matrix allowed the control of mechanical properties and the decrease of density. Previous work, by Al Rim et al.[14] has already investigated the mechanical properties of composite material based on sawdust or wood chips and a cement/clay matrix. They have evidenced a decrease of density and mechanical properties. However materials reinforced with fibers present a better deformability and insulating character[10].

The aim of the study is to investigate the impact of three parameters on the mechanical properties of geopolymers based on potassium or sodium alkaline elements. Various additives are introduced as reinforcements in the geopolymer matrix (silica, sand or fibers). As previously enounced geopolymer are not especially stable in time, so mechanical tests are performed on samples at 7 days and 28 days aged. The evolution of mechanical properties is investigated using a four-point flexural test.

## 2. EXPERIMENTAL PART

### 2.1. Samples preparation

Materials are prepared using metakaolin (from AGS[1]), alkali silicate (for potassium silicate: $Si/K = 1.7$, density 1.20, and for sodium silicate: $Si/Na = 1.71$, density 1.33) and alkali hydroxide (respectively 85.7% and 99.0% of purity for potassium and sodium). As shown in Figure 1, the mixture was prepared by mixing alkali hydroxide pellets in alkali silicate. Metakaolin was then added under magnetic stirring. After stirring, the mixture was put in a closed mold at 70°C for 24 hours. After this time the material is removed from the mold and kept in an oven at 50°C for aging. The mold used during this study was a rectangular mold allowing directly the obtaining of formatted sample 10x10x110 $mm^3$. Compositions of geopolymer matrix are presented Table 1.

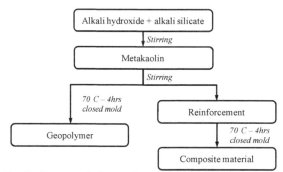

Figure 1. Synthesis protocol of geopolymer and composite geopolymer materials

Table 1. Compositions of the reactive mixture geopolymer (compositions are identical for 7 or 28 days aging).

|  | $n_{Si}/n_{Al}$ | $n_{Si}/n_K$ | $n_{Si}/n_{Na}$ | wt% water |
|---|---|---|---|---|
| $G_{K/Y}$ | 1.6 | 2.1 | - | 37.1 |
| $G_{Na/Y}$ | 1.8 | - | 1.9 | 37.1 |

[1] AGS France, 17270 Clerac, France.

In order to obtain composite materials four kinds of additives where introduced into the mixture before casting. Theses additives are mixed in the reactive mixture before placement at 70°C according to the flow chart in Figure 1. The four additives are silica[2] (denoted Si400, 99.0% $SiO_2$, $d_{50}$=10 µm), Fontainebleau Sand[2] (denoted SF), which has a high purity in silica and a particle size lower than 200 µm, normalized sand[3] (denoted SN, $d_{50}$=700 µm) and linen fibers[4] (denoted LF, diameter around 1.45mm). In order to increase the interaction between fibers and geopolymer matrix, these first are functionalized by immersion in a potash solution at 3M (LFT1) or 6M (LFT2) during 24hrs, rinsed with osmosis water and dry at 70°C. Additives amount in geopolymer matrix whatever the alkali cation are presented Table 2. After 24 hours of synthesis at 70°C, all the samples are unmolded and kept in an oven at 50°C and they are mechanically tested after 7 days or 28 days in order to study the influence of aging.

Table 2. Amount of additives added to geopolymer matrix. Percentages are by weight for Si400, SF and SN and by volume for linen fibers.

|  | Si400 | SF | SN | LF | LFT1 | LFT2 |
|---|---|---|---|---|---|---|
| Amount / % | 15 / 37 / 75 / 112 / 150 | 150 | 150 | 3 | 3 | 3 |

The nomenclature used to define samples is $G_{X/Y/Z}$:
-index X is relative to the alkaline element used for synthesis (potassium (K) or sodium (Na)),
-index Y is relative to the aging of samples (7 days (7) or 28 days (28))
-and index Z is relative to the additional reinforcement
So for example, sample $G_{K/28/SF}$ is a geopolymer synthesized with potassium and Fontainebleau sand aging 28 days at 50°C.

## 2.2. Characterization

### 2.2.1. Structural and physic-chemical properties

FTIR spectra were obtained from a ThermoFischer Scientific 380 infrared spectrometer (Nicolet) using the attenuated total reflection (ATR) method. The IR spectra were gathered between 500 and 4000 $cm^{-1}$ with a resolution of 4 $cm^{-1}$. OMNIC (Nicolet Instruments), the commercial software, was used for data acquisition and spectral analysis.

X-ray patterns were acquired via X-ray diffraction (XRD) experiments on a Brucker-AXS D 5005 powder diffractometer using $Cu_{K\alpha}$ radiation ($\lambda_{K\alpha}$ = 0.154186 nm) and a graphite back-monochromator. XRD patterns were obtained using the following conditions: dwell time: 2 s; step: 0.04° (2θ). Crystalline phases were identified by comparison with PDF standards (Powder Diffraction Files) from ICDD.

---

[2] Ceradel, 19 à 25 rue Fréderic Bastiat, BP 1598, 87022 Limoges Cedex 9, France.
[3] Société Nouvelle du Littoral, Z.A. – B.P. 9, 11370 Leucate, France.
[4] Sicomin, B.P 23 – 31 Avenue de la Lardière 13161 Châteauneuf les Martigues FRANCE

2.2.2. Mechanical properties

Mechanical properties were evaluated using the four points bend testing (Figure 2). This test is commonly used for brittle material. The sample is placed on the two lower supports and a load F is applied at two symmetrical points. During bend testing, the neutral axis generated by the mean curve is not subject to any solicitation and therefor has a zero elongation; fibers located on the upper side are subjected to compression while those located on the underside are subject to tension.

In this type of test, the maximum stress develops throughout the bottom fiber in the central part of the specimen so that the critical crack can occur in any section of this central part. Equation [2], based on the theory of elasticity, calculates the stress at break.

$$\sigma_r = \frac{3F(L-l)}{2bh^2}$$

[2]

With: F: load applied on the two symmetrical loading points (N); B: width of the specimen (mm); H: height of the specimen (mm); L: distance between two lower supports (mm); L: distance between the two loading points (mm). Specimens of rectangular cross section $10 \times 9 \times 110$ mm$^3$ were tested on a Lloyd EZ 20 apparatus with a sensor of 5 kN. The speed of movement of the span is 0.2 mm/min. For each type of material, 5 specimens were tested.

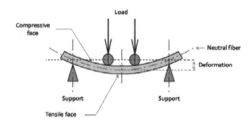

Figure 2. Deformation schematic of a sample in four point bend.

3. RESULT AND DISCUSSION

3.1. Geopolymer characterization

Two geopolymers matrices were first synthesized corresponding to the formulation $K_{0.10}^+\{(SiO_2)_{1.33}, AlO_2\}_{0.15}$, 0.071 $H_2O$ and $Na_{0.10}^+\{(SiO_2)_{1.58}, AlO_2\}_{0.12}$, 0.34 $H_2O$. Regardless of the alkaline element used, the geopolymers were strengthened by polycondensation reactions. This phenomenon was previously reported by monitoring dense or porous geopolymer formation using infrared spectroscopy[15]. Geopolymers are generally amorphous materials, and their characterization by X-ray diffraction (Figure 3 (a)) does not provide much information about their amorphous nature. The two geopolymers displayed an amorphous pattern with maximum diffraction intensity around 27 (°2θ). However, the diffraction pattern of metakaolin, which is an initial component of all synthesized geomaterials, shows some crystalline phases, like quartz, illite and orthoclase ($KAlSi_3O_8$), which are less present in the diffraction patterns of the final products, reflecting the alteration of these raw materials. This alteration and the restructuring of the materials are also evidenced by an obvious shift in the scattering diffraction peak of each geomaterial in relation to that of the raw materials. In fact, when the starting materials were activated with an alkaline solution, the scattering diffraction peak shifted from ~21 to ~27–28 (°2θ) (Figure 3 (a)), suggesting that the local bonding environment changed during the polymerization process. This mechanism could be observed in all geopolymer XRD patterns, corresponding to the exchanges between sodium or potassium elements. However, the maximum diffraction intensity was not the same for the geopolymers. This phenomenon can be explained by the difference in bond energy due to the alkaline cation[16].

(a)

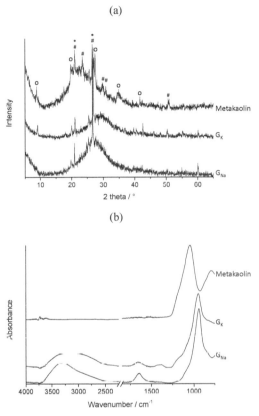

(b)

Figure 3. (a) XRD pattern and (b) infrared spectroscopy of metakaolin, potassium-based geopolymer and sodium-based geopolymer. PDF files: *, Quartz 01-086-1560; °, Illite 00-002-0056; #, KAlSi$_3$O$_8$ 00-025-0618.

The dissolution of the raw materials can also be examined by infrared spectroscopy. Figure 3 (b) allows for the comparison between metakaolin and the two kinds of geomaterials synthesized. The spectra of the geopolymers corresponded to the end of formation. The bands at 3255 and 1620 cm$^{-1}$, respectively, were attributed to the Si–O–H bond and water bending[17]. The main band detected for the geomaterial was due to Si–O–M bonds[18,19] (M=K or Na), located in the 1100–950 cm$^{-1}$ range. Their specific positions depend on the length and bending of the Si–O–M bond, which explains the small shift of the main peak among the two samples. Other peaks were located at 1200, 880, 800 and 670 cm$^{-1}$ corresponding to Si-O-Si in the asymmetric stretching mode[20,21], Al–OH, Si–O–Si[22], Si–O–Al[18], and O–Si–O[16], respectively.

3.2. Mechanical properties

Mechanical properties of the various samples are evaluated by the four point bend testing. This test presents the advantages to clearly evidence the different behaviors of composite material and to evidence, for example, the load take-over by fibers. Generally speaking, geopolymer materials are considered as brittle material due to their type of failure that only presents elastic deformation. Figure 4 shows the evolution of the flexural strength as a function of displacement for two kinds of sample, one without reinforcement and one with linen fibers. The first present an evolution really closed to a straight line reflecting the brittle character of this material. In comparison, the second curve relative to the same material reinforced with fibers presents a broken curve with different zones of failure. During zone 1, the matrix is principally solicited and the material behaves as a geopolymer matrix. At the end of zone 1, the matrix breaks inducing a loss of strength, which is taken up by the linen fiber. Solicitation continues with displacement increase, and fibers are principally solicited during zone 2 before the final failure. This final failure appears at a higher value than the matrix. This kind of curve evidences the role of fiber on the mechanical properties of materials. Indeed, even if they embrittle the material (lower maximal breaking strength for matrix at the end of zone 1), the load is taken up by fibers allowing higher strength and a less violent failure. This kind of phenomenon evidences also the presence of an interface between fibers and matrix.

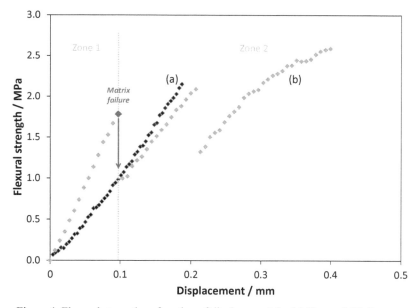

Figure 4. Flexural strength as function of displacement for (a) $G_{K/7}$ and (b) $G_{K/7/LF}$.

3.2.1. Study of geopolymer matrix

In a first approach, tests were performed on geopolymer matrix based on sodium or potassium alkali element without reinforcement. In order to evidence the impact of aging on the mechanical properties tests were realized after a cure of 7 days or 28 days at 50°C. Figure 5 shows the maximal flexural strength for the two different matrices age 7 or 28 days. At 7 days aging, $G_{Na/7}$ sample had higher maximal flexural strength than did the $G_{K/7}$ sample with a flexural strength of respectively 4.7 MPa and 1.2 MPa at 7 days.

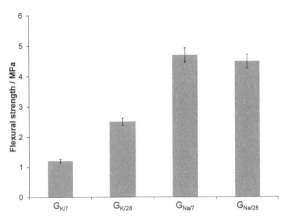

Figure 5. Evolution of the flexural strength as a function of aging and alkali element.

With aging the mechanical properties are constant for sodium-based samples. The difference of 0.2 MPa is included in the measurement error. At the opposite $G_K$ samples evidence a clear properties evolution with aging from 1.2 to 2.5 MPa. However the performance is still lower than $G_{Na}$. This difference can be explained by the difference in reactivity of these two alkali elements. Indeed, sodium is a high reactivity alkali element with its small ionic radius ($r_{Na}$=0.95 Å, $r_K$=1.33 Å). The alteration of raw material, the integration of sodium into network and then the structuration of material appears to be faster than for the potassium-based material. The system is then rapidly blocked. Alternatively, corresponding to the bigger size of potassium, reactions are slower. The network forms progressively. This hypothesis could be validated by testing material at longer aging times and could be correlated with the size of colloids.

3.2.2. Introduction of additive

In order to modify the mechanical properties of geopolymer materials, additives were introduced into the geopolymer reactive mixture and mechanical properties were evaluated by four point bend testing at 7 days and 28 days of aging. In a first time three additives were introduced, two sands (SF and SN) and linen fibers in the proportions given Table 2. Results are reported Figure 6.

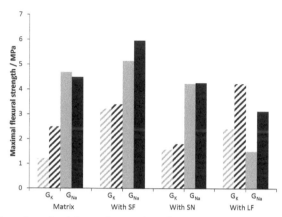

Figure 6. Evolution of maximal flexural strength as function of additives, aging (■ 7 days and ■ 28 days) and alkali element (■ potassium and ■ sodium).

Generally speaking, the addition of SN sand induces a decrease of mechanical properties compared to the matrix performances. Indeed, this normalized sand presents an important heterogeneity of size distribution and is constituted of fine (0.080 mm) and huge (1.60 mm) particles. This heterogeneity is relatively important compared to the sample size ($10\times10\times110$ mm$^3$). Moreover the presence of huge particles induces particle sedimentation and therefore the formation of different layers. Their cohesion can then be bad and induces low mechanical properties. On the other hand the addition of SF sand induces an increase of mechanical properties at 7 days of aging. This sand is more homogeneous in size, with particles size lower than 0.2 mm. The high concentration of sand in the composite (150 wt%) allows for a homogenous material, without formation of various layers. According to the work of Dove et al.[11], interfaces between geopolymer matrix and sand occur, which combined with the material homogeneity increase the mechanical properties. However, whatever the sand added, the mechanical properties do not evolve with aging time. For the sodium-based sample, it is in accordance with the behavior of the geopolymer matrix. In the case of potassium the matrix performance normally develops with aging time. With the addition of sand there is no performance evolution with aging time. The sand seems then to block the network evolution. The hypothesis can be formed that a part of potassium element was used to form an interface[23] and then did not participate to the formation of the geopolymer network.

The addition of linen fibers is a particular case due to their character. Indeed, linen fibers present a multitude of channels which permit the circulation of liquid containing distinct elements. In terms of mechanical properties, their introduction leads to an increase in the case of potassium-based material and a decrease in the case of sodium-based material (Figure 6). In fact in vegetable compound, the potassium element is one of the main constituent and has the ability to move into the cells through channels[5]. Consequently, it is easy for the potassium geopolymer matrix to establish an interphase at the interface between linen and Si-Al-K-O network enhancing mechanical properties.

Moreover for both alkali elements, the mechanical properties increase with aging time. This phenomenon was observed for the potassium geopolymer matrix but not for sodium geopolymer matrix. In this case, the very weak interphase create in the presence of sodium was not sufficient and some carbonate species could be formed involving a weakening of mechanical properties.

### 3.2.3. Impact of fiber treatment by potash solution

Previous results (§3.2.2.) demonstrated the increase of mechanical properties with aging for geopolymer reinforced with linen fibers. Whatever the alkali element used for synthesis, maximal flexural strengths were almost doubled with aging. This behavior can evidence the progressive development of interaction between matrix and fibers. This leads progressively to formation of interfaces. In order to accelerate this phenomenon the choice was taken to realize a treatment on linen fibers by immersion in potash solution. The choice of potassium is justified by its facility to circulate in vegetable matter[24]. Figure 7 shows the evolution of the flexural strength as a function of linen treatment for potassium or sodium-based matrix.

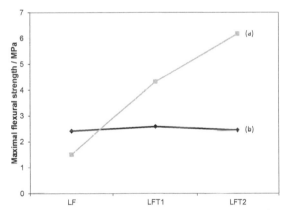

Figure 7. Flexural strength of (a) potassium and (b) sodium-based geopolymer reinforced by the various kinds of linen fibers at 7 days of aging.

In the case of the sodium-based sample the effect of linen fiber was useless for mechanical properties. Indeed this was constant at 2.4 MPa. In the case of potassium-based samples the maximal flexural strength increased with the concentration of potash solution treatment and shifted from 1.5 MPa for $G_{K/7/LF}$ to 4.3 MPa for $G_{K/7/LFT1}$ and to 6.5 MPa for $G_{K/7/LFT2}$. Then it can be assumed that a part of potassium contents of immersion solutions is absorbed by fibers and allowed the formation of a real interface between matrix and fibers.

This phenomenon does not appear with sodium due to the difference in cation. The interface was not improved since vegetable do not contain sodium element.

3.2.4. Impact of Si400 content

For all of these geopolymer compounds, mechanical properties highly depend on the nature of the alkali as well as the nature of the additive. Hence to understand only the effect of sand, we have chosen to investigate this effect in a potassium-based matrix. The amount of Si400 in the composite was varied from 0 to 150% by weight and the impact on mechanical properties was investigated (Figure 8). Generally, the evolution of mechanical properties can be described as a Gaussian with an optimum around 37 wt% of Si400 D addition. Flexural strength was then increased up to 5 MPa. In the presence of contents up to 20 wt%, it seems that the interaction between sand and matrix are isolated in the composite limiting the mechanical properties. On the other hand, in the presence of a high level of sand the interactions are too strong with certainly the creation of a sand coating which challenges the geopolymer matrix. The maximum in mechanical properties correspond to similar data obtained in the laboratory with another potassium silicate[25].

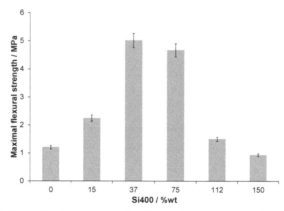

Figure 8. Evolution of the flexural strength of potassium-based geopolymer reinforced by different amount of Si400 at 7 days of aging.

4. CONCLUSION

The aim of composite materials is to combine materials with complementary properties to optimize their performances and to impart unique properties. In the case of geopolymers, the goal is to find a solution to their brittle nature. Metakaolin-based geopolymer materials reinforced with mineral or vegetable additives were developed, and the evolutions of the mechanical properties were investigated. Three parameters are investigated in this study; the alkali element used for synthesis (Na or K), the role of additives (sands, fine silica or fibers) and the impact of aging. The importance of the potassium geopolymer matrix in the presence of linen as fiber has been demonstrated. An optimal of 37 wt% content of sand was highlighted to favor the encapsulation of sand without a weak coating by silicate solution.

6. REFERENCES

[1] D. Gay, Matériaux composites, Hermès Edition, Paris, France, 1987.
[2] J. Davidovits, Chemistry and Applications, Institut Géopolymère, USA, 2008.
[3] H. Xu, Geopolymerisation of Aluminosilicate Minerals, PhD thesis, Department of Chemical Engineering, University of Melbourne, Australia, 2001.

[4] M.W. Grutzeck, D.D. Siemer, Zeolites Synthesized from Class F Fly Ash and Sodium Aluminate Slurry, *J. Am. Ceram. Soc.*, **80** (9), 2449-2458 (1997).

[5] P. Duxson, S.W. Mallicoat, G.C. Lukey, W.M. Kriven, J.S.J. Van Deventer, The effect of alkali and Si/Al ratio on the development of mechanical properties of metakaolin-based geopolymers, *Colloids Surf. A*, **292**, 8-20 (2007).

[6] P. Duxson, G.C. Lukey, F. Separovic, J.S.J. Van Deventer, Effect of alkali cations on aluminum incorporation in geopolymeric gels, *Ind. Eng. Chem. Res.*, **44** (4), 832-839 (2005).

[7] P. Duxson, J.L. Provis, G.C. Lukey, J.S.J. Van Deventer, F. Separovic, $^{29}$Si-NMR Study of Structural Ordering in Aluminosilicate Geopolymer Gels, *Langmuir*, **21** (7), 3028-3036 (2005).

[8] J. Hammell, P. Balaguru, R. Lyon, Influence of reinforcement types on the flexural properties of geopolymer composites, *International SAMPE Symposium and Exhibition* (Proceedings) **43** (2), 1600-1608 (1998).

[9] T.D. Hung, D. Pernica, D. Kroisová, O. Bortnovsky, P. Louda, V. Rylichova, Composites base on geopolymer matrices: Preliminary fabrication, mechanical properties and future applications, *Adv. Mater. Res.*, **55-57**, 477-480 (2008).

[10] Z. Li, Y. Zhang, X. Zhou, Short fiber reinforced geopolymer composites manufactured by extrusion, *J. Mater. Civ. Eng.*, **17** (6), 624-631 (2005).

[11] P.M. Dove, N. Han, A.F. Wallace, J.J. De Yoreo, Kinetics of amorphous silica dissolution and the paradox of the silica polymorphs, *Proceedings of the National Academy of Sciences of the United States of America*, **105** (29), 9903–9908 (2008).

[12] Y. Xiao, A.C. Lasaga, Ab initio quantum mechanical studies of the kinetics and mechanisms of quartz dissolution: OH⁻ catalysis, *Geochim. Cosmochim. Acta*, **60**, 2283–2295 (1996).

[13] M.A. Arsene, K. Bilba, A. Ouensanga, Etude d'un matériau composite : ciment/fibre de bagasse de canne à sucre, *Revue des Composites et Matériaux Avancés*, **1** (11), 7-20 (2001).

[14] K. Al Rim, A. Ledhem, O. Douzane, R.M. Dheilly, M. Queneudec, Influence of the proportion of wood on the thermal and mechanical performances of clay-cement-wood composites, *Cem. Concr. Compos.*, 1999, **21** (4), 269-276 (1999).

[15] E. Prud'homme, P. Michaud, E. Joussein, C. Peyratout, A. Smith, S. Arrii-Clacens, J.M. Clacens, S. Rossignol, Silica fume as porogent agent in geo-materials at low temperature, *J. Eur. Ceram. Soc.*, **30** (7) 1641–1648 (2010).

[16] D. Koloušek, J. Brus, M. Urbanova, J. Andertova, V. Hulinsky, J. Vorel, Preparation, Structure and Hydrothermal Stability of Alternative (Sodium Silicate-free) Geopolymers, *J. Non-Cryst. Solids*, **42** (2007) 967–9275.

[17] P. Innocenzi, Infrared spectroscopy of sol–gel derived silica-based films: a spectra-microstructure overview, *J. NonCryst. Solids*, **316**, 309–319 (2003).

[18] M. Criado, A. Palomo, A. Fernandez-Jiménez, Alkali activation of fly ashes. Part 1: Effect of curing conditions on the carbonation of the reaction products, *Fuel*, **84**, 2048–2054 (2005).

[19] J. Davidovits, Scientific Tools, X-rays, FTIR, NMR, Geopolymer: Chemistry and Applications, second ed, St-Quentin, France, 2008, 61–76.

[20] M.T. Tognonvi, S. Rossignol, J.-P. Bonnet, Effect of alkali cation on irreversible gel formation in basic medium, *J. NonCryst. Solids*, **357**, 43–49 (2011).

[21] C.A. Rees, J.L. Provis, G.C. Luckey, J.S.J. Van Deventer, Attenuated total reflectance Fourier transform infrared analysis of fly ash geopolymer gel aging, *Langmuir*, **23**, 8170–8179 (2007).

[22]T. Uchino, T. Sakka, K. Hotta, M. Iwasaki, Attenuated total reflectance Fourier transform infrared spectra of a hydrated sodium silicate glass, *J. Am. Ceram. Soc.*, **72** (11), 2173–2175 (1989).

[23]M. Tognonvi, J. Soro, S. Rossignol, Physical-chemistry of silica/alkaline silicate interactions during consolidation. Part 1: Effect of cation size, *J. Non-Cryst. Solids*, **358**, 81–87 (2012).

[24]F. Gouny, F. Fouchal, P. Maillard, S. Rossignol, A geopolymer mortar for wood and earth structures, Construction and Building Materials, **36**, 188–195 (2012).

[25]Private communication.

# EVALUATION OF GEOPOLYMER CONCRETES AT ELEVATED TEMPERATURE

Kunal Kupwade-Patil, Md. Sufian Badar, Milap Dhakal and Erez N. Allouche
Alternative Cementitious Binders Laboratory (ACBL),
Louisiana Tech University, Ruston, LA, USA

## ABSTRACT

Geopolymers are an emerging class of cementitious binders which possess a potential for high temperature resistance that could possibly be utilized in applications such as nozzles, aspirators and refractory linings. This study reports on the results of an investigation into the performance of fly ash based geopolymer binder in high temperature environments. Geopolymer concrete (GPC) was prepared using eleven types of fly ashes obtained from four different countries. High content alumina and silica sand were used in the mix for preparing GPC. GPC was subjected to thermal shock tests using ASTM C 1100-88. The GPC samples prepared with tabular alumina were kept in at 1093°C and immediately quenched in water. GPC specimens prepared with certain fly ashes exhibited signs of expansion along with cracking and spalling, while GPC prepared with specific Class "F" fly ash showed superior resistance to thermal shock. Microstructural analysis revealed that the resistance of GPC at elevated temperatures was dependent on the type of fly ash used, particle size distribution of fly ash, formation of zeolitic phases such as sodalite, analcime and nepheline and the overall pore structure of geopolymer concrete. The work indicates that the chemical composition and particle size distribution of the fly ash, type of fly ash (Class C & F) and the geopolymerization process plays a vital role in determining the suitability of geopolymer concrete for high temperature applications.

## INTRODUCTION

Ordinary Portland cement is the most widely used cementitious binder but It has a major limitation when subjected to elevated temperature. Traditional OPC based structures, when subjected to elevated temperature, suffer from loss of mechanical strength leading to a catastrophic failure [1]. The primary reason for OPC-based materials to fail after firing is the destruction of the calcium silicate hydrate (C-S-H) gel along with various crystalline hydrates [2-4]. A conventional approach to enhance the thermal properties of OPC is to use pozzolanic additives to binding calcium hydroxide to C-S-H gel; although this method could extend the temperature of application up to 700-800°C, it is associated with initial loss of mechanical strength and tends to lose strength further after firing. The pore structure of OPC concrete indicates that gel porosity increases significantly with increase in temperature. The gel and capillary water evaporate at 100-150°C, while this is accompanied by cracking and shrinkage between the temperature range of ~150°C-250°C. At 250-300°C, the compressive strength of the concrete decreases, due to the evaporation of chemically bound water from aluminum and ferrous constituents. An additional strength decrease was observed with the increase in temperature from 300-400°C, as the calcium hydroxide dehydrates to calcium oxide, while decomposition of C-S-H is completed at (400-650°C), exhibiting a significant strength reduction.

Alternative cements used for high temperature applications are costly and have disadvantages such as variations in mechanical strength, high viscosity, the short setting time finally leading to degradation and failure [5]. Studies have shown that alkali activated slag cements have exhibited higher resistance when subjected to elevated temperatures. The reasons for this superior behavior could be attributed to the formation of crystalline phases called anhydrous alumino silicates such as sodalite, analcime, and chabazite. These phases improve the crystallinity during heating up to 200-400°C, maintaining the structure up to approximately 800°C, and then recrystallize to new zeolite phases such as nepheline or albite. These contribute to enhancement of mechanical strength [6, 7]. Variables such as the type of fly ash (Class C or F), activation mechanism, silica to alumina ratio of the sodium alumino silicate hydrate (N-A-S-H), which affect the resistance of geopolymer concrete, are further investigated in the current study.

A comparison of advantages and disadvantages associated with alternative cementitious binders is shown in Table 1 [1]. Research in recent years has shown dramatic improvements in the performance of alternative cementitious binders, although a deep understanding is required of their chemistry, reaction mechanisms and property development. Geopolymer concrete is the next generation binder which is green in nature, sustainable, has a low carbon foot print, environmentally friendly and possesses high durability when compared to Ordinary Portland Cement [2-3]. Although the material shows superior durability in terms of high temperature, acid resistance and corrosion, a comprehensive study will provide quality guidelines for utilization of this product for public construction. The current study is intended to help characterize the various degradation mechanisms of geopolymer concrete when subjected to an aggressive environment. The proposed study will relate the durability testing to the microstructure when subject to aggressive environment, so that the overall understanding of durability aspect of GPC is well understood at elevated temperatures. This will also help to mitigate the risk for public construction and will provide guidelines on the limitations of geopolymer concrete at elevated temperatures.

Table 1. Comparison of alternative binders to Portland cement [1]

| | Portland cement | Calcium Aluminate Cement | Calcium Sulfoaluminate Cement | Alkali-activated binder | Super sulfated cement |
|---|---|---|---|---|---|
| Primary phases/materials | $C_3S$ | CA | $C_4A_3S$, $C_4AF$ | Aluminosilicate alkali | Slag, $CS/CSH_2$ |
| Hydrates | C-S-H, CH, Aft | $CAH_{10}$, $C_2AH_8$, C-S-H | Aft, $AH_3$, AFm, $C_2ASH_8$ | Gel (N-A-S-(H) and C-(A)-S-H, zeolytes | Aft, C-S-H, AFm |
| Advantages | • Long History<br>• Standard Composition | • Rapid Strength<br>• Sulfate Resistant<br>• Abrasion Resistant | • Low energy<br>• Rapid Strength<br>• Shrinkage | • Low heat of reaction<br>• Heat and acid resistant | • Low heat of hydration<br>• Durable in aggressive environment |
| Disadvantages | • High Energy<br>• High $CO_2$<br>• Limited early strength<br>• Poor in aggressive environment | • Strength loss on conversion of metastable to stable hydrates | • Durability unproven<br>• Costly | • Challenging rheology<br>• Durability unproven | • Slow strength gain |

Geopolymer Concrete Cement (GPC) is an up and coming class of cementitious binder, which can provide a 100% replacement to Ordinary Portland Cement for building green civil engineering infrastructures [8, 9]. Thus, GPC has the potential to be at the leading edge of a measurable shift in the construction industry towards the development of highly sustainable, durable, and minimum energy consuming cementitious binders featuring an almost negligible carbon footprint. Geopolymer cements offer an intriguing combination of characteristics such as higher mechanical strength, excellent chemical durability, variety of environmental benefits, and strong potential for commercial applications [10-12]. The field of geopolymer cements also provides significant scientific challenges associated with the need for better understanding of the polymerization reactions, kinetics and the precursors involved in this reaction, the relationships between mix design and the mechanical properties of the resulting cementitious matrix, and in general the durability mechanism when subject to extreme environments [1, 13].

EXPERIMENTAL PROCEDURE

Geopolymer concrete (GPC) was prepared by using eleven different types of fly ashes obtained from three different countries (USA, Israel and China). The specimens were cubes of 50 mm in length, breadth and height. White fused alumina with a nominal size of 5 mm was used as coarse aggregate. Silica sand and commercially available fine alumina aggregate (tabular alumina of nominal size of 2.36 mm) was used as fine aggregate in the preparation of GPC specimens as shown in Table 2. The chemical composition of the fly ashes is shown in Table 3. This study examines geopolymer concrete when subjected to elevated temperature prepared from both Class C and F fly ash stockpiles. Sodium hydroxide (14 M NaOH) and sodium silicate obtained from PQ Corporation (Valley Forge, PA, USA) was used as an activator for preparation of fly ash based geopolymer concrete. Sodium silicate composed of 45 % by weight and $SiO_2$ to $Na_2O$ ratio of 2:1, was used in the preparation of GPC. Sodium silicate to sodium hydroxide ratio was 1:1 and the activator (sodium hydroxide + sodium silicate) to binder ratio was 0.45. After 24 hours of batching, the fly ash based geopolymer concrete specimens were removed from the molds and cured at a temperature of 80°C for 72 hours.

Samples were subject to thermal shock testing by keeping in the oven at 1093°C and quenching it in water after one hour. Samples prepared from silica sand and commercially available fine alumina aggregate were then subjected to 5 cycles of thermal shock as shown in Figure 1. Performance evaluation of the samples after each cycle was evaluated for major or minor cracks, expansion and total failure. Visual analysis was conducted after each cycle and digital micrographs of each sample was recorded. The chemical composition of GPC specimens (controls and thermal shock) was conducted via Energy Dispersive-X-Ray fluorescence (XRF) spectroscopy (ARL QUANT'X EDXRF Spectrometer). In addition, microstructure characterization was conducted using scanning electron microscopy (SEM) and X-Ray diffraction analysis was done using the D8 Advanced Bruker AXS spectrometer. In addition, X-ray micro tomography was conducted to analyze the pore structure of geopolymer concrete when subjected to thermal shock treatment.

Measurements were carried out using hard X-ray synchrotron radiation (25 kev) in a parallel beam configuration, with 0.25° rotation per step with 2 second exposure time per step. X-ray detection was achieved with (Ce) YAG X-ray scintillation and CCD camera, capturing 2048 × 512 pixels with voxel resolution of 2.5 μm.

Table 2. Sample designation, fly ash and aggregate type used in preparation of geopolymer concrete

| Sample Name | Fly Ash type used to prepare GPC | Country of origin of fly ash | Fine Aggregate | Test of exposure | Sample Name | Fly Ash type used to prepare GPC | Country of origin of fly ash | Fine Aggregate | Test of exposure |
|---|---|---|---|---|---|---|---|---|---|
| TS-W-1 | Class F | USA | "P" Gravel | TS | TS-W-7 | Class C | USA | "P" Gravel | TS |
| C-W-1 | Class F | USA | "P" Gravel | Control | C-W-7 | Class C | USA | "P" Gravel | Control |
| TS-WO-1 | Class F | USA | Alumina | TS | TS-WO-7 | Class C | USA | Alumina | TS |
| C-WO-1 | Class F | USA | Alunina | Control | C-WO-7 | Class C | USA | Alumina | Control |
| TS-W-2 | Class F | USA | "P" Gravel | TS | TS-W-8 | Class F | Israel | "P" Gravel | TS |
| C-W-2 | Class F | USA | "P" Gravel | Control | C-W-8 | Class F | Israel | "P" Gravel | Control |
| TS-WO-2 | Class F | USA | Alumina | TS | TS-WO-8 | Class F | Israel | Alumina | TS |
| C-WO-2 | Class F | USA | Alumina | Control | C-WO-8 | Class F | Israel | Alumina | Control |
| TS-W-3 | Class F | China | "P" Gravel | TS | TS-W-9 | Class F | Israel | "P" Gravel | TS |
| C-W-3 | Class F | China | "P" Gravel | Control | C-W-9 | Class F | Israel | "P" Gravel | Control |
| TS-WO-3 | Class F | China | Alumina | TS | TS-WO-9 | Class F | Israel | Alumina | TS |
| C-WO-3 | Class F | China | Alumina | Control | C-WO-9 | Class F | Israel | Alumina | Control |
| TS-W-4 | Class F | China | "P" Gravel | TS | TS-W-10 | Class F | China | "P" Gravel | TS |
| C-W-4 | Class F | China | "P" Gravel | Control | C-W-10 | Class F | China | "P" Gravel | Control |
| TS-WO-4 | Class F | China | Alumina | TS | TS-WO-10 | Class F | China | Alumina | TS |
| C-WO-4 | Class F | China | Alumina | Control | C-WO-10 | Class F | China | Alumina | Control |
| TS-W-5 | Class F | China | "P" Gravel | TS | TS-W-11 | Class C | USA | "P" Gravel | TS |
| C-W-5 | Class F | China | "P" Gravel | Control | C-W-11 | Class C | USA | "P" Gravel | Control |
| TS-WO-5 | Class F | China | Alumina | TS | TS-WO-11 | Class C | USA | Alumina | TS |
| C-WO-5 | Class F | China | Alumina | Control | C-WO-11 | Class C | USA | Alumina | Control |
| TS-W-6 | Class F | China | "P" Gravel | TS | | | | | |
| C-W-6 | Class F | China | "P" Gravel | Control | | | | | |
| TS-WO-6 | Class F | China | Alumina | TS | | | | | |
| C-WO-6 | Class F | China | Alumina | Control | | | | | |

TS: Thermal Shock

RESULTS AND DISCUSSION

The performance evaluation of GPC specimens prepared with silica sand and alumina as fine aggregate was subjected to five thermal shock cycles as shown in Table 4. GPC specimens [TS-WO-1 (Class F) and TS-WO-11 (Class C)] prepared with fine alumina aggregate did not suffer any physical damage nor show signs of cracking or expansion as compared to other samples. GPC specimen [TS-W-5 (class F)] prepared with silica sand did not suffer any mechanical damage until four cycles, signs of major cracking and failure were observed at the end of the fifth cycle. In contrast, certain GPC specimens [ TS-W-4, TS-W-6 and TS-W-8 (Class F), TS-W-7 (Class C)] suffered severe damage after the first cycle of thermal shock cycle, while the rest of the samples suffered moderate deterioration in the form of cracking and expansion.

Table 3. Chemical composition of fly ash stockpiles

| SI. NO | Fly Ash Type | Country of Origin | $SiO_2$ | $Al_2O_3$ | $SiO_2/Al_2O_3$ | CaO | $Fe_2O_3$ | MgO | $SO_3$ | $Na_2O$ | $K_2O$ | LOI |
|---|---|---|---|---|---|---|---|---|---|---|---|---|
| 1 | Class F | USA | 55.07 | 28.61 | 1.92 | 1.97 | 6.22 | 1.08 | 0.19 | 0.38 | 2.63 | 1.82 |
| 2 | Class F | USA | 58.52 | 20.61 | 2.84 | 5.00 | 9.43 | 1.86 | 0.49 | 0.52 | - | 0.05 |
| 3 | Class F | CHINA | 47.98 | 31.17 | 1.54 | 8.14 | 6.50 | 1.06 | 0.44 | 0.25 | 0.89 | 1.11 |
| 4 | Class F | CHINA | 48.14 | 27.12 | 1.78 | 8.51 | 9.14 | 2.07 | 1.22 | 0.28 | 1.19 | 0.54 |
| 5 | Class F | CHINA | 55.65 | 20.93 | 2.66 | 7.25 | 5.55 | 2.93 | 0.16 | 3.39 | 1.35 | 0.45 |
| 6 | Class F | CHINA | 56.41 | 21.47 | 2.54 | 11.2 | 7.3 | 0.73 | 0.24 | 0.87 | 1.28 | 0.24 |
| 7 | Class C | USA | 55.61 | 19.87 | 2.80 | 12.93 | 4.52 | 2.49 | 0.49 | 0.67 | 0.86 | 0.22 |
| 8 | Class F | ISREAL | 52.48 | 25.63 | 2.05 | 3.30 | 9.36 | 1.69 | 0.20 | 0.70 | 2.20 | 2.10 |
| 9 | Class F | ISREAL | 55.05 | 24.58 | 2.24 | 3.46 | 8.52 | 0.95 | 0.18 | 0.73 | 1.27 | 2.36 |
| 10 | Class F | CHINA | 45.96 | 37.00 | 1.24 | 2.74 | 8.49 | 0.79 | 0.25 | 0.33 | 0.99 | 0.82 |
| 11 | Class C | USA | 37.77 | 19.33 | 1.97 | 22.45 | 7.33 | 4.81 | 1.56 | 1.80 | 0.41 | 0.17 |

LOI: Loss of ignition

Digital micrographs after each cycle for the GPC specimens (TS-WO-3,5,7) prepared with fine alumina aggregate and with silica sand (TS-W-3,5,7) are shown in Figures 1 and 2, respectively. The GPC's prepared with fine alumina aggregate from Class F Fly ash (TS-WO-5) and (TS-WO-3) did not show any major signs of deterioration until the last cycle, while GPC prepared with Class C fly ash showed signs of cracking and deterioration after one cycle of thermal shock followed by complete failure at the end of cycle five. In general, this indicates that the GPC batched with class "F" fly ash performed better as compared to Class "C" fly ash, when subjected to thermal shock testing.

Digital micrographs of the GPC specimens (TS-W-3, 5 and 8) prepared with silica sand and Class F fly ash is shown in Figure 2. Specimen (TS-W-5) did not suffer any signs of deterioration after 5 thermal shock cycles while GPC specimen (TS-W-3, TS-W-8) prepared with Class F fly ash exhibited major cracking and complete failure, after one thermal shock cycle as shown in Figure 2. The GPC specimens prepared with Class C fly ash showed complete deterioration at the end of one thermal shock cycle (See Table 4). This shows that all Class F fly ash are not stable at elevated temperature and are not suitable for high temperature applications. In addition, the GPC specimens were studied for chemical analysis via XRF and microstructure characterization using XRD and SEM.

Table 4: Performance evaluation of geopolymer concrete subjected to 5 thermal shock cycles

| Sample No | Fly Ash Origin | Fly Ash Type | Fine Aggregate | Cycles | | | | |
|---|---|---|---|---|---|---|---|---|
| | | | | 1 | 2 | 3 | 4 | 5 |
| TS-W-1 | USA | Class F | Silica sand | ● | ▲ | ▲ | ▲ | X |
| TS-WO-1 | USA | Class F | Alumina | ● | ● | ● | ● | ● |
| TS-W-2 | USA | Class F | Silica sand | ▲ | ▲ | ◆ | ◆, X | ◆ X |
| TS-WO-2 | USA | Class F | Alumina | ● | ▼ | ▲ | X | X |
| TS-W-3 | China | Class F | Silica sand | ◆,▼ | ◆▼ | ◆,▼ | ◆, ▼ | ◆ ▼ |
| TS-WO-3 | China | Class F | Alumina | ● | ● | ● | ▲ | ▲ |
| TS-W-4 | China | Class F | Silica sand | X | | | | |
| TS-WO-4 | China | Class F | Alumina | ● | ▲ | ▲ | ▲ | ▲ |
| TS-W-5 | China | Class F | Silica sand | ● | ● | ● | ● | ▼ |
| TS-WO-5 | China | Class F | Alumina | ● | ● | ● | ● | X |
| TS-W-6 | China | Class F | Silica sand | ▼, ◆, X | | | | |
| TS-WO-6 | China | Class F | Alumina | ▲ | ▲ | ▲ | ▲ X | X |
| TS-W-7 | USA | Class C | Silica sand | X | | | | |
| TS-WO-7 | USA | Class C | Alumina | ▲ | ▲ | ◆ | X | X |
| TS-W-8 | Israel | Class F | Silica sand | X | | | | |
| TS-WO-8 | Israel | Class F | Alumina | X | ▼X | | | |
| TS-W-9 | Israel | Class F | Silica sand | ▼ | ◆, ▼ | ◆, ▼ | ◆, ▼ | ◆, ▼ |
| TS-WO-9 | Israel | Class F | Alumina | ▼ | ▼ | X | | |
| TS-W-10 | China | Class F | Silica sand | ● | ▲ | X | | |
| TS-WO-10 | China | Class F | Alumina | ▲ | ▲ | X | | |
| TS-W-11 | USA | Class C | Silica sand | X | | | | |
| TS-WO-10 | USA | Class C | Alumina | ● | ● | ● | ● | ● |

**No Cracking: ●   Minor Crack: ▲   Major Cracking:   ◆Expansion:   ▼ Total Failure: X**

XRF analysis of all the specimens prepared with alumina aggregate and silica sand as a fine aggregate is shown in Figures 3 and 4, respectively. Figure 3 exhibits the control and thermal shock specimens batched with alumina aggregate. The $Al_2O_3$ increased for most of the specimens after thermal shock treatment as compared to the controls, expect for certain specimens (TS-WO-1, TS-WO-2 and TS-W-4) as shown in Figure 3. The alumina from the fine aggregate contributed in the formation of additional $Al_2O_3$ when subjected to elevated temperature. In contrast, for certain samples (TS-WO-1, TS-WO-2 and TS-W-4) the $Al_2O_3$ decreased by ~ 50% when subject to 5 cycles of thermal shock treatment causing the $SiO_2/Al_2O_3$ ratio to increase for these samples 8-9%. These samples (TS-WO-1, TS-WO-2 and TS-WO-4) did not exhibit any severe mechanical damage till when subjected to thermal shock treatment (See Table 4). This may indicate that the aluminum oxide may be involved in the formation of an amorphous zone of geopolymer. Further studies using the Nuclear Magnetic Resonance (NMR) technique are required to quantify this process [9, 14].

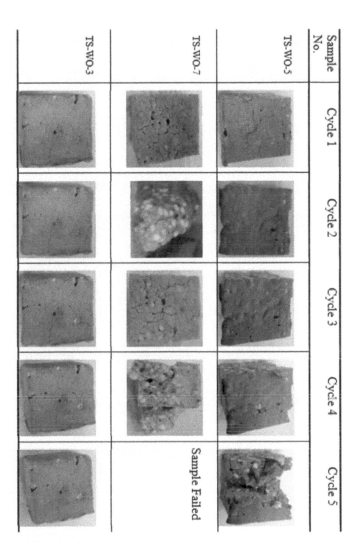

Figure 1. Geopolymer concrete cubes with alumina aggregate subject to 5 cycles thermal shock.

The XRF analysis for GPC specimens prepared with silica sand is shown in Figure 4. The GPC specimens suffered severe mechanical damage as compared to specimens prepared with alumina aggregate. The $SiO_2/Al_2O_3$ ratio was much higher in specimens prepared with silica sand as compared to alumina aggregate (See Figure 5). The GPC prepared with silica sand did not have additional alumina and had greater content of un-reacted silica, therefore sufficient formation of an amorphous zone in the form of sodium aluminosilicate hydrate (N-A-S-H) was not formed [13]. This shows that additional alumina is required to form an amorphous zone of N-A-S-H, which plays a vital role in durability resistance and mechanical performance of the binder at elevated temperatures [15, 16].

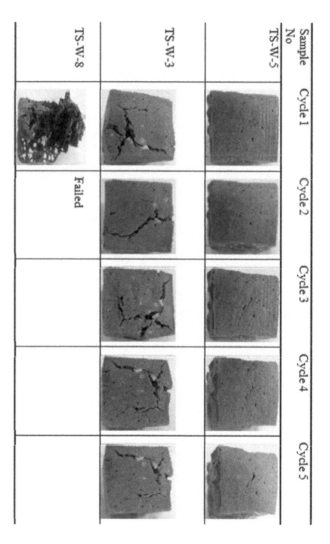

Figure 2. GPC with silica sand subjected to 5 thermal shock cycles

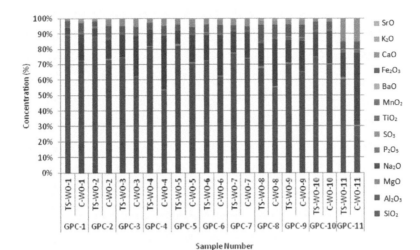

A. GPC prepared with alumina aggregate

Figure 3. XRF analysis of GPC prepared with fine alumina aggregate

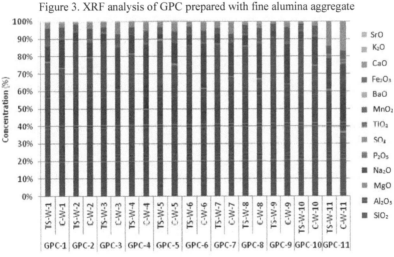

Figure 4. XRF analysis on Geopolymer concrete with silica sand

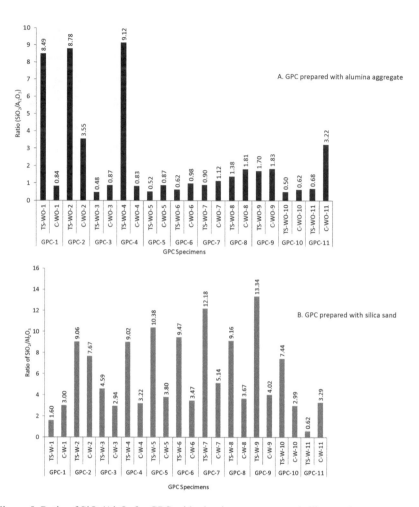

Figure 5. Ratio of $SiO_2/Al_2O_3$ for GPC with alumina aggregate and silica sand

XRD analysis of six specimens (two class "F" and one class "C") GPC's are shown in Figure 6. XRD studies of GPC control (C-WO-07) prepared with Class "C" fly ash shows phases such as quartz, albite, nepheline and gehlenite. The thermal shock treated specimens showed strong peaks of analcime and sodalite in addition to nepheline. XRD analysis of GPC with Class F ash (TS-WO-3) exhibited similar crystalline zeolitic phases as GPC with Class C fly ash, in addition to fayalite and mullite. The control specimens exhibited mullite and after thermal shock treatment the mullite phase disappeared as it was involved in the geopolymerization. Later it may have formed amorphous geopolymer. Thermally stable phases such as sodalite and analcime were detected after the thermal shock treatment. These phases possess similar structures as does N-A-S-H gel and recovered their crystallinity during 204-426°C, then retained their structure up to ~ 815°C. The formation of the amorphous phase due to geopolymerization is initiated by crystallization of the zeolite precursor such as albite, stalbite, microcline and labradorite [17, 18]. The precursor plays a crucial role in the formation of crystallization of stable phases which leads to amorphization of geopolymeric gels [9, 19].

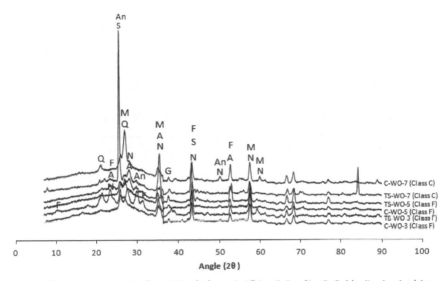

S: Sodalite Q: Quartz M: Mullite N: Nepheline A: Albite Γ: Fayalite G: Gehlenite An: Analcime

Figure 6. XRD analysis of Geopolymer Concrete (control and thermal shocked) with Class "C" and "F" fly ashes prepared with fine alumina aggregate

Studies have shown that at temperatures below 500°C, the primary reaction products of amorphous aluminosilicate semicrystalline gels such as N-A-S-H are formed, along with zeolite crystals such as mullite. The formation of zeolite crystals depends on the composition of the fly ash and the chemical activator used for akali activation of the fly ash. Zeolite products such as analcime and chabazite are formed up to 572°C. Upon increasing the temperature to 752°C promotes the recrystallization forming high silica stable structures (crystalline feldspathoid) such as nepheline, leucite and labradorite. Thermally stable phases such as sodalite detected via XRD in the GPC exposed to thermal shock cycles indicates reduced contraction after exposing to thermal shock cycles. This phase (sodalite) then re-crystallizes to nepheline and albite without destruction of the alumino silicate framework which is responsible for the formation of the N-A-S-H geopolymer gel.

SEM analyses for GPC specimens (C-WO-3, TS-WO-3) prepared with Class "F" fly ash are shown in Figure 7. The control specimens (C-WO-3) showed un-reacted fly ash crystals along with crystals of mullite. The specimen subjected to thermal shock treatment exhibited crystals of nepheline along with the amorphous zone, which indicates that the thermal shock treatment leads to the crystallization of unreacted fly ash which was not involved in the geopolymerization. In addition, microcracks were observed in this specimen after the thermal shock treatment. The performance evaluation after 5 cycles also indicated minor cracking for three cycles along with major cracking for the fourth and the fifth cycles.

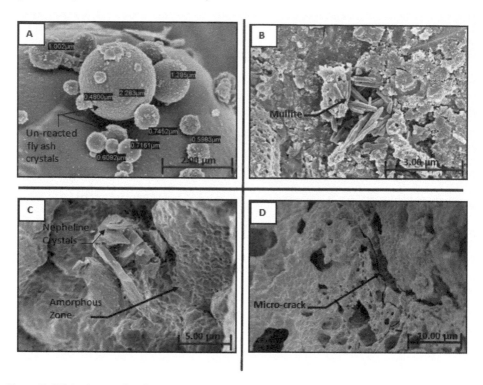

Figure 7. SEM micrographs of control sample (C-WO-3) exhibiting unreacted fly ash crystals and zeolite crystals (A and B), Fig 7 C and D shows amorphous zone with nepheline crystals on the specimens subjected to thermal shock.

SEM analysis of GPC (TS-WO-5) prepared with a Class "F" fly ash procured from China is shown in Figure 8. The control specimen (C-WO-5) exhibited un-reacted crystals along with some zeolite crystallization (Figure 8A and B). Upon thermal shock treatment the unreacted fly ash underwent geopolymerization forming an amorphous zone. The specimen (TS-WO-5) exhibited superior performance when subjected to 5 cycles of thermal shock treatment as shown in Table 4 and Figure 2. The superior performance of this specimen could be attributed to the formation of an amorphous zone and almost a full geopolymerization of the fly ash crystals which were not involved in the initial geopolymerization [9, 20, 21]. Related research has shown that geopolymer concrete when subject to elevated temperatures, retain amorphous nature while exhibiting some changes in crystalline phase composition. Sodium-based geopolymer concretes showed crystalline phases such as nepheline, albite and tridymite. These phases have been reported to be responsible for the improvement of thermal resistance of geopolymer concretes [20].

SEM analysis of Class C fly ash is shown in Figure 9. Figure 9A exhibits crystallization in the form of zeolite T crystals; in addition un-reacted fly ash crystals were observed as shown in Figure 9B. Amorphization was observed in the un-reacted fly ash crystal, this indicates that the size of the fly crystals plays an important role in the geopolymerization. The fly ash crystals which were not involved in the geopolymerization was 11.3 μm. Further study is required to examine the effect on particle size on geopolymerization, which will lead to the successful formation of the amorphous phase. The thermal shock specimen led to the crystallization of geopolymeric gel, leading to the formation of analcime crystals in the form of plates, as shown in Figure 9. The analcime phase indicates that the thermally stable zeolite structures were developed when subjected to elevated temperature. The formation of hydrated sodium alumino silicate depends on the cation. Analcime and nepheline depend on sodium cation leading to zeolite formation. This indicates that thermal shock treatment led to additional formation of N-A-S-H phase, which is responsible for higher strength and durability of geopolymer concrete.

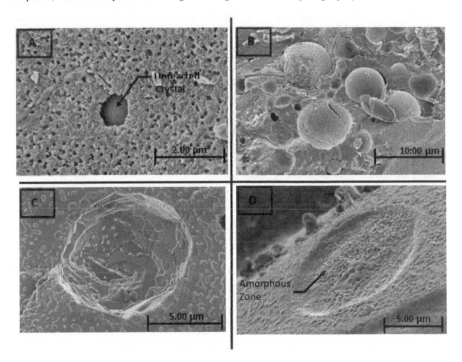

Figure 8. SEM micrographs of control sample (C-WO-5) exhibiting crystallization over unreacted crystals, which (C and D) shows amorphous zone on (TS-WO-5).

X-ray micro tomographs, exhibiting the slice through image of GPC prepared with Class F (TS-WO-2) and Class C fly ash (TS-WO-7) after thermal shock treatment are shown in Figure 10. Both GPC's prepared from Class F and C fly ash showed micro-cracks after 5 cycles of thermal shock testing. The corresponding cubic images of GPC specimens are shown in Figures 10 B and D, respectively. These images exhibited a 3D porous view of the specimens, when exposed to elevated temperature.

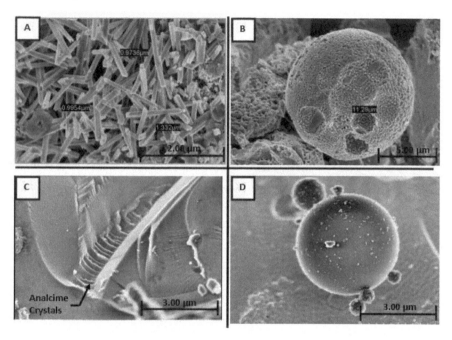

Figure 9. SEM micrograph of zeolite-T crystals (A) and unreacted reacted crystals in the control samples, while plates of analcime on the thermal shock along with unreacted crystals after thermal shock

The maximum pore diameter determined via X-ray micro tomography for class F GPC (TS-WO-02) and Class C GPC (TS-WO-07) was 2000 μm and 2500 μm, respectively. The GPC prepared with Class C fly ash exhibited an increase in pore diameter by a factor of 1.5. The pore connectivity network of the GPC's was examined using the ortho-slice view as shown in Figures 11 A and B. Class F specimens showed pore connectivity after the thermal shock treatment while GPC prepared with Class C fly ash did not exhibit signs of pore connectivity. This shows that due to elevated temperature exposure the pores were expanding and connecting to form a pore connectivity network. Further studies are required to quantify the pore connectivity network and to examine the tortuosity of the pore network, which plays a critical role in controlling the strength and preventing the ingress of deleterious species such as chlorides and sulfates, which lead to the degradation of concrete structures.

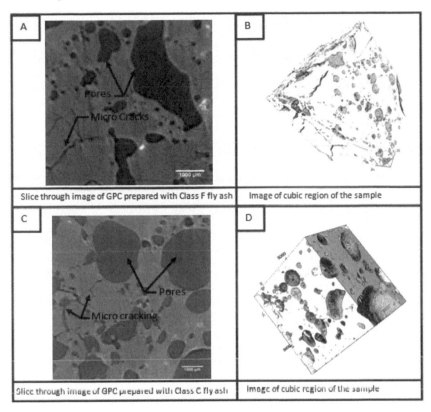

Figure 10. X-ray μC tomography of Class F fly ash (TS-WO-2) and Class C fly ash (TS-WO-07)

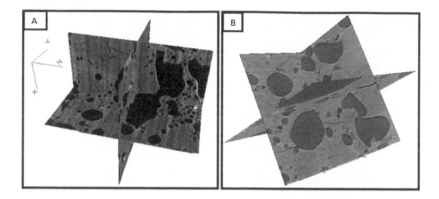

Figure 11. Ortho-slice view of Class F and Class C geopolymer concrete showing the pore connectivity network.

CONCLUSIONS

Geopolymer concrete prepared using eleven different types of fly ashes obtained from three different countries were subject to thermal shock treatment. The specimens prepared with alumina filler as fine aggregate exhibited superior performance as compared to the specimens made with silica sand. This indicates that thermal shock treatment leads to additional formation of N-A-S-H phase, which is responsible for higher strength and durability of geopolymer concrete. Thermally stable phases such as sodalite and analcime were detected after the thermal shock treatment. The formation of amorphous phase of geopolymerization as initiated by crystallization of the zeolite precursor plays a crucial role in the formation of crystallization of stable phases which leads to amorphization of gepoloymeric gels. This shows that additional alumina is required to form the amorphous zonc of N-A-S-H, which plays a vital role in durability, resistance and mechanical performance of the binder at elevated temperature.

REFERENCES

[1] M.C.G. Juenger, F. Winnefeld, J.L. Provis, J.H. Ideker, Advances in alternative cementitious binders, Cement and Concrete Research, **41** (2011) 1232-1243.
[2] A.M. Rashad, Y. Bai, P.A.M. Basheer, N.C. Collier, N.B. Milestone, Chemical and mechanical stability of sodium sulfate activated slag after exposure to elevated temperature, Cement and Concrete Research, **42** (2012) 333-343.
[3] W. Khaliq, V. Kodur, Behavior of high strength fly ash concrete columns under fire conditions, Materials and Structures/Materiaux et Constructions, **46** (2013) 857-867.
[4] W. Khaliq, V.K.R. Kodur, Effect of high temperature on tensile strength of different types of high-strength concrete, ACI Materials Journal, **108** (2011) 394-402.
[5] M.J. DeJong, F.-J. Ulm, The nanogranular behavior of C-S-H at elevated temperatures (up to 700 °C), Cement and Concrete Research, **37** (2007) 1-12.
[6] A. Fernández-Jiménez, J.Y. Pastor, A. Martín, A. Palomo, High-Temperature Resistance in Alkali-Activated Cement, Journal of the American Ceramic Society, **93** (2010) 3411-3417.

[7] T. Bakharev, J.G. Sanjayan, Y.B. Cheng, Effect of elevated temperature curing on properties of alkali-activated slag concrete, Cement and Concrete Research, **29** (1999) 1619-1625.
[8] J. Davidovits, Geopolymer Chemistry and Applications, Institut Géopolymère, 2011.
[9] J.L. Provis, J.S.J. Van Deventer, Geopolymers: Structure, Processing, Properties and Industrial Applications, Woodhead Publishing Limited, 2009.
[10] P. Duxson, J.L. Provis, G.C. Lukey, J.S.J. van Deventer, The role of inorganic polymer technology in the development of 'green concrete', Cement and Concrete Research, **37** (2007) 1590-1597.
[11] P. Duxson, A. Fernández-Jiménez, J. Provis, G. Lukey, A. Palomo, J. van Deventer, Geopolymer technology: the current state of the art, Journal of Materials Science, **42** (2007) 2917-2933.
[12] E.I. Diaz-Loya, E.N. Allouche, S. Vaidya, Mechanical properties of fly-ash-based geopolymer concrete, ACI Materials Journal, **108** (2011) 300-306.
[13] C. Shi, A.F. Jiménez, A. Palomo, New cements for the 21st century: The pursuit of an alternative to Portland cement, Cement and Concrete Research, **41** (2011) 750-763.
[14] P. Duxson, J.L. Provis, G.C. Lukey, F. Separovic, J.S.J. van Deventer, 29Si NMR Study of Structural Ordering in Aluminosilicate Geopolymer Gels, Langmuir, **21** (2005) 3028-3036.
[15] A. Fernández-Jiménez, J.G. Palomo, F. Puertas, Alkali-activated slag mortars: Mechanical strength behaviour, Cement and Concrete Research, **29** (1999) 1313-1321.
[16] I. Garcia-Lodeiro, A. Palomo, A. Fernández-Jiménez, D.E. Macphee, Compatibility studies between N-A-S-H and C-A-S-H gels. Study in the ternary diagram $Na_2O-CaO-Al_2O_3-SiO_2-H_2O$, Cement and Concrete Research, **41** (2011) 923-931.
[17] A. Fernández-Jimenez, A.G. de la Torre, A. Palomo, G. López-Olmo, M.M. Alonso, M.A.G. Aranda, Quantitative determination of phases in the alkali activation of fly ash. Part I. Potential ash reactivity, Fuel, **85** (2006) 625-634.
[18] G. Kovalchuk, A. Fernández-Jiménez, A. Palomo, Alkali-activated fly ash: Effect of thermal curing conditions on mechanical and microstructural development – Part II, Fuel, **86** (2007) 315-322.
[19] P. Duxson, J.L. Provis, Designing Precursors for Geopolymer Cements, Journal of the American Ceramic Society, **91** (2008) 3864-3869.
[20] W.D.A. Rickard, J. Temuujin, A. van Riessen, Thermal analysis of geopolymer pastes synthesised from five fly ashes of variable composition, Journal of Non-Crystalline Solids, **358** (2012) 1830-1839.
[21] J. Temuujin, W. Rickard, M. Lee, A. van Riessen, Preparation and thermal properties of fire resistant metakaolin-based geopolymer-type coatings, Journal of Non-Crystalline Solids, **357** (2011) 1399-1404.

# BASIC RESEARCH ON GEOPOLYMER GELS FOR PRODUCTION OF GREEN BINDERS AND HYDROGEN STORAGE

C. H. Rüscher, L. Schomborg, A. Schulz, J. C. Buhl

ABSTRACT

Geopolymer gels could be optimized for binders, possessing high mechanical strength. Two further applications will be reported. Firstly NaBH₄ crystals can be recrystallized within a geopolymer type matrix. The compositions and solidification conditions could be optimized, obtaining an hydrogen release of about 72 wt% of pure NaBH₄. The hydrogen release is obtained by acid titration without the requirement of any catalyst. Secondly sodalites of micro- and nano-crystalline sizes could be synthesized, enclosing the BH₄-anion in the cages. The crystals share extended contributions of a geopolymer type matrix. The matrix stores and transports the water molecules for the hydrogen release reactions, revealing new anion species $[H_{4-n}B(OH)_n]^-$, n = 1, 2, 3 in the sodalite cages.

## INTRODUCTION

The syntheses of aluminosilicate gels - also called geopolymer gels - can be modified in many aspects. One route concerns the optimization for revealing rigid bodies of high mechanical strength (geopolymer cements, green binders) which generally relies on quasi X-ray amorphous or short range ordered poly-sialate and poly-siloxo bondings [1,2]. It could be shown that such binders consist of long poly-siloxo chains, crosslinked via sialate bondings [3-6]. Other routes are devoted to the production of crystalline microporous and mesoporous systems (zeolites) with well known applications, starting from about 50 years ago [7,8]. Zeolites have also been discussed for about 30 years for potential hydrogen storage, among other new materials like metal hydrides, carbon nanotubes or clathrates [9-13], as well as new metal-organic framework compounds (MOFs) [14-17]. Sodalites (code SOD) possess the highest density of sodalite cages among the zeolites in a closed packed structure. Theoretically sodalites could enclose 8 H₂ per cage [18]. For zeolites RHO, FAU, KFI, LTA and CHA theoretical values of maximal capacities range between 2.6 and 2.9 wt% H₂ [19]. In an earlier investigation Weitkamp et al. [20] reported an uptake of about 1.6 wt% H₂ by sodalite, loaded at 300°C and 10 MPa. More recently [21,22], it could be shown that zeolite X provides storage capacity of about up to 3 wt% optimized at -200°C and 100 MPa. However, using zeolites in this way, the hydrogen storage may not have any advantage compared to hydrogen storage in ordinary gas cylinders. According to the U.S. DOE (U.S. Department of Energy) the hydrogen content should presently not be much smaller than about 6-7 wt% of the storage capacity considered for any portable application. In this respect, it is interesting to consider hydrides, e.g. NaBH₄ covering in total 5.2 wt% H₂ which can effectively be doubled to about 10.4 wt% H₂ upon reaction with water. Barrer [8] proposed an impregnation of boronhydride salts like Al(BH₄)₃ or NaBH₄ into preformed zeolites like A, X and Y, which could not be realized in so far. Here we report two other possibilities of inclusion of NaBH₄. These are the inclusion of NaBH₄ in a geopolymer matrix, i.e. the mother compounds of the zeolites, and by direct synthesis in the sodalite structure with formula Na₈[Al₆Si₆O₂₄](BH₄)₂ per unit cell.

Our recent studies succeeded in an easy and safe way of handling NaBH₄ salt in strong alkaline aluminate and silicate solutions. Brought together gelation occurs, which could be further solidified by drying and handled for hydrogen release as first reported in a conference contribution [23]. The gelation reveals a heterogeneous solid, composed of NaBH₄ crystals and sodalite type nanocrystals which are glued together in a geopolymer type matrix as has already been noted in a book chapter [24]. Geopolymer synthesis, using similar solutions without $NaBH_4$, has also been reported, i.e. sodium aluminate solution brought together with sodium waterglass solution (silica fume solved in sodium hydroxide solution) [25].

This contribution describes in a first step properties of geopolymer binders which could be helpful for a better understanding of the geopolymer contribution in the new material. Secondly details about synthesis of the gel, alteration during solidification and optimization of hydrogen storage capabilities and results concerning the geopolymer type matrix are given. Thirdly it is shown that a new way of investigating the properties of the $BH_4$-anion could be realized for $NaBH_4$ enclosed in the sodalite cage. Such material could be synthesized in a direct way revealing sodalite as micro-crystalline [26] and nano-crystalline powders [27]. The disadvantage of much lower stored hydrogen content in $BH_4$-sodalites of only one-tenth compared to $NaBH_4$ could at the moment, probably, disregarded due to the benefit of a better understanding of the hydrogen release reactions. This will be discussed with respect to the role of the geopolymer type matrix present in all types of $BH_4$-aluminosilicate sodalite samples prepared so far.

EXPERIMENTAL

Geopolymer cements were prepared using metakaolin (Metastar 501) and potassium waterglass solution with mass ratio MKM/KWG = 1.7 resulting in nominal molar Si/Al ratio of 1.9 and K/Al of about 1.1 (Cem.1). Further details of characterizations of the starting material, also including compositional variation and mechanical property relationship are given in another contribution in this issue [28]. The cement considered here was prepared by mixing MKM and KWG for 10 minutes forming an homogeneous slurry, which was filled into cylindrical PE-containers (18 mm diameter, 10 mm height) and closed. These samples were aged at 22°C in the laboratory for certain times for the experiments described below.

The new $NaBH_4$-geopolymer composite material was prepared from commercially available sodium-tetrahydroborate ($NaBH_4$), sodium-silicate ($Na_2SiO_3$), sodium-aluminate ($NaAlO_2$) and water. It is prepared in a two step process under open conditions. In the first step a synthesis batch is prepared by dissolving $NaAlO_2$ in water. A certain amount of $NaBH_4$ is added and dissolved under stirring until clear (solution I). A second solution is prepared from $Na_2SiO_3$ and water, which is added with $NaBH_4$ until clear, too (solution II). Afterwards gel precipitation starts by the dropwise addition of solution II to solution I. A pastey liquid results from this mixture. In a second step, the gel is dried typically between 80 and 110°C under open conditions, too. Typical amounts taken are 310 mg $Na_2SiO_3$ solved in 1.5 ml $H_2O$ and added with 250 mg $NaBH_4$ (solution I) and 250 mg $NaAlO_2$ solved in 1.5 ml $H_2O$ and added with 250 mg $NaBH_4$ (solution II). Such a batch encloses a $NaBH_4$/solid ratio of 0.47 (by weight) and a Si/Al ratio of 1.03 (molar). In NaBH4/solid ratio solid means the total amount of $NaBH_4$, $NaAlO_2$ and $Na_2SiO_3$ used.

Hydrogen release experiments of the new composite material were carried out using ordinary glassware. A 100 ml two necked bulb was used. One neck was attached to a 100 ml gas syringe to measure the volume of the formed gas. The second neck was used for injection of acid solution to the powder sample in the bulb. The bulb was also used for temperature dependent stability tests of the powder with injected water. Further details of the apparatus and reproducibility tests are described elsewhere [24]. A proof that hydrogen is released has been given by the sound of the well known hydrogen-oxygen reaction (Knallgas-reaction). The concentration of $CO_2$ released from the samples was below the detection limit of 1000 ppm, as could be determined by a gas detector (ION Science GasChek).

The hydrothermal synthesis of micro- and nano-crystalline sodium aluminosilicate tetrahydroborate sodalites (micro- and nano-NaBH$_4$-SOD) has been developed [26, 27]. A successful enclathration of tetrahydroborate in the sodalite cages requires low temperatures, high concentrations of NaOH, and a low liquid/solid ratio in closed Teflon cups for the creation of autogeneous water pressure. The Teflon cups were filled with the hydroborate salt, kaolinite and sodium hydroxide solution. After this filling procedure the Teflon cups were placed in autoclaves and typically heated up to temperatures of 120°C, using reaction periods up to 24 hours for obtaining micro-BH$_4$-SOD. In a similar way, nano-BH$_4$-SOD could be prepared using lower temperatures (60°C) and shorter reaction times (up to 12 hours).

Thermogravimetric (TG) and differential thermo analysis (DTA) were carried out using a Setaram equipment (Setsys Evolution). The measurements were conducted with heating and cooling rates between 2 and 10°C/min conducting a flow of 20 ml/min of He. XRD pattern were recorded on a Bruker AXS D4 ENDEAVOR diffractometer (Ni filtered Cu $K\alpha$ radiation) with a step width of 0.03° $2\theta$ and 1 second per step. The powder data were analyzed with the Stoe WinXPOW software package and also with the program TOPAS (Bruker). FTIR analysis were performed on a Bruker IFS66v FTIR spectrometer by using the pellet technique in the 370–4000 cm$^{-1}$ range, with a resolution 2 cm$^{-1}$ (1 mg sample diluted in 200 mg of KBr or NaCl). In selected experiments the pellets were used for temperature dependent IR (TIR) investigations in an in house made device as described in detail elsewhere [29]. SEM/EDX investigations were carried out on crushed and gold sputtered samples on machines JEOL SM-6390A (in our own Laboratory) and on JEOL 7000F (at Frederic Seitz Materials research Laboratory, University of Illinois at Urbana-Champaign, USA).

RESULTS AND DISCUSSION

Structural changes of geopolymer cements during prolonged ageing

Depending on the Si/Al ratios the structure of geopolymers could be described via poly-sialate M$_n$-(Si-O-Al-O-)$_n$, poly-sialate siloxo M$_n$-(Si-O-Al-O-Si-O-)$_n$, and poly-sialate disiloxo M$_n$-(Si-O-Al-O-Si-O-Si-O-)$_n$, units, revealing Si/Al ratios of 1/1, 2/1, 3/1, and so on. For Si/Al $>> 3$, cross linking of poly-siloxo chains including a varying number of sialate links becomes relevant [1]. Similar to aluminosilicate zeolites, considering faujasite type zeolites A, X, Y, the geopolymer network contains mesopores (2-50 nm) possessing a significant distribution in sizes. The maximum in the distribution of pore sizes varies with the nominal Si/Al ratio, e.g. at 5 nm (Si/Al = 2.15) and 2.5 nm (Si/Al = 1.15) [30]. Fig. 1 shows some TG/DTA results of Cem.1.

Cem.1 possesses nominal Si/Al of about 1.9. Aged between 24 h and 800 h at room temperature a mass loss of about 25 % due to evaporation of H$_2$O could be obtained during heating up to 600°C. About 90 % of the loss occurs already below 120°C, peaked at about 80°C, whereas the rest gradually dehydrates up to about 400°C (Fig. 1 a). The DTA curves reveal endothermic contributions peaked at about 90°C (Fig. 1 b). The same sample rehydrates under atmospheric conditions, revealing now a mass loss of about 10 % during heating again up to 600°C (Fig. 1 c). Now the main loss is peaked at about 60°C, followed by a more gradual loss up to about 400°C, showing the endothermic peak shifted to the lower temperature (Fig. 1 d). Some small but significant variations in mass loss and heat flow may also be noted for samples aged for shortest and longest time given as marked in Fig. 1. These changes as well as changes between first and second heating runs are related to changes in the mesopores due to further condensations and modifications in the structure of the Si-O-Al network. Such changes are relevant during ageing and accelerated ageing during heating. Small changes could be detected during prolonged ageing in IR absorption spectra in the line shape of the density of states (DOS) of asymmetric Si-O vibrations [5, 6]. Similar changes could be observed in the heat treated sample (Fig. 2). A broadening occurs in the DOS of asymmetric Si-O vibrations

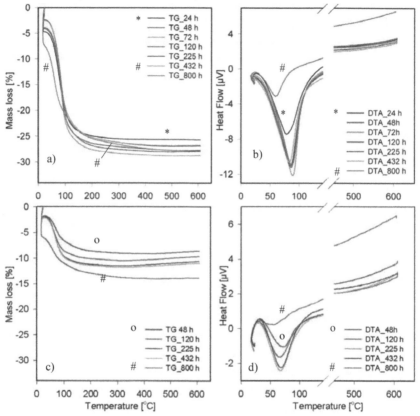

Fig. 1 Mass loss (a) and heat flow (b) of geopolymer Cem.1 aged between 24 and 800 h during heating up to 600°C. Mass loss (c) and heat flow (d) during second heating up to 600°C. Note that samples were exposed to air under ambient conditions between the first and second heating run. Curves of samples aged 24, 48 and 800 h are marked by *, # and o, respectively (see text).

most significantly during the first heating up to 600°C. The second heating run reveals much less changes in DOS. This indicates a significant change in the network structure of the geopolymer in first heat treatment gaining some thermal stability for further heat treatments. The related reactions may dramatically slow down in time and are hard to be followed using IR, XRD or NMR methods . Significant changes could, however, be observed using the Molybdate method as described previously [3-5]. The typical XRD pattern of a "fully reacted" geopolymer-cement shows a broad peak with a maximum at about 28-30°2Theta using CuKα radiation, which does not show significant variations either during prolonged ageing or in heating experiments up to 1000°C (Fig. 3). Above 1000°C crystallization takes place.

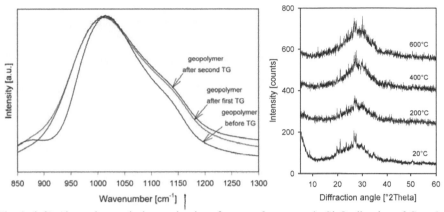

Fig. 2 (left) Absorption peak due to density of states of asymmetric Si-O-vibration of Cem.1, aged 225 h before, after first, and after second heating to 600°C. Samples taken from the experiments used for Fig. 1.

Fig. 3 (right) XRD pattern of Cem.1, aged 200 h and after heating to 200, 400, and 600°C under open conditions. Sharp diffraction peaks are related to minor amounts of impurity phases (muscovite, quartz, feldspar) of in total less than 10 wt% in the as received metakaolin [28], which remain unaltered during alkali activation, ageing and heat treatment.

NaBH₄-geopolymer composites

Directly after precipitation of the $NaBH_4$ gel the product has the state of an amorphous or very short range ordered sodium aluminosilicate, containing the whole amount of $BH_4$-anions from the inserted $NaBH_4$-salt. As a result of separation of NaOH during gel precipitation and the alteration process, the alkalinity remains very high, thus preventing the tetrahydroborate from decomposition by hydrolysis. The subsequent process of drying at 80°C up to 110°C for times between 0.5 h - 4.0 h causes rapid gel hardening. Gel precipitation and alteration during drying at 110°C up to two hours was followed by X-ray powder diffraction (XRD) as shown in Fig. 4. The powder diagram of the pure $NaBH_4$-salt is also shown for comparison. Without further drying only a very broad peak with a maximum in average at about 30°2 theta (d = 2.97 Å) and an extended shoulder towards higher 2 Theta. This peak-structure reveals some similarities to the peak-structure also obtained for the geopolymer, Cem.1 in Fig. 3. However, Cem.1, may be described with an average d value of about 3.1 Å, which is close to typical Si-O-Si bonding distances. Therefore, it may be concluded that shorter bonding distances, probably due to NaOH bondings may have a strong influence here. A significant crystallization can already be seen after 30 minutes of drying, reducing significantly the intensity of the broad X-ray peak. The five most intensive peaks can uniquely be indexed as (110), (211), (310), (222) and (330) and consistently refined for a cubic lattice with parameter of about 9.09(5) Å (0.5 h drying), 8.941(22) Å (1 h drying) and 8.924(22) Å (2 h drying). The pattern and lattice parameter could be related to basic hydro-sodalite $Na_8(Al_6Si_6O_{24})(OH)_2(H_2O)_2$ (ICSD 36050: 8.89 Å, ICSD 72059: 8.875 Å), hydro-sodalite $(Na_8(Al_6Si_6O_{24})(H_2O)$ (ICSD 85512: 8.848 Å) and $NaBH_4$-sodalite $Na_8(Al_6Si_6O_{24})(BH_4)_2$ (ICSD 153255: 8.916 Å). Ageing at 110°C does not alter the sodalite type peaks (Fig. 4 b, c, d). The rather broad peaks of the sodalite show its typical nano-crystalline character. The crystallization of NaBH$_4$ follows the formation of nano-crystalline sodalites as can be distinct in Fig. 4 c, d by comparison to the pattern for single phase NaBH$_4$ (Fig. 4 e). Pattern and refined lattice parameter, 6.1612(7) Å, are consistent with that of NaBH$_4$ (ICSD 165707: 6.118 Å). An additional sharp diffraction peak is observed at 6.5°2 theta (d = 13.66 Å). This peak does not appear quite general in every batch we prepared and also disappears with furth⋅ The other

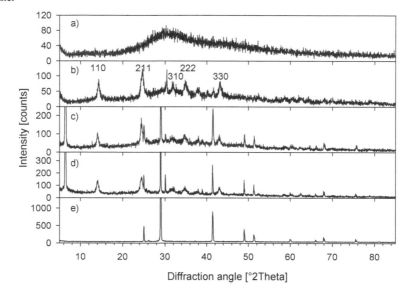

Fig. 4. X-ray powder patterns of NaBH$_4$-gel in dependence of the drying period during drying at 110°C: directly after gel precipitation (a) and after drying for 0.5 h (b), 1.0 h (c) and 2.0 h (d). The pattern of the pure NaBH$_4$-salt, dried at 110°C is inserted for comparison (e). Peaks corresponding to NaBH$_4$ can also be identified in the new composite in pattern c and d. The peaks as indexed in b and also observed in c and d are related to sodalite phase (compare also below Fig. 9, and text for further details).

SEM micrographs of the NaBH$_4$-gel shows the typical appearance of a sheeted-like gel similar to photographs also obtained for Cem.1 (Fig. 5). In few cases rather big cube-like crystals (up to 5 μm) were observed which could depict NaBH$_4$ covered in the gel. EDX-area analyses always revealed very strong Na-signals. The signals of Si and Al indicate molar ratios of the gel composition in agreement with the nominal input.

Advantages of handling of NaBH$_4$ in the geopolymer gel could be related to an effect of protection against moisture, an increased thermal stability and a simple handling for hydrogen release. Moreover the total capacity of the hydrogen storage should not become too much diluted compared to the pure NaBH$_4$-salt. All these requirements are well justified. It may be noted that below 40°C NaBH$_4$ creates a stable hydrated form NaBH$_4$·2H$_2$O in contact with water. This species dehydrates at about 40°C to water and NaBH$_4$, which leads to the uncontrollable reaction [31]. In contrast, water added to the NaBH$_4$-gels does not lead to a significant reaction. This was proven for example in an experiment at 70°C with the addition of water [24], using the glassware apparatus (see Experimental). The $NaBH_4$ in the composite remains protected, whereas about 60% was lost in the same experiment for unprotected $NaBH_4$.

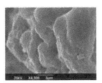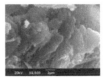

Fig. 5 SEM micrographs of geopolymer samples: Cem.1 (a, b) and the $NaBH_4$ containing Gel (c, d). (Taken at 20 kV, magnification with scale units/bars are given).

A controlled hydrogen release from the $NaBH_4$ enclosed in the geopolymer gels can be achieved by adding diluted hydrochloric acid (1%) as described under Experimental. It could be shown that the hydrogen content of the solidified geopolymer gels can be varied with the amount of added $NaBH_4$ during the synthesis. In a first step a series of syntheses were carried out with a molar ratio Si/Al = 0.83 and drying temperature of 110°C. The $NaBH_4$/solid weight ratio was varied between 0.26 and 0.75. The resulting hydrogen release increased approximately linearly with the $NaBH_4$ input up to about 72% of the obtained hydrogen release from the pure $NaBH_4$ (Fig. 6 a). Higher amounts of $NaBH_4$ could not be dissolved in the solutions. In a second step the drying temperature was varied. Generally lower temperatures requires longer drying times for solidification. At 110°C samples could be dried in 2 hours. Samples dried at 80°C and 95°C could be used after 4 hours whereas at 20°C and 40°C it takes

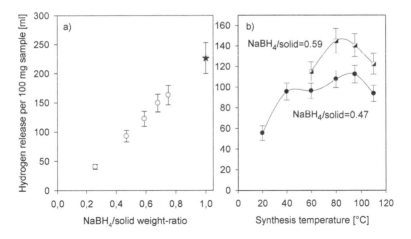

Fig. 6 Hydrogen release from $NaBH_4$-composites with a) different $NaBH_4$/solid ratios for samples dried at 110°C and b) for two $NaBH_4$/solid ratios (by weight) as function of drying temperature.

four days under open conditions. The effect of drying temperature is shown in Fig. 6 b for two examples of x = NaBH$_4$/solid ratios (by weight). An optimum for the amount of hydrogen release could be seen at about 80 and 90°C for 0.59 and 0.47, respectively. In a third step the nominal Si/Al ratio of the geopolymer was varied in series of drying experiments at 110°C (2 h). No significant variations in the absolute content of hydrogen release could be obtained for nominal Si/Al ratios between 0.4 and 1. A slight increase in content could be seen for Si/Al = 2 and Si/Al = 2.6. IR spectra taken along this series also indicate that the formed sodalite contribution in the aluminosilcate gels becomes most significant for Si/Al around 1.

Temperature dependent infrared (TIR) absorption of two examples of the new material (denoted composite I and II) at temperatures between 20°C and 400°C in comparison to the behavior of NaBH$_4$-salt are shown in Fig. 7. KBr was used as diluting matrix for the samples. The tetrahedral BH$_4$-anion group in the NaBH$_4$-salt is identified by vibration modes at 1123 cm$^{-1}$ ($\upsilon_4$), 2222 cm$^{-1}$ ($\upsilon_{Fl}$), 2291 cm$^{-1}$ ($\upsilon_{Fh}$) and 2383 cm$^{-1}$ ($\upsilon_2+\upsilon_4$) [24,25]. Two further vibrations in addition to the IR active normal vibrations ($\upsilon_3$, $\upsilon_4$) of the BH$_4$-tetrahedra are seen. One is explained by the combination mode ($\upsilon_2+\upsilon_4$). The other one is due to the Fermi-Resonance effect according to $2\upsilon_4 \approx \upsilon_3$. Therefore, both modes share their intensities and become shifted towards higher ($\upsilon_{Fh}$) and lower ($\upsilon_{Fl}$) frequency [31]. A high frequency shoulder in each peak, which becomes better resolved in spectra below room temperature, is due to the isotope distribution $^{10}$B/$^{11}$B [32]. With increasing temperature the BH$_4$-related peaks become very much broadened, which appears completely reversible during cooling. Some new peak structures appear in the spectrum of NaBH$_4$ for a sample heated to 400°C and cooled down in the KBr pellet. These peaks indicate some destruction products (dp) of NaBH$_4$ which become visible for the first time at temperatures above 300°C. A peak around 700 cm$^{-1}$ may also be noted. An assignment of these peaks to some polymeric borate species could be possible [32]. Further investigations are required for an assignment of these peaks.

Nearly the same characteristic peak positions, intensity relations and temperature dependencies as observed for the NaBH$_4$-salt are also obtained for the NaBH$_4$ in the composite material, i.e. vibration modes are at 1127 ($\upsilon_4$), 2225 ($\upsilon_{Fl}$), 2293 ($\upsilon_{Fh}$) and 2388 ($\upsilon_2+\upsilon_4$) (in cm$^{-1}$). For composite II some very small indications of destruction products (dp) could be seen in the spectra of the pellet heated at 400°C, which is however completely absent for composite I thermally treated in the same way. These two composites could further be distinct in their infrared absorption below 1100 cm$^{-1}$. For composite II typical peaks related to fundamental lattice vibrations of sodalite could be identified at 436, 469, 668, 705, 733 and a contribution at 990 (in cm$^{-1}$). These peaks are superimposed on the broader peaks of the geopolymer matrix. The main peak positions are therefore denoted as "sod+geop" in Fig. 7. For composite I the three main peaks (denoted as "geop" in Fig. 7) correspond to the geopolymer. There is not any significant indication of sodalite. It may be suggested that the formation of sodalite makes no favorable contribution in the composite and might be avoided for further optimization of storage capacities.

There are indications of CO$_3$-anions, H$_2$O molecules and also of OH-groups in the spectra of composites. The CO$_3$-anions are related to the formation of Na$_2$CO$_3$ during preparation of the composites due to handling under open conditions (attraction of CO$_2$ of the alkaline solutions). The main absorption peak at around 1450 cm$^{-1}$ and also the corresponding smaller one at about 880 cm$^{-1}$ remain stable in intensity during heat treatment in composite I. Composite II shows some changes which could be related to decrease in distortion of the CO$_3$-anions with increasing temperature related to dehydration. The absorption peak at 1630 cm$^{-1}$ indicates the presence of H$_2$O molecules via their bending modes. Their corresponding symmetric and asymmetric stretching vibrations show up in the broad distribution of absorption intensities in the range

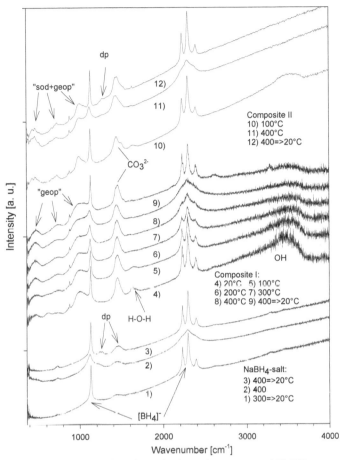

Fig. 7 Temperature dependent infrared (TIR) absorption spectra of NaBH$_4$, composite I, and composite II samples at temperatures and cooled down from temperatures as denoted (details are described in the text).

between 2800 up to 3600 cm$^{-1}$, covering propably also some specific OH-groups (NaOH). In the 20°C spectrum of composite I spurious amounts of NaBr•2H$_2$O could be identified by a double peak structure (maxima at 1632 and 1650 cm$^{-1}$) and a more significant peak at about 620 cm$^{-1}$. These features and some additional peaks observed in the range between 3200 and 3500 cm$^{-1}$ [31] could be identified in further room temperature spectra of various composites but could not be seen in the 20°C spectra of composite I in Fig. 7 due to a too small signal to noise ratio. It can be seen, that with increasing temperature, the peaks related to NaBr•2H$_2$O disappear due to dehydration. The formation of NaBr•2H$_2$O is related to the use of KBr as diluting matrix for the IR measurement due to some Na/K exchange reaction between the KBr and Na-ions from the composite in the presence of significant water adsorption in the composite.

It is interesting to note that significant amounts of water can be enclosed in the geopolymer part of the composite. During drying, e.g. at 120°C, most of the water enclosed in the pores of the geopolymer becomes desorbed. A similar amount of water is adsorbed, upon holding the sample under open conditions. The water attack to $NaBH_4$ is, however, negligible. The residual water present above 200°C could also not lead to any significant reaction with the $NaBH_4$ in the composite material (Fig. 7). The "water effect" could be visualized by some preliminary TG/DTA measurements as shown in Fig. 8. For composite II samples the TG curves show a mass loss of about 2 wt% up to 210°C and a further loss of about 1 % between 325 and 375°C, the highest temperature used. Corresponding DTA curves show a pronounced endothermic signal at about 200°C which could be related to a rather well defined dehydration of hydrosodalite. Another explanation can also not be ruled out at the moment, since such a sharp signal in the DTA curve and also such a sharp drop in mass loss seems to be unusual compared to ordinary hydrosodalite [33,34]. The much broader endothermic signals, which could be realized with respect to a hypothetical baseline at around 100 and 150°C, are related to different dehydration effects of the geopolymer matrix in composite II. The IR spectra of the sample taken before the TG experiment and afterwards show almost no loss in the $BH_4$-vibration intensity, reproducing well the behavior also observed in TIR experiments carried out for composite II shown in Fig. 7. It may be noted that for the TG/DTA measurements at 120°C dried samples were used. Longer holding times under open conditions and at 20°C could lead to more significant rehydration, which leads to much higher losses and changes in the related DTA curves. Further experiments are necessary to better understand details of the hydration and dehydration effects of the geopolymer contribution in the composites.

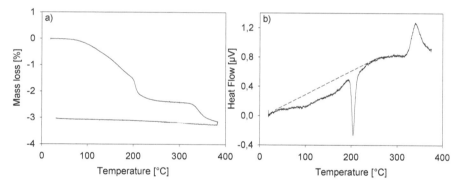

Fig. 8 Mass loss (a) and heat flow (b) of composite type II sample.

The geopolymer-type residual in micro- and nano-$NaBH_4$-SOD
The XRD pattern of nano-crystalline $NaBH_4$-sodalites (nano-$NaBH_4$-SOD) obtained for reaction times of 4 h, 6 h and 12 h at 60°C are compared to a micro-crystalline sample in Fig. 9. With decreasing reaction time the broad peak with maximum around 30°2 theta in the XRD pattern strongly increases in intensity (Fig. 9). This peak shows close similarities with the broad peak typically observed for the geopolymer Cem.1 (Fig. 3) and also with the peak obtained for the geopolymer gel which encloses the $NaBH_4$-crystals (Fig. 4). It is important to note that a similar peak is observed for the micro-$NaBH_4$-SOD, too, but with much lower intensity (see inset in Fig. 9 d). It can be related here to the precursor material for the crystallization of the sodalites and remains to a more or less extent in the samples and cannot be removed, e.g. in washing

procedures (washing is typically carried out to remove residual $Na_2CO_3$ from the samples). This "partially crystalline", short ranged ordered, or - we may call it geopolymer-type - aluminiosilicate phase stores and transports the water molecules for the thermally controlled hydrogen release r̶ ̶ ̶ ̶ ̶ ̶ ̶ (̶H̶ ̶)̶P̶(̶O̶H̶)̶ ̶ ̶ ̶ ̶ ̶ ̶ 3̶)̶ species in all the so far pre

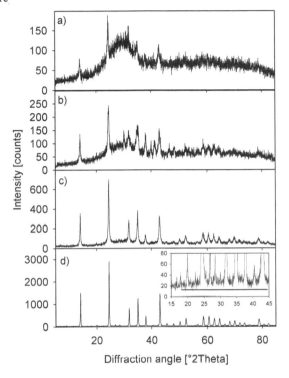

Fig. 9 XRD pattern of nano-NaBH₄-SOD as synthesized hydrothermally at 60°C with heating for 4 h (a), 6 h (b) and 12 h (c) and of micro-NaBH₄-SOD (d) heated at 120°C for 24 h. (Insert in d: enlarged scale for selected part of diffraction angle). The sample (c) possesses about 20 nm average crystal size according to X-ray refinement. The pattern of micro-NaBH₄-SOD is consistent with ICSD 153255. The refined lattice parameter is 8.916 A.

Spectra as obtained in heating/cooling runs up to 500°C in TIR experiments of a micro-NaBH₄-SOD in NaCl pellet are shown in Fig. 10. Shown are the sodalite specific triplicate lattice vibrations (Fig. 10 a). The non-framework species show up in the spectral range above 1000 cm$^{-1}$ (Fig. 10 b). The as-synthesized sample (starting sample) reveals the four peaks of the BH₄-anion as also seen for pure NaBH₄ and the NaBH₄ enclosed in the geopolymer (compare Fig. 7). The positions of the peak maxima appear at about 1131, 2239, 2287 and 2391 (all in cm$^{-1}$) and are obviously shifted compared to NaBH₄ in a characteristic way due to the framework-BH₄ interaction. The broad peak denoted by D corresponds to O-H vibrations of water molecules. Their presence can also be seen by the weak peak at around 1630 cm$^{-1}$. These peaks become much more intensive for the nano-NaBH$_4$-SOD (see below, Fig. 11, denoted OH and HOH) due to the higher amount of the geopolymer type part in the sample. Therefore, the water content can be related to the geopolymer type phase. The sharper peak D′ could correspond to some B(OH)$_4$-anion groups or basic OH-groups in the sodalite [26,27], but may be relatively small in intensity. An increase in intensity at around D′ can be seen with increasing heating temperature of the sample whereas the H$_2$O related intensity decreases in the TIR experiment. In the same way the intensity related to the BH$_4$-anion decreases and new peaks in the range as denoted by A, B, B′, C and C′ app

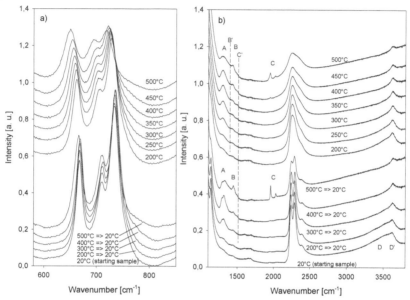

Fig. 10 Absorption of micro-NaBH$_4$-SOD at 20°C (starting sample) and as obtained in temperature dependent infrared (TIR) experiments in the range of characteristic sodalite framework vibrations (a) and of non-framework species (b). Spectra are taken in heating/cooling runs at temperatures as denoted. Further details are described in the text.

$$[BH_4]^-_{SOD} + H_2O \Rightarrow [BH_3(OH)]^-_{SOD} + H_2 \qquad (1)$$

$$[BH_3(OH)]^-_{SOD} + H_2O \Rightarrow [BH_2(OH)_2]^-_{SOD} + H_2 \qquad (2)$$

$$[BH_2(OH)_2]^-_{SOD} + H_2O \Rightarrow [BH(OH)_3]^-_{SOD} + H_2 \qquad (3)$$

$$[BH(OH)_3]^-_{SOD} + H_2O \Rightarrow [B(OH)_4]^-_{SOD} + H_2 \qquad (4)$$

According to this, peaks A, B, B′ could be assigned to the species [BH$_3$(OH)]$^-_{SOD}$, [BH$_2$(OH)$_2$]$^-_{SOD}$, [BH(OH)$_3$]$^-_{SOD}$, respectively [24, 26, 27]. Further investigations [24] have shown that peaks C′ and C are related to [BO(OH)$_2$]$^-_{SOD}$ and [BO$_2$]$^-_{SOD}$ according to the dehydration steps

$$[B(OH)_4]^-{}_{SOD} \iff [BO(OH)_2]^-{}_{SOD} + H_2O \qquad (5)$$

$$[BO(OH)_2]^-{}_{SOD} \iff [BO_2]^-{}_{SOD} + H_2O. \qquad (6)$$

It may be noted that reaction 5 and 6 occur in consecutive reaction steps above about 250°C and 400°C, respectively. Therefore, at least in spectra above 400°C, the appearance of OH-groups with increasing temperature cannot be related to $[B(OH)_4]^-{}_{SOD}$ species or even to the intermediate $[BO(OH)_2]^-{}_{SOD}$ species. This conclusion is supported since peak C' related to the $[BO(OH)_2]_{SOD}$-species just increase and decrease in intensity between 200 and 500°C. Moreover peak C is related to the $[BO_2]_{SOD}$-species appears, indicating dehydration. The presence of significant amounts of "basic OH" groups as known for hydroxy-sodalites, can also be ruled out for temperatures above 400°C [33, 34]. Therefore, the pronounced OH-related peak observed in the TIR spectra when cooled down from 400 and 500°C could mainly be related to OH groups formed in reactions 1 to 3. Accordingly, it can be suggested that anion species $[H_{4-n}B(OH)_n]^-$ with n = 1, 2, 3 could be stabilized in the sodalite cages. There are slight but significant changes in the sodalite typical triplicate lattice vibrations (Fig. 10 a), which correspond to the change in cage fillings. These changes could correspond to not more than 20% of the cages. Since all the available water for the reaction in part is consumed and in another part simply dehydrated, the reaction necessarily stops or becomes very slow against two end-members, namely $NaBO_2$- and $NaBH_4$-sodalite. In separated experiments it could be shown that the sample can be rehydrated under atmospheric conditions and that the hydrogen release reaction could be continued in a further heating run. Typically, in seven of such cycles with heating to 500°C the $BH_4$-content decreases below 10% of the initial value [24]. The micro-$NaBH_4$-SOD together with the new $H_{4-n}B(OH)_n$-anions remains stable up to about 600°C. Above this temperature the sodalite framework cannot withstand the reduction potential of $BH_4$-anion anymore.

Similar effects as observed for the micro-$NaBH_4$-SOD can be depicted for nano-$NaBH_4$-SOD samples (Fig. 11). Shown is the as-synthesized sample at 20°C and at 20°C after heat treatment up to 200, 300, 400 and 500°C in the TIR experiment, using a NaCl pellet as above. Here already the starting sample shows peak A intensity related to some pre-reaction according to reaction 1. An increased intensity of this peak and a decrease of $BH_4$-intensity is detectable for heating at 200°C, which becomes more significant at 300°C. Additionally peak B is observed, as has been related to $H_2B(OH)_2$-anions in accordance with reaction 2 in the micro-$NaBH_4$-SOD. At 400 and 500°C peak B' could be seen positioned between A and B, which could be related to $HB(OH)_3$-species. There is only very low intensity of peaks C indicating only very small concentration of $BO_2$-species. On the other hand the occurrence of C' related to $BO(OH)_2$-species can well be identified in the spectra taken at 400°C and 500°C (7, 9) in Fig. 11, as well as cooled down to 20°C (4, 5). The OH-related peaks in these spectra may also be assigned to $H_{4-n}B(OH)_n$ groups in all these species. There is, however a larger distribution in intensity compared to the spectra obtained for the microcrystalline sample at the same temperatures. It can presently not be ruled out that a type B-peak observed in the spectra 5 (cooled from 500°C) as marked with B in brackets could be related to the intensification of $CO_3^{2-}$-group with increasing temperature. A similar effect could also be observed in hydro-sodalite samples due to the dehydration of hydrogen-carbonate groups, formed during synthesis which could not completely be removed by washing procedures. However, it can be seen that the total amount of $BH_4$-anions has almost disappeared at 500°C without any significant destruction of the sodalite. This could be concluded as the sodalite-lattice vibrations remain nearly unchanged from the heat treatment. It may also show that $H_{4-n}B(OH)_n$-species may not show significant contribution to the absorption intensity in the range between 2000 to 2400 $cm^{-1}$. The change seen in the spectral range of the lattice vibrations are mainly related to the dehydration effect and is, therefore, related to the geopolymer part of the san

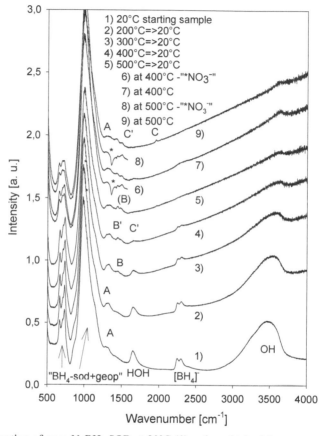

Fig. 11 Absorption of nano-$NaBH_4$-SOD at 20°C (1) and as obtained in temperature dependent infrared (TIR) experiments (2-9) in the range of characteristic sodalite framework vibration and of non-framework species. Spectra are taken in heating/cooling runs at temperatures as denoted. Further details are described in the text.

Two further spectra taken at 400 and 500°C (spectra 6 and 8) are shown in Fig. 11 drawn from 500 cm$^{-1}$ only up to 1500 cm$^{-1}$. These spectra were taken in exactly the same way as spectra 7 and 9, however with the addition of some amount of $NaNO_3$ in the starting pellet and as referenced to a NaCl pellet with the same addition of $NaNO_3$. It could be shown that the hydrogen released from the sodalite immediately reduces $NO_3$-anions which leads, therefore, to the "negative peak" in the spectra (denoted by *). This "tracer-reaction" shows that hydrogen does not leave the sodalite as ordinary $H_2$-molecule but in an activated form. Similar results were also obtained for the microcrystalline sample [24].

Typical TG curves of micro- and nano-NaBH$_4$-SOD are shown in Fig. 12. A total mass loss of about 9 wt% is observed for nano-NaBH$_4$-SOD. The change in slopes could be related with the occurrence of two broad endothermic peaks centered around about 80°C and 175°C observed in the DTA curves (not shown). Thus the main content of water dehydrates. Above about 200°C a mass loss of about 1 % could be seen. This loss is related to the combined effect of dehydration and release of hydrogen. For the micro-NaBH$_4$-SOD there is a total mass loss of about 1-1.5 wt%. Here also the main dehydration occurs below about 250 °C, which is also related to two peaks observed in the DTA curves centered around 80°C and 200°C. The interaction for hydrogen release occurs above about 250°C leading to a very small mass loss, which could also still be accompanied by a small contribution of dehydration.

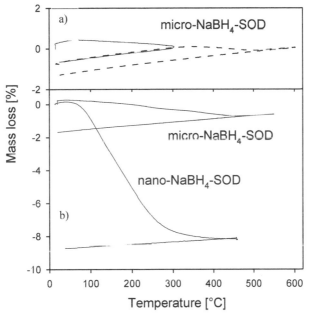

Fig. 12 Mass loss of micro- and nano-NaBH$_4$-SOD during various heating/cooling cycles: a) cycles to 300°C followed by cycle to 600°C for micro-NaBH$_4$-SOD. b) cycle of micro-NaBH$_4$-SOD to 600°C compared to nano-NaBH$_4$-SOD to 450°C.

CONCLUSION

A key point for the understanding of hydrogen release form the BH$_4$-sodalite concerns the supply of water at elevated temperature, i.e. at temperatures between 200°C up to 500°C. The supply of this water is related to the geopolymer type part of the samples. As shown for the geopolymer binder, most of the water is dehydrated below 200°C (Fig. 1 a, 1 c), which was similarly also observed for the geopolymer part in the composite (Fig. 8 a). However, there is still enough water supply above 200°C in the sodalite samples for the hydrogen release reaction. This water is related to residual amounts of water as could also be seen in the narrow range around D′ in the TIR spectra of micro-NaBH$_4$-SOD (Fig. 10) and nano-BH$_4$-SOD (Fig. 11) as well as in the TG-experiments (Fig. 12). A similar hydrogen release reaction does not occur in the composite material. This is explained by the presence of a strongly alkaline layer around $NaBH_4$ crystals which grows in a secondary step in the drying sequence (Fig. 4) from the residual $BH_4$-bearing alkaline solution. Such a layer is absent for $BH_4$-anions in the sodalite. Here the protection is related to the size of the sodalite windows, i.e. their closure via Na-ions. The entrance windows become more opened with the increasing mobility of the Na-ions with increasing temperature [24, 29]. The comparison of the nano- and micro-$NaBH_4$-SOD samples strongly indicates a diffusion controlled mechanism of the hydrogen release reaction. Further investigations may go in this direction. Based on this understanding preliminary success in hydrogen reloading could be reported [24].

Finally, working systems using $NaBH_4$ as hydrogen storage for fuel cell applications could be considered [35,36]. In fact this work has shown that the effective amount of hydrogen in the new geopolymer-$NaBH_4$ composite reaches about 72 wt% of pure $NaBH_4$. Presently available fuel cell applications use about 30 wt% $NaBH_4$ in alkaline solutions [36]. It remains an open question if suitable technical solutions could also be available for direct applications with the new composite. On the other hand a breakthrough for large scale application for $NaBH_4$ as a fuel has not come although there have been many investigations since the discovery of $NaBH_4$ more than 60 years ago, and intensified efforts during the last ten years [37]. This could basically be related to the "no way of any reloading property" of consumed $NaBH_4$. Further investigations in the field of $NaBH_4$-enclosure in geopolymers and the transformation dynamics of the sodalite $NaH_{4-n}B(OH)_n$-anion species concerning hydrogen release and reloading properties could possibly open new ways to overcome these difficulties.

ACKNOWLEDGEMENTS

Some results were obtained by LS given prior to publication in his Ph. D. thesis work. We thank the "Land Niedersachsen" and the Leibniz University Hannover for some financial support in these investigations. Some SEM investigations could be carried out at Frederic Seitz Materials research Laboratory, University of Illinois at Urbana-Champaign, USA, by Dr. E. Mielcarek during her stay abroad, with technical assistance by Dr. W. Swiech, which is gratefully acknowledged. Finally we thank an unknown referee and the editor for helpful comments for improving the manuscript.

REFERENCES
[1] J. Davidovits, Chemistry of geopolymeric systems, terminology, in Geopolymere '99 International Conference, J. Davidovits, R. Davidovits, C. James, eds., Geopolymer Institute, Saint Quentin, France, 9-40, 1999.
[2] J. L. Provis, J. S. J. van Deventer, eds. "Geopolymers-Structure, Processing, Properties and Industrial Applications", Woodhead Publishing Limited, Oxford, 2009.
[3] C. H. Rüscher, E. M. Mielcarek, W. Lutz, A. Ritzmann, W. M. Kriven, Weakening of alkali activated metakaolin during ageing investigated by molybdate method and infrared absorption spectroscopy. J. Am. Ceram. Soc., 93, 2585-2590 (2010).
[4] C. H. Rüscher, E. M. Mielcarek, W. Lutz, A. Ritzmann, W. M. Kriven, The ageing process of alkali activated metakaolin. Ceram. Trans., 215, 1-10 (2010).
[5] C. H. Rüscher, E. M. Mielcarek, W. Lutz, F. Jirasit, J. Wongpa, New insights on geopolymerisation using molybdate, Raman, and infrared spectroscopy, Ceramic Engineering and Science Proceedings, 31, issue 10, Strategic Materials and Computational Design, W. M. Kriven, Y. Zhou, M. Radovic, eds., 19-35 (2010).

[6]C. H. Rüscher, E. M. Mielcarek, J. Wongpa, C. Jaturapitakkul, F. Jirasit, L. Lohaus, Silicate-, aluminosilicate and calciumsilicate gels for building materials: chemical and mechanical properties during ageing, Eur. J. Mineral. **23**, 111-124 (2011).

[7]D. W. Breck, Zeolite Molecular Sieves, Wiley, New York, 1974.

[8]R. M. Barrer, Hydrothermal Chemistry of Zeolites, Academic Press, London, 1982.

[9]F. Schüth, Mobile Wasserstoffspeicher mit Hydriden der leichten Elemente. Nachrichten aus der Chemie **54**, 24-28 (2006).

[10]L. Schlapbach, A. Züttel, Hydrogen storage materials for mobile applications. Nature **414**, 353-358 (2001).

[11]B. Bogdanivic, M. Schwickardi, Ti-doped alkali metal aluminium hydrides as potential novel reversible hydrogen storage material, J. Alloys Compd., **253**, 1-9 (1997).

[12]B. K. Pradhan, A. R. Harutyunyan, D. Stojkovic, J. C. Grossman, P. Zhang, M. W. Cole, V. H. Crespi, H. Goto, J. Fujiwara, P. C. Eklund, Large cryogenic storage of hydrogen in carbon nanotubes at low pressures. J. Mat. Res. **17**, 2209-2216 (2002).

[13]Y. H. Hu, E. Ruckenstein, Clathrate hydrogen hydrate - A promising material for hydrogen storage, Angew. Chem. Int. Ed. **45**, 2011-2013 (2006).

[14]J. Sculley, D. Yuan, H. C. Zhou, The current status of hydrogen storage in metal-organic frameworks updated. Energy Environ. Sci. **4**, 2721-2735 (2011).

[15]R. J. Kuppler, D. J. Timmons, Q. R. Fang, J. R. Li, T. A. Makal, M. D. Young, D. Yuan, D. Zhao, W. Zhuang, H. C. Zhou, Potential applications of metal-organic frameworks. Coordination Chemistry Reviews, **253**, 3042-3066 (2009).

[16]L. J. Murray, M. Dinca, J. R. Long, Hydrogen in metal-organic frameworks. Chem. Soc. Rev. **38**, 1294-1314 (2009).

[17]C. Jia, X. Yuan, Z. Ma, Metal-organic frameworks (MOFs) as hydrogen storage materials. Prog. in Chem. **21**, 1954-1962 (2009).

[18]A. W. C. van den Berg, S. T. Bromley, J. C. Jansen, Thermodynamic limits on hydrogen storage in sodalite: a molecular mechanics investigation, Microp. Mesop. Mat. **78**, 63-71 (2005).

[19]J. G. Vitillo, G. Ricchiardi, G. Spoto, A. Zecchina, Theoretical maximal storage of hydrogen in zeolitic frameworks, Phys. Chem. Chem. Phys. **7**, 3948-3954 (2005).

[20]J. Weitkamp, M. Fritz, St. Ernst, Zeolites as media for hydrogen storage, J. Hydrogen Energy **20**, 967-970 (1995).

[21]D. P. Boom, Hydrogen Storage Materials, Green Energy and Technology, Springer Verlag London Limited, 2011.

[22]H. W. Langmi, A. Walton, M. M. Al-Mamouri, S. R. Johnson, D. Book, J. D. Speight, P. P. Edwards, I. Gameson, P. A. Anderson, I. R. Harris, Hydrogen adsorption in zeolites A, X, Y and RHO, J. Alloys Comp. 356-357, 710-715 (2003).

[23]J. C. Buhl, L. Schomborg, C. H. Rüscher, NaBH$_4$ in solidified aluminosilicate gel: a new hydrogen storage with interesting properties. NSTI-Nanotech 2012, www.nsti.org. ISBN 978-1-4665-6276-9, **Vol. 3**, 559-562 (2012).

[24]J. C. Buhl, L. Schomborg, C. H. Rüscher, Enclosure of sodium tetrahydroborate (NaBH$_4$) in solidified aluminosilicate gels and microporous crystalline solids for fuel processing, in Hydrogen Storage, J. Liu ed., INTECH ISBN 978-953-51-0371-6, free online editons www.intechopen.com, **Chapter 3**, 49-90, (2012).

[25]D. R. M. Brew, K.J. D. MacKenzie, Geopolymer synthesis using silica fume and sodium aluminate, J- Mater. Sci. **42**, 3990-3993 (2007).

[26]J. C. Buhl, Th. M. Gesing, C. H. Rüscher, Synthesis, crystal structure and thermal stability of tetrahydroborate sodalite Na$_8$[AlSiO$_4$]$_6$(BH$_4$)$_2$, Microp. Mesop. Mat. **80**, 57-63 (2005).

[27]J. C. Buhl, L. Schomborg, C. H. Rüscher, Tetrahydroborate sodalite nanocrystals: Low temperature synthesis and thermally controlled intra cage reactions for hydrogen release of nano- and micro crystals. Micro. Meso. Mater. 132 (2010) 210-218.

[28]C. H. Rüscher, A. Schulz, M. H. Gougazeh, A. Ritzmann, Mechanical strength development of geopolymer binder and the effect of quartz content, to be published in Ceramic Engineering and Science Proceedings **34** (2013).

[29]C. H. Rüscher, Chemical reactions and structural phase transitions of sodalites and cancrinites in temperature dependent infrared (TIR) experiments, Microp. Mesop. Mat. **86**, 58-68 (2005).

[30] P. Duxson, J.L. Provis, G.C. Lukey, S.W. Mallicoat, W.M. Kriven, J.S.J. van Deventer, Understanding the relationship between geopolymer composition, microstructure and mechanical properties, Colloids and Surfaces A: Physicochem. Eng. Aspects 269, 47-58 (2005).

[31]Y. Filinchuk, H. Hagemann, Structure and properties of $NaBH_4 \cdot 2H_2O$ and $NaBH_4$, Eur. J. Inorg. Chem. 3127-3133 (2008).

[32]Nakamoto, K., Infrared and Raman Spectra of Inorganic and Coordination Compounds, John Wiley & Sons, New York, 1978.

[33]J. Felsche, S. Luger, Phases and thermal characteristics of hydro-sodalites $Na_{6+x}[AlSiO_4]_6(OH)_x \cdot nH_2O$, Thermochimica Acta **118**, 35-55 (1987).

[34]J. Felsche, S. Luger. Structural collapse or expansion of the hydro-sodalite series $Na_8[AlSiO_4]_6(OH)_2 \cdot nH_2O$ and $Na_6[AlSiO_4]_6 \cdot nH_2O$ upon dehydration. Ber. Bunsengesellschaft Phys. Chem. **90**, 731-736 (1986)

[35]C. Ponce de Leon, F. C. Walsh, D. Pletcher, D. J. Browning, J. B. Lakeman, Review: Direct borohydride fuel cells. J. Power Sources **155**, 172-181 (2006).

[36]Z. P. Li, B. H. Liu, K. Arai, K. Asaba, S. Suda. Evaluation of alkaline borohydride solutions as the fuel for fuel cell. J. Power Sources **126**, 24-33 (2004).

[37]U. B. Demirci, O. Akdim, P. Miele, Ten years efforts and no-go recommendation for sodium borohydride for on-board automotive hydrogen storage. Intern. J. Hydrogen Energy **34**, 2638-2645 (2009).

MECHANICAL CHARACTERISTICS OF COTTON FIBRE REINFORCED GEOPOLYMER
COMPOSITES

T. Alomayri and I.M. Low
Centre for Materials Research, Department of Imaging & Applied Physics, Curtin University of
Technology, GPO Box U1987, Perth, WA 6845, Australia

ABSTRACT
    This research is concerned with the mechanical properties of cotton fibre reinforced geopolymer
composites (CFRGC). The effect of cotton fibre content on the flexural strength, the flexural modulus,
the impact strength and the compressive strength were investigated. The density and porosity of cotton
fibre reinforced geopolymer composites were also measured. The results indicate that cotton fibre
content has a strong influence on the flexural strength, flexural modulus, impact strength and
compressive strength of CFRGC. These cotton fibres reinforced geopolymer composites may be useful
in civil engineering applications.

INTRODUCTION
    Geopolymers are inorganic compounds can be cured and hardened at near-ambient temperatures
to form materials that are effectively low-temperature ceramics with the typical temperature resistance
and strength of ceramics.[1] In recent years, geopolymers have emerged as an alternative to cements.[2, 3]
However, despite their many desirable attributes such as relatively high strength, elastic modulus and
low shrinkage, geopolymers suffer from brittle failure like most ceramics. This limitation may be
readily overcome through fibre reinforcement.
    The concept of using natural fibres to reinforce materials is not new. Cements and polymers have
been reinforced with natural fibres for many years, particularly in developing countries that have used
local materials, such as bamboo, sisal, jute and coir with some success.[4-8] These natural materials are
not only cheap, but their low density and favourable mechanical properties make them attractive
alternatives to synthetic fibres.[7, 8] It is known that certain types of natural fibres tend to produce greater
mechanical properties in concrete. For instance, Rahmann et al.[9] found that bamboo fibres improved
the flexural strength of concrete, and Lin et al.[10] also observed a similar improvement in wood-fibre
reinforced concrete. Similarly, the use of hemp fibres has been found to improve the fracture toughness
of natural fibre-reinforced concrete.[11]
    In this paper, cotton fibre reinforced geopolymer composites with different fibre contents (0.3,
0.5, 0.7 and 1.0 wt%) have been successfully fabricated. The effect of fibre content on the mechanical
properties has been investigated in terms of flexural strength, flexural modulus, impact strength and
compressive strength. Scanning electron microscopy (SEM) has been used to investigate the
microstructure of the cotton fibre reinforced geopolymer composites.

EXPERIMENTAL METHODS
Specimens Preparation
    The fly-ash used in the production of geopolymers was class-F and it was sourced from the coal
fired power station in Western Australia. The results of X-ray fluorescence testing (XRF) are shown in
Table 1 for the fly-ash used. Sodium silicate grade D solution was used for the geopolymer production.
The chemical composition of sodium silicate is 14.7 wt% $Na_2O$, 29.4 wt% $SiO_2$ and 55.9 wt% water.
Sodium hydroxide solution was prepared by dissolving sodium hydroxide flakes in water. The flakes
utilized were commercial grade with 98% purity. An 8 molar sodium hydroxide solution was prepared

and combined with the sodium silicate solution one day before mixing. In this work, cotton fibres with an average length of 10 mm, average diameter of 0.2 mm and density of 1.54 g/cm$^3$were used to reinforce the geopolymer.

In the design of geopolymer composites, an alkaline solution to fly ash ratio of 0.35 was used and the ratio of sodium silicate solution to sodium hydroxide solution was fixed at 2.5. The fibres were added slowly to the dry fly ash in a Hobart mixer at a low speed until the mix became homogeneous at which time the alkaline solution was added. This was mixed for ten minutes at low speed and another ten minutes at high speed. The walls of the mixing container were scraped down to ensure the consistency of the mixture. This procedure was followed for all of the test specimens. Each mix was cast in 25 rectangular silicon moulds of 80 mm × 20 mm × 10 mm and placed on a vibration table for five minutes. The specimens were covered with a plastic film and cured at 105 °C for three hours, then rested for 24 hours at room temperature before de-moulding. They were then dried under ambient conditions for 28 days. [12]

Table1: Chemical composition of class-F fly-ash.

| SiO$_2$ | Al$_2$O$_3$ | Fe$_2$O$_3$ | CaO | MgO | SO$_3$ | Na$_2$O | K$_2$O | LOI |
|---|---|---|---|---|---|---|---|---|
| 50% | 28.25% | 13.5% | 1.78% | 0.89% | 0.38% | 0.32% | 0.46% | 1.64% |

**Density and Porosity**

The thickness, width, length and weight were measured in order to determine the bulk density using the following equation:

$$P = M/V \tag{1}$$

where P = density in kg/m$^3$, M = mass of the test specimen (kg) and V = volume of the test specimen (m$^3$).

The value of apparent porosity ($P_s$) was determined using Archimedes' principle according to ASTM Standard (C-20) and tap water was used as the immersion water. The apparent porosity ($P_s$) was calculated using the following equation:[13]

$$P_s = \frac{m_s - m_d}{m_s - m_i} \times 100 \tag{2}$$

where $m_d$ = mass of the dried sample, $m_i$ = mass of the sample saturated with and suspended in water and $m_s$ = mass of the sample saturated in air.

**Mechanical Properties**

Rectangular bars of geopolymer composites were prepared for three-point bend and Charpy impact tests. Three-point bend tests were performed on A LLOYD Material Testing Machines Twin Column Bench Mounted (5–50 kN) with a span of 40 mm and displacement rate of 1.0 mm/min to determine the flexural strength and flexural modulus. The Charpy impact test was performed on un-notched samples using a Zwick Charpy impact machine with a 1.0 J pendulum hammer. In addition, cylindrical samples with a 2:1 height to diameter ratio were fabricated to determine the compressive strength using a Lloyds EZ50 Universal Tester (Lloyd UK) with a 0.5% grade 50 kN load cell and 150 mm diameter compression platens. A preload of 50 N was applied followed by an extension rate of 5 mm/min.

The flexural strength ($\sigma_F$) and flexural modulus ($E_F$) were computed using the following equations:

$$\sigma_F = \frac{3}{2}\frac{P_m S}{WD^2} \tag{3}$$

$$E_F = \frac{S^3}{4WD^3}\left(\frac{\Delta P}{\Delta X}\right) \tag{4}$$

where $P_m$ is the maximum load at crack extension (N), $S$ is the span of the sample (40 mm), $D$ is the specimen thickness (mm), $W$ is the specimen width (mm). $\Delta P/\Delta X$ is the initial slope of the load displacement curve. The impact strength ($\sigma_i$) was calculated using the following formula:

$$\sigma_i = \frac{E}{A} \tag{5}$$

where $E$ is the impact energy to break a sample with a ligament of area $A$.

The compressive strength ($\sigma_c$) was calculated using the following formula:

$$\sigma_c = \frac{P}{A} \tag{6}$$

where $P$ = total load on the sample at failure (N) and $A$ = calculated area of the bearing surface of the specimen ($m^2$).

**Microstructure Analysis**
        Images of scanning electron microscopy were obtained using a Zeiss Evo 40XVP. SEM imaging was used to examine the microstructure of geopolymer composites. Subsequently, the specimens were coated with a thin layer of platinum before observation to avoid charging.

RESULTS AND DISCUSSION
**Density and Porosity**
        Fig. 1 illustrates the effect of cotton fibre (CF) content on the density and porosity of CF/geopolymer composites after 28 days of curing at ambient temperature. Fig. 1(a) shows that density decreases as the content of cotton fibres increases. However, the porosity also increases with increases in the content of cotton fibres as shown in Fig. 1 (b). The increase in porosity with rising cotton fibre content could be attributed to the tendency of fibres clumping together during mixing, which resulted in entrapped water-filled spaces that subsequently turned into voids.

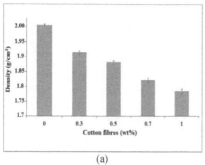
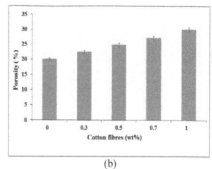

(a)                                          (b)

Fig. 1: Effect of cotton fibre content on (a) density and (b) porosity.

**Flexural Strength and Modulus**

Geopolymer composites were prepared with increasing fibre content and the mechanical properties at 28-days were measured to determine the optimum fibre content. The effect of fibre content on flexural strength of cotton fibre reinforced geopolymer composites is plotted in Fig. 2 (a). The flexural strength is found to increase up to 0.5 wt% cotton fibre content after which it decreases. This enhancement in flexural strength is due to the good dispersion of cotton fibres throughout the matrix which helps to increase the interaction or adhesion at the matrix/fibre interface. The lower flexural strength at higher fibre content may be attributed to agglomeration of fibres and thus poor dispersion of fibres within the matrix. As a result, lower loads were transferred from the matrix to the fibres, thus resulting in lower flexural strengths.

The flexural modulus versus fibre content for the geopolymer composites is shown in Fig. 2 (b). The flexural modulus shows a similar trend to flexural strength. As fibre content increases up to 0.5 wt%, the flexural modulus increases. However, the flexural modulus decreases as a result of fibre agglomeration, and this reduction is most pronounced for composites with high cotton fibre content.

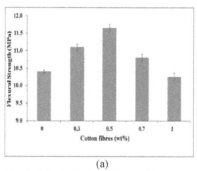
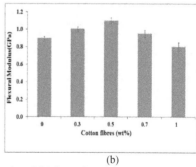

(a)                                          (b)

Fig. 2: Effect of cotton fibre addition on (a) flexural strength and (b) flexural modulus.

**Impact and Compressive Strength**

The impact strengths of geopolymer composites recorded during the impact tests are shown in Fig. 3 (a), which indicates similar trends to the flexural strength results. The addition of 0.5 wt% cotton fibres in the geopolymer matrix increases the impact strength over plain geopolymer, but this trend reverses with the addition of 0.7 and 1 wt% cotton fibres. This reduction in impact strength at higher content of cotton fibre was again due to the formation of fibre agglomerates and voids as a result of increased matrix viscosity. In addition, the presence of the cotton fibre caused a reduction in the fibre-matrix adhesion and the concomitant efficiency in stress-transfer.

Fig. 3 (b) shows that the compressive strength of geopolymer reinforced with cotton fibres follows the same trend to impact strength. The compressive strength is observed to decrease with increasing cotton fibre content greater than 0.5 wt%. This is believed to be due in part to the fact that as the amount of cotton fibres increases, the workability of the mixture decreases thus preventing thorough mixing, which may lead to reduced bond strength between the cotton fibre and the geopolymer paste.

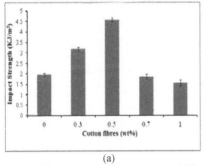 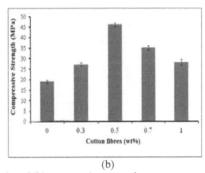

(a)               (b)

Fig. 3: Effect of cotton fibre content on (a) impact strength and (b) compressive strength.

**Microstructural Features**

The microstructures of cotton fibre reinforced geoplymer composites with 0.5 and 1.0 wt% have been shown in Fig. 4 (a) and (b). In Fig. 4 (a), it can be seen that at 0.5 wt% the fibres are distributed relatively homogeneously within the matrix. The effective dispersion of cotton fibres throughout the matrix helps to increase the interaction or adhesion at the matrix/cotton fibre interface. Hence, this facilitates optimal efficiency in stress-transfer from the matrix to the cotton fibres, thus resulting in the improvement of mechanical properties. In contrast, the microstructure of geoplymer composites with cotton fibre content equal or greater than 1.0 wt% shows poor homogeneity due to fibre agglomerations (see Fig. 4 (b)). As a result, these samples showed inferior mechanical properties because of the poor interface bonding or stress-transfer between the geopolymer matrix and the cotton fibres. The presence of agglomerations in geoplymer composites is very undesirable because they deteriorate the mechanical properties. Therefore, up to 0.5 wt % short cotton fibres can be used as optimum reinforcement in geopolymer composites for use in structural applications.

(a)                                                     (b)

Fig. 4: Scanning electron micrographs showing the microstructures of geopolymer composites reinforced with (a) 0.5 % and (b) 1.0 wt% cotton fibres.

CONCLUSIONS

The effect of fibre content on the mechanical properties of cotton fibre reinforced geopolymer composites has been investigated. The mechanical properties (i.e., flexural strength, flexural modulus, impact strength and compressive strength) were found to increase at an optimum fibre content of 0.5 wt%.The porosity of geopolymer composites increased with an increase in fibre content by virtue of increasing fibre agglomerations. Geopolymer composites containing lower fibre content tend to show better fibre homogeneity and stronger fibre-matrix interfacial bonding than those with higher fibre content.

ACKNOWLEDGEMENTS

The authors would like to thank Ms E. Miller for her assistance with SEM. The authors would also like to acknowledge Dr. W. Rickard and Mr. L. Vickers for their assistance with the data collection. Finally, the authors wish to express their sincere thanks to the Physics Department of Umm Alqura University for their generous financial support in the form of a PhD scholarship.

REFERENCES

[1]K.C. Goretta, N. Chen, F. Gutierrez-Mora, J.L. Routbort, G.C. Lukey and J. S. J. Van Deventer, Solid-particles erosion of a geoplymer containing fly ash and blast furnace slag, *Wear*, **256**, 714-719 (2004).

[2]J. Davidovits, Geopolymers: Inorganic polymeric new materials, *J. Therm. Anal.* **37**, 1633-1656 (1991).

[3]J. Temuujin, A. Van Riessen and K.J.D. MacKenzie, Preparation and characterisation of fly ash based geopolymer mortars, *Constr. Build. Mater.* **24**, 1906-1910 (2010).

[4]H. Baluch, Y.N. Ziraba and A. K, Azad, Fracture characteristics of sisal fibre reinforced concrete, *J. Cem. Compos. Concr.* **9**, 157-168 (1987).

[5]B. Pakotiprapha, R.P. Pama and S.L. Lee, Behaviour of a bamboo fibre-cement paste composite, *J. Ferro. Cem.* **13**, 235-248 (1983).

[6]H. S. Ramaswamy, B.M. Ahuja and S. Krishnamoorthy, Behaviour of concrete reinforced with jute, coir and bamboo fibres, *J. Cem. Compos. Concr.* **5**, 3-13 (1983).

[7]H. Alamri, J.M. Low and Z. Alothman, Mechanical, thermal and microstructural characteristics of cellulose fibre reinforced epoxy/organoclay nanocomposites, *Compos B: Eng*, **43**, 2762-71, (2012).

[8]P. Wambua, J. Ivens and I. Verpoest, Natural fibres: can they replace glass in fibre reinforced plastics, *J. Compos. Sci. Technol.* **63**, 1256-1264 (2003).

[9]M.M Rahman, M.H. Rashid, M.A. Hossain and M.T. Hasan, Performance evaluation of bamboo reinforced concrete beam, *J. Eng. Technol.* **11**, 142-146 (2011).

[10]X. Lin , M.R. Silsbee, D.M. Roy, K. Kessler and P.R. Blankenhorn, Approaches to improve the properties of wood fibre reinforced cementitious composites, *Cem. Concr. Res.* **24**, 1558-1566 (1994).

[11]Z. Li, L. Wang and X. Wang, Compressive and flexural properties of hemp fibre reinforced concrete, *Fibres Polymer.* **5**, 187-197 (2004).

[12]T. Bakharev, Thermal behaviour of geopolymers prepared using class F fly-ash and elevated temperature curing, *Cem. Concr. Res.* **36**, 1134-47(2006).

[13]Standard test methods for apparent porosity, water absorption, apparent specific gravity, and bulk density of burned refractory brick and shapes by boiling water 2000: Pennsylvania, ASTM C-20.

# GREEN COMPOSITE: SODIUM-BASED GEOPOLYMER REINFORCED WITH CHEMICALLY EXTRACTED CORN HUSK FIBERS

Sean S. Musil
Department of Materials Science and Engineering, University of Illinois at Urbana-Champaign, Urbana, IL, USA

P. F. Keane and W. M. Kriven
Keanetech LLC, Champaign, IL, USA

ABSTRACT
Geopolymers are an inorganic polymeric material composed of alumina, silica, and alkali metal oxides. Geopolymers have shown promise as a low cost, environmentally friendly structural material. The addition of a reinforcing phase vastly improves the strength and toughness of the composite. For this study, sodium-based geopolymer is reinforced with chemically extracted corn husk fibers via a paste and weave approach followed by cold press and 50°C curing. Corn husk fibers (CHF) are a low cost, abundant and sustainable resource. They show moderate strength and high elongation to failure equating to high work of rupture for a natural fiber which acts as a toughening mechanism for this biocomposite. CHF are chemically extracted from corn husks using a room temperature alkali bath (10:1 molar ratio of $H_2O$:NaOH) with manual agitation for 30 minutes and are rinsed with water to neutralize the pH. Composite tensile and flexural strength are determined for both quasi-aligned and randomly oriented fibers. Composite microstructure is also evaluated with the SEM focusing on fiber continuity and fiber/matrix interface.

INTRODUCTION
Geopolymers are synthesized as a liquid and do not shrink upon curing, which allows for net shape casting as well as easy addition of reinforcing materials. The abundance of aluminosilicate raw materials makes geopolymers a low cost option for use as a building material. Mechanical properties of the geopolymer can be tailored through the addition of a reinforcing phase to either add strength or toughness depending on the application. Reinforcements can range from high cost, high strength metal oxide or silicon carbide fibers to low cost, abundant and renewable natural fibers. For this study, low cost and readily available corn husk fibers were chosen as the reinforcement to determine their viability as a low cost room temperature building material when integrated into a geopolymer matrix. Sodium based geopolymer was also utilized for its low cost.

World corn production is estimated at 849 million tons this fiscal year, with approximately 274 million tons being produced in the U.S. alone.[1] Corn husks comprise approximately 10% of the lignocellulosic cornstover that remains after harvesting, equating to around 27.4 million tons of husks available, of which over 5 million tons of fibers can be extracted.[2] The use of these agricultural byproducts as a potential reinforcement for building materials has great potential in developing countries where more expensive and more rare materials are infeasible or cost prohibitive. Natural cellulosic fibers have been used primarily in paper making, but have more recently been integrated into various forms of composites including polymer matrix composites as performed by Yang et. al.[3,2]

Corn husks can be chemically extracted from the husk through an alkalization technique which removes most of the non-cellulose webbing material (lignin and ash). What remains are lignocellulosic fiber bundles comprised of fine cellulose fibers held together by hemicellulose and the remaining lignin. Further enzymatic treatment can be performed to depolymerize the hemicellulose and break lignin covalent bonding to further extract the fine cellulose fibers, but was not done for this study.[2]

As compared to another natural fiber, jute, CHF has lower strength, but has significantly higher elongation and work of rupture, which can be utilized to improve the toughness of the geopolymer composite.[2] Other properties compared to similar cellulosic fibers can be seen in Table I below. Reddy, et al. provide cellulose content in %, which is assumed to be weight% based on the stated Norman-Jenkins method for determining cellulose content.

Table I. Comparison of cellulosic fiber properties[4]

| Fiber/Bundle | Cellulose (wt%) | Color | Bundle Length (cm) | Fiber Length (mm) | Fiber Width (μm) |
|---|---|---|---|---|---|
| Corn | 80–87 | Yellowish white | 2–20 | 0.5–1.5 | 10.0–20.0 |
| Cotton | 88–95 | Off white | 1.5–5.5 | 15.0–56.0 | 12.0–25.0 |
| Linen | 72–82 | Creamy white | 20–140 | 4.0–77.0 | 5.0–76.0 |
| Jute | 62–64 | Brownish | 150–360 | 0.8–6.0 | 15.0–25.0 |

EXPERIMENTAL PROCEDURES

Composite Processing

Corn husk fiber bundles were chemically extracted from commercially purchased dried corn husks (Figure 1) using a 10:1 molar ratio of $H_2O$:NaOH alkali solution (5.5 N, pH=11). This ratio was chosen as twice the stoichiometric ratio or half the alkalinity used in geopolymer synthesis (10 $H_2O$:2 NaOH). This differs from the process of Yang et. al., who used a less caustic 0.5 N NaOH solution, but elevated the temperature to 85°C.[2] Due to the exothermic nature of mixing sodium hydroxide and water, the alkali solution was cooled to ambient temperature before use. Intact corn husks were immersed in the alkali bath and manually agitated for 30 minutes until dissolution of the husk webbing occurred. The separated fiber bundles were then rinsed with deionized water to remove dissolved substances and residual oils from the chemical separation process and to equilibrate the pH back to neutral. A magnified view of a single dried CHF bundle can be seen in Figure 2. The fiber bundles were then evenly laid out in either quasi-aligned or random orientations and allowed to dry at ambient conditions. The condition of the CHF during each step can be seen in Figure 3 and Figure 4 below.

Figure 1. As received corn husk                    Figure 2. Single extracted CHF bundle

Figure 3.  Chemically separated CHF bundles     Figure 4.  Dried, randomly oriented CHF

Sodium geopolymer binder was developed by mixing a solution of sodium hydroxide, deionized water and fumed silica (sodium water-glass) with Metamax metakaolin clay (1.3 μm particle size). The slurry was first mixed using an IKA mixer (Model RW20DZM, IKA, Germany) with high shear mixing blade to break up and evenly disseminate the metakaolin particles. The high shear mixing was accomplished at 1800 rpm for 5 minutes to produce a low viscosity, homogeneous slurry. The slurry was then vibrated on a FMC Syntron vibrating table (FMC Technologies, Houston, Texas) to remove trapped air bubbles introduced during the high shear mixing. Additional mixing and degassing was then performed on a Thinky ARE-250 planetary conditioning mixer (Intertronics, Kidlington, Oxfordshire, England). The Thinky mixer was run at 1200 rpm for 3 minutes to further mix and 1400 rpm for 3 minutes to centrifugally de-gas the slurry. The resulting geopolymer chemical composition was $Na_2O \cdot Al_2O_3 \cdot 4SiO_2 \cdot 11H_2O$.

The final sodium geopolymer slurry was then used along with the prepared fibers to create a total of four composite panels. Panels were created using a 6 inch by 8 inch rectangular Delrin mold. The panels were created with fibers in both quasi-aligned and random fiber orientations. For the quasi-aligned panels, previously aligned and dried fibers were cut into 6 inch by 8 inch plies (Figure 5) and then laid up along with geopolymer slurry in a paste and weave method to guaranteed maximum infiltration of matrix material within fiber bundles. Randomly oriented panels utilized a single ply of randomly oriented, but evenly distributed CHF, which was then infiltrated with the geopolymer slurry. Each variant of fiber orientation was also molded with/without the fibers pre-wet prior to molding. Due to the hydrophilic nature of CHF, a fiber pre-wetting variable was introduced to determine if the amount of water absorbed by dried fibers affected the geopolymerization reaction and thus the overall mechanical properties of the biocomposite. Wetted fibers were moistened with deionized water and tamped to remove excess water before being laid up in the panel mold.

Figure 5.  Quasi-aligned CHF plies

Figure 6.  Na-geopolymer with CHF cured panel

The uncured panel was then covered with a Delrin top piece and placed in a hydraulic press at 50 psi for 24 hours to compress the panel and remove excess geopolymer matrix material. Following the cold press process, the panel mold was sealed and placing in the oven at 50°C for 24 hours for final curing. An example of an as-cured panel can be seen in Figure 6 above. Panel properties are depicted in Table II below, where $M_f$ denotes mass fraction or weight%.

Table II.  Na-geopolymer/CHF composite panel properties

|  | $M_f$ | Fiber Orientation | Wetted/Unwetted |
|---|---|---|---|
| Panel 1 | 13.28% | Quasi-aligned | Wetted |
| Panel 2 | 12.99% | Quasi-aligned | Unwetted |
| Panel 3 | 14.45% | Random | Wetted |
| Panel 4 | 12.01% | Random | Unwetted |

Mechanical Testing

Cured panels were demolded, trimmed and cut into mechanical testing samples using a wet tile saw with high speed diamond abrasive cutting wheel. Flexural test specimens were cut to approximately 10 mm by 60 mm, while straight-sided tensile test specimens were approximately 25 mm by 150 mm.

The flexure samples were subjected to room temperature four-point bend testing using an Instron Universal Testing Frame, following ASTM standard C78/C78M-10. Samples were mounted in a SiC high-temperature 4-point bending apparatus. The span length between the lower supports was 40 mm and the upper supports were 20 mm apart and equidistant from the point of load application. Loading rate was determined using ASTM C78/C78M-10 guidelines to maintain a 1.0 MPa/min target rate of stress increase on the tension face of the specimen.

Tensile coupons where tabbed at each end using Garolite glass/epoxy composite in order to reduce stress induced by gripping and avoid damage to the composite prior to testing. Tensile test samples were tested to failure in uniaxial tension at room temperature also on an Instron Universal Testing Frame with manual wedge grips. Specimens were gripped at the tabs using light manually applied pressure. Testing followed ASTM C1275-10 guidelines. Specimens were loaded utilizing a constant displacement rate of 0.02 mm/sec. Strains were too high to measure with an extensometer, so cross-head displacement was recorded instead.

RESULTS AND DISCUSSION

Flexure and monotonic tensile tests were performed. Monolithic sodium geopolymer (NaGP) was too brittle and flaw insensitive to directly test the static tensile strength, therefore the tensile strength $(\sigma_T)$ was estimated using the flexural strength $(\sigma_{4B})$ and the Weibull modulus $(m)$ as derived from the flexural test data using equation (1) below.[5] Weibull parameters were calculated using the procedure outlined later in this paper.

$$\frac{\sigma_{4B}}{\sigma_T} = \left[\frac{4(m+1)^2}{m+2}\right]^{\frac{1}{m}} \tag{1}$$

A summary of the mechanical testing results is shown is Table III below. Flexure tests were not performed on the quasi-aligned and wetted fiber composite due to a lack of usable material for testing. Tensile testing was not performed on the randomly aligned fiber samples because strength would be even lower than that of the quasi-aligned fiber composite. Average flexure strength values for reinforced samples could not be explicitly determined due to the fact that nearly half of the test samples elongated to the point of impact with the lower portion of the test fixture. The reported average strength is therefore the average of the stress at the point of impact and test termination, for samples that made contact, and the actual strength values for those samples that failed before contact was made. Those instances are indicated by an asterisk in the table below. Standard deviation on those sample batches was deemed irrelevant based on the aforementioned results.

Table III. Summary of mechanical testing results

| Test Type | Fiber Orientation | Wetted/Unwetted | Average Strength (MPa) | Standard Deviation (MPa) |
|---|---|---|---|---|
| 4-pt Bend | | | 14.139 | 1.865 |
| 4-pt Bend | Quasi-aligned | Unwetted | 8.788* | - |
| 4-pt Bend | Random | Wetted | 7.735* | - |
| 4-pt Bend | Random | Wetted | 6.701* | |
| Tensile | | | 9.027 (Est) | 1.191 (Est) |
| Tensile | Quasi-aligned | Wetted | 3.158 | 0.278 |
| Tensile | Quasi-aligned | Unwetted | 2.882 | 0.334 |

Flexural Testing

Four point flexure testing was performed on both the monolithic sodium geopolymer and the CHF composite specimen. Flexure tests were not performed on the quasi-aligned and wetted fiber composite due to a lack of usable material for testing. Figure 7 through Figure 10 depict the results of the flexure testing. Average flexure strength of the pure sodium was approximately 14 MPa. Few reinforced flexure samples actually failed completely, most strained until contact with the test fixture was made and the test was terminated. This can be seen in Figure 11 which shows the large amounts of strain produced during testing which caused impact with the test fixture prior to actual failure of the composite. Actual flexure strength for reinforced samples could therefore not be determined, but strain values were significant (>10x that of pure NaGP). Actual flexure strength values would only be greater than those listed in Table III above. Visually, as seen in Figure 11, there is a large loss in flexure strength due to reinforcement with cornhusk fibers. However, gains in toughness are substantial in that such high values of failure (or impact prior to failure) are seen.

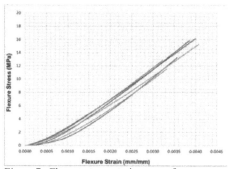

Figure 7. Flexure stress-strain curves for pure NaGP

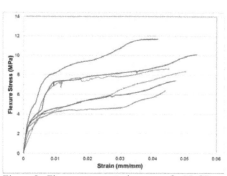

Figure 8. Flexure stress-strain curves for NaGP with CHF (quasi-aligned, wetted)

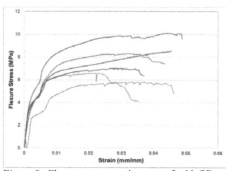

Figure 9. Flexure stress-strain curves for NaGP with CHF (random, unwetted)

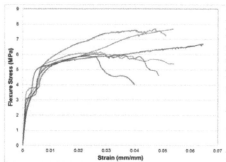

Figure 10. Flexure stress-strain curves for NaGP with CHF (random, wetted)

Figure 11. As-tested flexure samples of NaGP with CHF

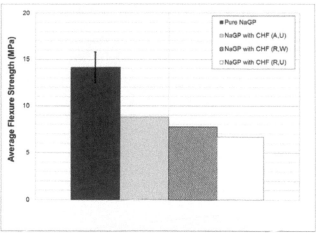

Figure 12. Flexure strength comparison chart

Tensile Testing

While tensile testing is non-standard for brittle, ceramic-like materials, when reinforced the resulting composite can be successfully tested in static tension with much reduced flaw sensitivity due to the fibrous reinforcement acting as a crack bridge against growing flaws.

Tensile test results for the composite panels with quasi-aligned and both the wetted and unwetted fibers are show below in Figure 13 and Figure 14. Tensile behavior remains nearly linear, in both cases, until the point of initial matrix failure at approximately 2.5 MPa. From there all curves exhibit a non-linear response indicating progressive matrix failure. During the non-linear portion strength is still increasing due to the straightening and elongation of the fibers before initial fiber failure at the peak. Failure following ultimate tensile strength is graceful and shows high amounts of fiber elongation and pullout. This is consistent with the 15.3% average elongation of CHF reported by Huda and Yang.[2] High amounts of fiber pullout are indicative of a weak fiber/matrix interface and results is high amounts of energy absorption due to friction with the matrix material during pullout.

SEM micrographs Figure 16 through Figure 19 show gaps between the matrix and the fiber representative of a weak interface. SEM samples were cut using a high speed diamond abrasive saw and were not polished due to the hydrophilic nature of geopolymer and the CHF. The striations seen in the images are a result of the cutting process. Since the CHF bundles are hydrophilic, during pressing or as a result of pre-wetting they absorb water and expand slightly, then during the curing process, they dry out and shrink away from the geopolymer creating gaps. Additionally, cracks can be seen propagating through the matrix and deflecting around the fiber bundles also representative of a weak interface. From the micrographs the wide variation in diameter and shape of the CHF bundles is also apparent.

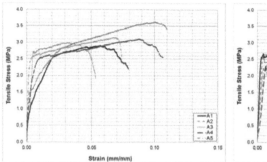 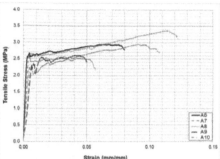

Figure 13. Tensile stress-strain curves for NaGP with quasi-aligned and wetted fibers

Figure 14. Tensile stress-strain curves for NaGP with quasi-aligned and unwetted fibers

Figure 15. As-tested tensile samples of NaGP with CHF

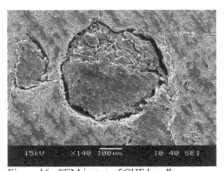

Figure 16. SEM image of CHF bundle embedded in NaGP matrix, M=140x

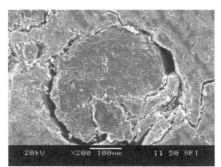

Figure 17. SEM image of CHF bundle embedded in NaGP matrix, M=200x

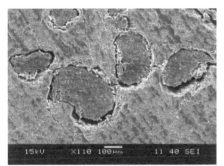

Figure 18. SEM image of CHF bundles in Na GP matrix with weak interface

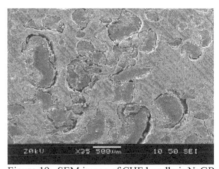

Figure 19. SEM image of CHF bundle is NaGP matrix show varying size and shape

While tensile strength is low, composite toughness is much higher than that of monolithic geopolymer. Stress-strain curves seem to follow a trend similar to thermoplastics in that there is an initial elastic region follow by yielding, in this case initial matrix failure, and then a period of high elongation or drawing.

The average ultimate tensile strength for each fiber type, as compared to the estimated tensile strength for monolithic sodium geopolymer is depicted in Figure 20. Error bars represent ± one standard deviation. Even without Weibull analysis, it can be seen from reduced magnitude of the deviations that reinforcement with CHF bundles resulted in higher flaw tolerance. The results showed that reinforcing with CHF bundles also reduced the ultimate tensile strength by approximately two thirds.

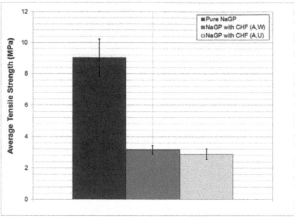

Figure 20. Tensile strength comparison chart

Weibull Statistical Analysis

Weibull statistics were only calculated for the pure sodium geoploymer (NaGP) flexure data in order to estimate the tensile strength of the monolithic material. A distribution function, $F$, was estimated based on the maximum failure stresses using the median rank method shown in equation (2). Where the maximum stress for each sample is given a rank, $i$, from 1 to $n$, which is the number of samples in the batch; with a rank of 1 given to the lowest stress value.[7]

$$F \cong \frac{i - 0.3}{n + 0.4} \qquad (2)$$

The Weibull distribution function is also given by equation (3) below, where $v$ is the non-dimensional volume, $\sigma$ is the stress, $\sigma_o$ is the scale parameter and $m$ is the shape parameter or Weibull modulus.

$$F = 1 - e^{\left[ -v \left( \frac{\sigma}{\sigma_o} \right)^m \right]} \qquad (3)$$

To determine the Weibull parameters, it was assumed that gage length is unchanged in the specimens, so $v = 1$, and equation (3) can be rearranged to get equation (4) below.

$$\ln \left[ \ln \left( \frac{1}{1-F} \right) \right] = m \ln(\sigma) - m \ln(\sigma_o) \qquad (4)$$

This fits the common slope-intercept form of a line, so $\ln \left[ \ln \left( \frac{1}{1-F} \right) \right]$ was then plotted versus $\ln(\sigma)$, using the approximation of $F$ given in equation (2), and a linear fit was applied for each data series. The slope of the fit line then corresponds to the Weibull modulus, $m$, and the y-intercept corresponds to $-m \ln(\sigma_o)$ which can be solved for $\sigma_o$. Table IV summarizes the resulting Weibull parameters for the pure sodium geopolymer test samples.

Table IV. Weibull parameters for pure NaGP flexure tests.

| $m$ | $\sigma_o$ |
|---|---|
| 7.654 | 9.564 |

CONCLUSIONS

Corn husk fiber (CHF) bundles were successfully chemically extracted from corn husks for use as a low cost, abundant reinforcement to sodium geopolymer. While a decrease in tensile and bend strength was found, vast improvements in toughness (ductility) were gained through the addition of the CHF. SEM imagery and test results indicate a weak fiber/matrix interface, evidenced by gaps between fiber and cured matrix as well as significant fiber pullout as a failure mechanism. Additional testing at lower fiber contents could determine an optimum configuration where matrix connectivity is improved to take advantage of NaGP strength and CHF ductility.

ACKNOWLEDGEMENTS

The authors would like to thank Keanetech, LLC, Champaign IL for supplying the corn husks and Na-waterglass for this study. SEM Images were taken on the JEOL 6060LV Scanning Electron Microscope in the Frederick Seitz Materials Research Laboratory Central Facilities, University of Illinois.

REFERENCES

[1]"US Grains Council: Global Analysis of Grain Supply," The Crop Site, http://www.thecropsite.com/news/11723/us-grains-council-global-analysis-of-grain-supply

[2]"Chemically Extracted Corn husk Fibers as Reinforcement in Light-Weight Poly(propylene) Composites," Shah Huda and Yiqi Yang, Macromolecular Materials and Engineering, **293,** 235-243 (2008)

[3]"Structure and Properties of Natural Cellulose Fibers Obtained from Corn husks, Cornstalks, Rice, Wheet, Soybean Straw and Sorghum Stalks and Leaves," Narendra Reddy, Dissertation (2006)

[4]"Properties and potential applications of natural cellulose fibers from corn husks," Narendra Reddy and Yiqi Yang, Green Chem., **7,** 190–195 (2005)

[5]"Engineering Composite Materials," Bryan Harris, The Institute for Materials, London (1999)

[6]"Mechanical Behavior of Materials", 2nd Ed, Marc Meyers and Krishan Chawla, Cambridge University Press, Cambridge, New York (2009)

[7]"Statistical Analysis of Fracture Strength of Composite Materials Using Weibull Distribution," M. Hüsnü Dirikolu, Alaattin Aktas and Burak Birgören, Turkish J. Eng. Env. Sci., **26,** 45-48 (2002).

# OPTIMIZATION AND CHARACTERIZATION OF GEOPOLYMER MORTARS USING RESPONSE SURFACE METHODOLOGY

Milap Dhakal, Kunal Kupwade-Patil, Erez N. Allouche, Charles Conner la Baume Johnson
Alternative Cementitious Binders Laboratory (ACBL),
Louisiana Tech University, Ruston, LA 71270, USA

Kyungmin Ham
Center for Advanced Microstructure and Devices (CAMD)
Louisiana State University, Baton Rouge, LA 70806, USA

ABSTRACT

The current study uses a statistical approach known as Response Surface Methodology (RSM) to optimize the constituents for the production of a cementitous binder referred as engineered geopolymer concrete (E-GPC). RSM method was used to optimize the influence on strength and porosity using three variables such as silica and alumina content of the fly ash and activator solution to binder ratio. Fifteen experimental runs were designed using Central Composite design as suggested by RSM and was compared with the experimental data. The performance of the models suggested a possible relation between the three variables to achieve a desired engineering E-GPC in terms of strength and porosity. In addition, microstructural characterization was carried out using scanning electron microscope (SEM), X-Ray Diffraction (XRD) and Fourier Transform Infrared Spectroscopy (FTIR). The characterization revealed information related to crystalline to amorphous content of E-GPC, zeolitic crystalline phases and bonds between Si and Al to examine the effect of hydrated sodium alumino silicate hydrate (N-A-S-H) on strength of geopolymeric gels. Optimization of geopolymer concrete, using RSM and microstructural study helped in designing an engineering E-GPC in terms of strength, porosity, workability and setting time for specific civil infrastructure applications.

INTRODUCTION

*1.1 Geopolymer Concrete*

Geopolymers are a novel and promising class of cementitious materials and are considered as the emerging next generation construction material targeting the utilization of an industrial by-product i.e. fly ash as a 100% substitute of Ordinary Portland Cement (OPC) [1, 2]. The concrete obtained from the resulting polymerization of solid aluminosilicates (e.g. fly ash or metakaolin) under the activation of highly alkaline medium (e.g. NaOH and $Na_2SiO_3$) has been named variedly by different authors as 'geopolymer concrete', 'inorganic polymer concrete' or 'alkali activated concrete' [2-5]

Evaluation of mechanical properties and the developed relationship by Diaz et al. [3] shows the similarity of those with respect to OPC and hints toward a viable application of GPC in the construction industry. In addition, by comparison to Ordinary Portland concrete (OPC), Geopolymer concrete (GPC) is virtually acid and sulfate resistant, offers high compressive and tensile strengths, rapid increase in strength gain and undergoes little shrinkage [6, 7]. Alkali-aggregate reactions do not occur in Geopolymer concrete and its nano-porous pore-system results in greatly reduced permeability and ionic diffusions. [7, 8]

## 1.2 Response Surface Methodology

Response surface methodology (RSM) is a combination of mathematical and statistical techniques used for the development, improvement, and optimization of processes [9]. RSM has been a powerful tool for the process evaluation and optimization dominantly in the industries but also presents strong applications in the research sectors in order to predict the results when they are function of controlled independent variables.

The fundamental equation relating the result of any experiment is stated in equation 1.

$$y = f(\xi_1, \xi_2, \ldots, \xi_k) + \varepsilon \tag{1}$$

where $y$ is the response of the experiment which is the function of various independent controllable variables $\xi_1$, $\xi_2$..., and $\xi_k$ and $\varepsilon$ is the representation of other uncontrollable sources of variations for the outcome $y$. The controllable variables in Equation 1 in the natural form with their respective physical units are difficult to handle. For convenience these variable are converted into coded variables $x_1, x_2, \ldots, x_k$ which are unitless and have zero mean and the same standard deviations. In term of the coded variables the Equation 1 gets simplified as given in Equation 2.

$$\eta = f(x_1, x_2, \ldots, x_k) \tag{2}$$

The function $f$ in Equation 2 has two models to fit the required response surface for any desired response, one being a first order model and other being the second order. Most of the times the second-order model is opted to predict the curvature of the response surface as the first-order model is not powerful enough to predict the strong curvature of the response surface. The general second-order model for the prediction of '$f$' is given in Equation 3.

$$\eta = \beta_0 + \sum_{j=1}^{k} \beta_j x_j + \sum_{j=1}^{k} \beta_{jj} x_j^2 + \sum_{i<j=2}^{k} \sum \beta_{ij} x_i x_j \tag{3}$$

where, $\beta$'s are the unknown coefficients to be determined and $x$'s the input parameters. To establish the unknown coefficients $\beta$'s, experimental points are to be determined. The number of experimental points required to establish these unknown coefficients is obtained using sampling methods or experimental designs. In this study Central Composite experimental design is used to determine the number of experimental points required to establish the coefficients for Equation 3. Fly ash being an industrial by-product, the quality assurance in its production is irrational unlike the case of Portland cement production. Therefore, the current study holds a significant role in establishing a mathematical correlation between the fly ash property and the mechanical characteristics of its derived GPC.

## MODEL DETERMINATION

In this model, three known parameters of the fly ash precursors namely, activator to fly ratio (AS/FA), particle size distribution (PSD) and, ratio of the sum of silica and alumina to calcium content ((S+A)/C) were taken as input parameters and compressive strength of the derived GP mortar was taken as the output parameter (PSD refers to the percentage of particle less than 45 μm in the fly ash precursor). Therefore, in this case using k = 3 in equation 3 we can expand it to get the following equation 4.

$$\eta = \beta_0 + \beta_1 x_1 + \beta_2 x_2 + \beta_3 x_3 + \beta_{11} x_1^2 + \beta_{22} x_2^2 + \beta_{33} x_3^2 + \beta_{12} x_1 x_2 + \beta_{13} x_1 x_3 + \beta_{23} x_2 x_3 \tag{4}$$

Fifteen sets of experimental points were set up as required in Central Composite design given by Equation 5.

$$N = 2^n + 2n + 1 \tag{5}$$

where, $N$ is the required experimental points, $n$ is the number of input parameters for each experimental point. In our case $n = 3$ and, hence the required number of experimental points is $N = 15$.

EXPERIMENTAL PROCEDURE

Geopolymer mortar samples were prepared using fly ashes of both the classes C and F collected from different thermal power stations located both within and outside the United States, as well as fine sand with fineness modulus (FM) of 2. Commercially available sodium silicate (45% w/w and $SiO_2$ to $Na_2O$ ratio of 2:1) and freshly prepared sodium hydroxide were used to serve the purpose of activator solution. The mortar samples were prepared as per ASTM C305-12. For each fly ash type three samples of cubes with dimension 2" were made following a mix design as given in Table 1.

**Table 1.** Mix Design Parameters

| Parameters | Levels |
|---|---|
| Aggregate | Sand |
| Binder | Fly Ash |
| Binder:Aggregate | 1:1 |
| Activator Solution:Binder | 0.4-0.94 (Table 3) |
| $Na_2SiO_3$:NaOH | 1:1 |
| Concentration of NaOH | 14M |

**Table 2.** Chemical composition of fly ash precursors

| FA ID | FA Code | $SiO_2$ | $Al_2O_3$ | CaO | $Fe_2O_3$ | MgO | $SO_3$ | $Na_2O$ | $K_2O$ |
|---|---|---|---|---|---|---|---|---|---|
| 1 | BY (%) | 37.77 | 19.13 | 22.45 | 7.33 | 4.81 | 1.56 | 1.80 | - |
| 2 | DH (%) | 58.52 | 20.61 | 5.00 | 9.43 | 1.86 | 0.49 | 0.52 | - |
| 3 | DH5 (%) | 59.32 | 19.72 | 6.90 | 7.22 | 2.23 | 0.36 | 1.11 | 1.27 |
| 4 | OH (%) | 55.07 | 28.61 | 1.97 | 6.22 | 1.08 | 0.19 | 0.38 | 2.63 |
| 5 | SJ2 (%) | 56.39 | 27.36 | 4.69 | 3.34 | 0.75 | 0.26 | 1.50 | 0.95 |
| 6 | SM (%) | 66.50 | 18.80 | 4.91 | 1.95 | 0.63 | 0.22 | 2.90 | 2.63 |
| 7 | CC (%) | 33.02 | 19.82 | 26.19 | 6.75 | 6.34 | 1.36 | 1.92 | 0.35 |
| 8 | PK (%) | 59.25 | 18.43 | 9.23 | 5.61 | 3.23 | 0.35 | 0.50 | 1.63 |
| 9 | NE (%) | 31.26 | 19.76 | 28.53 | 6.47 | 4.81 | 1.45 | 2.27 | 0.47 |
| 10 | HW (%) | 33.38 | 14.72 | 26.80 | 7.69 | 1.49 | 10.36 | 0.56 | 0.79 |
| 11 | MO (%) | 55.61 | 19.87 | 12.93 | 4.52 | 2.49 | 0.49 | 0.67 | 0.86 |
| 12 | AD (%) | 53.00 | 25.88 | 3.34 | 9.45 | 1.70 | 0.20 | 0.71 | 2.23 |
| 13 | DL-2 (%) | 60.40 | 23.75 | 12.36 | 8.07 | 0.80 | 0.27 | 0.96 | 1.42 |
| 14 | NH1 (%) | 40.75 | 22.79 | 4.64 | 17.76 | 1.23 | 1.29 | 1.33 | 2.19 |
| 15 | NH2 (%) | 36.18 | 17.70 | 2.26 | 10.59 | 1.20 | 0.20 | 0.73 | 1.59 |

The samples were subjected to a thermal curing at 140±5 °F for the period of 72 hours. The GP samples were then subjected to axial compression test as per ASTM C109/C109M-11b. The compression test data for all the geopolymer mortar were then used to construct a constitutive Response Surface model as shown in Table 3. All the input made in the model was first coded so as to obtain all unitless quantities.

**Table 3.** Input parameters for the RSM model

| FA ID | FA Code | Source | FA Class | AS/FA | PSD | (S+A)/C | Strength (MPa) |
|---|---|---|---|---|---|---|---|
| 1 | BY | USA | C | 0.45 | 83.01 | 2.535 | 23 |
| 2 | DH | USA | F | 0.4 | 63.5 | 15.826 | 58.27 |
| 3 | DH5 | USA | F | 0.4 | 62.97 | 11.455 | 56.41 |
| 4 | OH | USA | F | 0.5 | 71.26 | 42.477 | 47.44 |
| 5 | SJ2 | USA | F | 0.94 | 58.19 | 17.857 | 12.22 |
| 6 | SM | USA | F | 0.85 | 30.28 | 17.373 | 17.91 |
| 7 | CC | USA | C | 0.4 | 85.68 | 2.018 | 21.73 |
| 8 | PK | USA | C | 0.4 | 63.24 | 8.416 | 49.7 |
| 9 | NE | USA | C | 0.4 | 85.68 | 1.788 | 23.55 |
| 10 | HW | USA | C | 0.85 | 76.21 | 1.795 | 4.01 |
| 11 | MO | USA | C | 0.43 | 68.75 | 5.838 | 35.31 |
| 12 | IL1 | Israel | F | 0.52 | 70.57 | 23.636 | 68.93 |
| 13 | DL2 | China | C | 0.5 | 89.21 | 6.809 | 68.4 |
| 14 | NH1 | USA | F | 0.4 | 87.5 | 13.694 | 69.51 |
| 15 | NH2 | USA | F | 0.75 | 43.49 | 23.841 | 39.22 |

The RS regression analysis was carried out after coding the uncoded raw data. Coding was done to achieve all the input parameters in a dimensionless state. The raw output i.e. the regression coefficients in coded units were therefore converted into the uncoded units so as to carry out further RS analyses.

In addition to the statistical modeling, X-Ray fluorescence (XRF) using an ARL QUANTX EDRF spectrometer was used for the chemical analysis of all the used fly ashes and their derived geopolymer mortar. Further, micro X-ray tomography and X-Ray diffraction were applied to the selected geopolymer sample for their microstructural characterization. The X-ray microcomputed tomography was carried out using a parallel beam configuration on a hard X-ray synchrotron beamline (25 KeV), applying the X-Ray over the sample rotating half circle about its vertical axis at the rate of 0.25°/sec with total exposure time of 2.5 sec for each rotation. The X-ray imaging was done using camera of definition 2048×2048 pixels with pixel size of 13.5 μm and voxel size of 2.5 μm. X-Ray Diffraction analysis was carried out using diffraction spectra obtained from a commercial diffractometer (D8 Advance Bruker AXS Spectrometer). ATR-FTIR analysis was conducted using a Nicolete 6700 manufactured by Thermo-Scientific (West Palm Beach, FL).

RESULTS AND DISCUSSION

Fifteen different geopolymer mortars were made so as to satisfy the requirement of Central Composite experimental design using fly ashes obtained from fifteen different stock piles around the world. Fly ash stockpiles used in the current study were obtained from the USA, China and Israel. The experimental observation of the response (compressive strength) of all the 15 geopolymer mortars is given in Table 3. The regression coefficients obtained from the selected experimental design along with their corresponding t and p values are shown in Table 4. The p-value (<0.05) showing high significance of all the input parameters was for the construction of the RS model. The regression model obtained had an R-squared value of 99.74% and s-value of 7.56.

**Table 4.** Regression coefficients for the RS model

| Coefficients | Coef. (Coded) | SE Coef. (Coded) | t | p | Coef. (Uncoded) |
|---|---|---|---|---|---|
| Constant ($\beta_0$) | 54.51 | 1.828 | 29.818 | 0 | 893.703 |
| AS/FA ($\beta_1$) | -14.558 | 2.198 | -6.623 | 0.001 | -807.496 |
| PSD ($\beta_2$) | 38.499 | 3.187 | 12.078 | 0 | -18.529 |
| (S+A)/C ($\beta_3$) | 27.202 | 3.783 | 7.190 | 0.001 | -4.602 |
| AS/FA×AS/FA ($\beta_{11}$) | -4.891 | 2.959 | -1.653 | 0.159 | -67.096 |
| PSD×PSD ($\beta_{22}$) | 77.847 | 6.751 | 11.531 | 0 | 0.089 |
| $Al_2O_3$ x $Al_2O_3$ ($\beta_{33}$) | -36.327 | 1.699 | -21.382 | 0 | -0.088 |
| AS/FA x $SiO_2$ ($\beta_{12}$) | 88.708 | 8.063 | 11.001 | 0 | 11.151 |
| AS/FA x $Al_2O_3$ ($\beta_{13}$) | 44.003 | 5.641 | 7.800 | 0.001 | 8.012 |
| $SiO_2$ x $Al_2O_3$ ($\beta_{23}$) | 44.717 | 6.424 | 6.961 | 0.001 | 0.075 |

The regression diagrams exhibiting the fit and the experimental observation for the response (strength) is shown in Figure.1. The parity of the experimental and the fitted values of the response is shown in Figure. 2. The correlation between strength, (S+A)/C, PSD and, AS/FA is shown in Figure. 3-5. Using the RSM method, surface and contour plots were generated for AS/FA ratio for 0.4, 0.5 and, 0.6, which shows the compressive strength as a function of (S+A)/C and PSD at respective levels of AS/FA ratio. The surface and contour plot shows that increase in (S+A) content leads to higher values of strength. In this model the CaO is responsible for reduction in strength. It is well known that silica and alumina play a vital role in the formation of amorphous geopolymerization network known as the hydrated sodium alumino silicate (N-A-S-H) [10-13]. The current study examines the role of NASH system by using chemical analysis such as XRD, XRFS and FTIR along with microstructure studies (X-ray μ tomography).

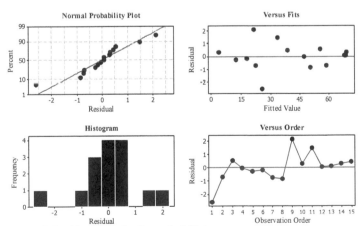

**Figure 1.** Residual Plots for Strength (MPa)

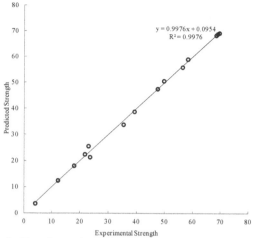

**Figure 2.** Parity Plots for Experimental and Fit Strengths (MPa)

Studies have been conducted on relating the effect of CaO content of the fly ash to the compressive strength [14]. In general, CaO content helps to increase the compressive strength while causing a major reduction in the setting time, hence affecting the workability of GPC. CaO content greater than 20% significantly affects the workability of GPC by shortening the setting and therefore, such GPC cannot be used for practical field application [15]. The study conducted by van Riessen et. al. [14] shows that GPC cured at elevated temperature demonstrates higher CaO content, which can hinder the formation of geopolymer and lead to significant reduction in mechanical strength. Further studies are required to clearly establish a bench mark range on the CaO content which provides an accurate relation to the compressive strength of GPC considering workability as a primary parameter.

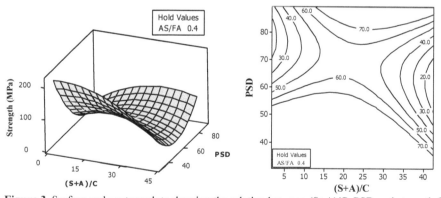

**Figure 3.** Surface and contour plots showing the relation between (S+A)/C, PSD and strength for

AS/FA = 0.4

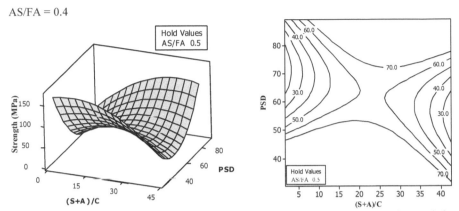

**Figure 4.** Surface and contour plots showing the relation between (S+A)/C, PSD and strength for AS/FA = 0.5

     The surface and contour plots show that with the increase in PSD and decrease (S+A)/CaO ratio leads to lower mechanical strength. PSD is an important parameter which plays a critical role in geopolymerization mechanism and controlling the overall mechanical strength of the concrete. Fly ashes containing larger particle size posses large specific area which requires higher demand for the activator solution to complete the alkali activation of the fly ash. Un-reacted fly ash crystals which are not involved in the amorphous geopolymer gel network could react with other impurities in the fly ash causing various crystalline phases. Hence, PSD could affect the packing characteristics along with the overall pore network of the GPC cement matrix system, which is responsible for strength and durability of the concrete[15, 16].

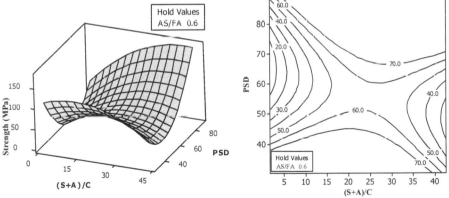

**Figure 5.** Surface and contour plots showing the relation between (S+A)/C, PSD and strength for AS/FA = 0.6

    The validation of the model was carried out using samples of geopolymer mortars developed using new fly ash precursors. The comparison of the actual experimental strength with the predicted model is shown in Table 5. Keeping in mind the complex chemistry of the precursors and its inter-relationship with its particle size distribution (PSD) (Figure 6), the model generated using the RS methodology helps to predict the characteristic strength of a fly ash within acceptable range. This helps us to predict the strength of geopolymer concrete derived from that particular fly ash with a satisfactory confidence limit.

**Table 5.** Validation of the developed RS regression model

| FA | AS/FA | PSD (%) | (S+A)/C | Predicted Strength (MPa) | Experimental strength (MPa) | ERROR (%) |
|---|---|---|---|---|---|---|
| IL2 | 0.50 | 73.48 | 23.02 | 71 | 65 | 9.23 |
| ZN1 | 0.50 | 80.54 | 9.96 | 57 | 77 | -25.97 |
| DH4 | 0.40 | 61.66 | 16.31 | 62 | 52 | 19.23 |
| WA | 0.50 | 81.23 | 1.68 | 21 | 17 | 23.53 |
| SE | 0.40 | 67.33 | 6.35 | 39 | 55 | -29.09 |

Chemical composition obtained via XRF for the fly ash and the GPC is shown in Figure 6. The model predicts that the increase in ratio of (S+A)/C would lead to an increase in mechanical strength of concrete. Experimental evaluation of the ratio of (S+A)/C on compressive strength is shown in Figure 7. The ratio of (S+A)/C on compressive strength increased for 9 specimens, while the rest of the specimens exhibited decrease in compressive strength.. The specimens which exhibited strength below 30 MPa did not follow this trend, while the specimens showing strength above 40 MPa mostly adhered to the model. This might be due to the fact that compressive strength of GPC is a function of complex correlation between the chemical composition and physical properties (e.g. PSD) of the fly ash. As shown in Figure 2, if the contribution of (S+A)/C, PSD and AS/FA is considered, the developed model is highly compatible with the experimental observations for strength. Also, it must be kept in mind that various factors like impurities in the fly ash, amount of fly ash actually involved in the geopolymerization which led to the formulation of the N-A-S-H network and the overall pore connectivity network of the specimen were not considered in the model. Therefore, microstructure analysis plays a vital role in examining the compressive strength and durability of the GPC.

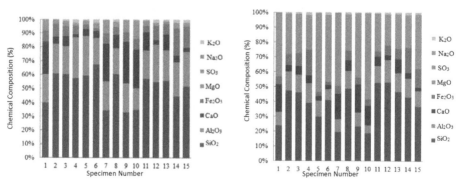

**a)** Chemical composition of Fly Ash samples      **b)** Chemical composition of the GPC samples
**Figure 6.** Chemical composition of Fly ash's and GPC specimens

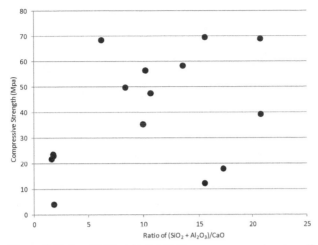

**Figure 7.** Ratio of $(SiO_2 + Al_2O_3)/CaO$ versus compressive strength

The XRD analysis of GPC specimens is shown in Figure 8. GPC which exhibited strength higher than 35 MPa showed significant phases such as quartz, albite, and mullite along with hydrosodalite. The albite and hydrosodalite could be related to the N-A-S-H system, which is one of the primary phases responsible for development of strength in geopolymer concrete [1]. In addition, mullite and quartz were also observed among the GPC specimens. This phase could have been observed from the unreacted fly ash which was not involved in the geopolymerization.

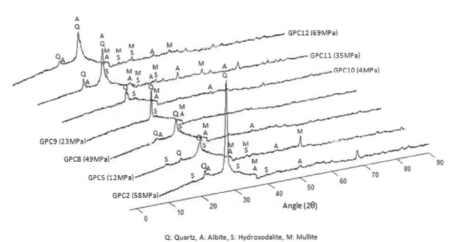

Q: Quartz, A: Albite, S: Hydrosodalite, M: Mullite

**Figure 8.** X-Ray diffraction spectra for selected GP mortar samples

ATR-FTIR analyses of the selected geopolymer concrete along with their corresponding fly ash are shown in Figure 9. The fly ash precursor showed geopolymer spectrum at approximately 1060 cm$^{-1}$, due to the Si-O-T(Al/Si) asymmetric stretching band which indicates unreacted fly ash. A significant reduction in the band was observed from 1060 cm$^{-1}$ to 960 cm$^{-1}$ in the GPC specimens. This band could be attributed to the formation of a new geopolymer gel network, and indicates that the unreacted fly ash was involved in the geopolymerization [2]

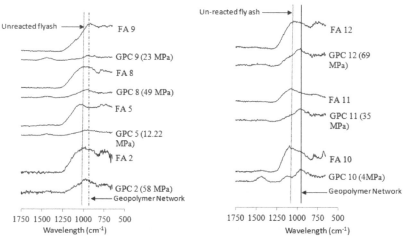

**Figure 9.** Absorbance spectra obtained via ATR-FTIR analysis for fly ash and geopolymer concrete

The density of GPC matrix along with the pore configuration was studied on low (4 MPa), medium (35 MPa) and high strength (69MPa) GPC specimens (GPC 10, 11 and 12) using X-Ray μ-CT. Specimen with high strength revealed the presence of dispersed and discrete micro pores (~600 μm). In contrast, large cracks (~5100 μm) interconnected with fully developed micro-cracks were observed in the low strength GPC, whereas large but discrete pores (4800 μm) were observed in the case of medium strength GPC. The pore size distribution is shown in Figure 11.

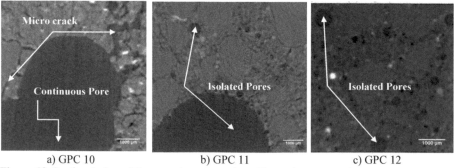

a) GPC 10          b) GPC 11          c) GPC 12

**Figure 10.** Sectional view of the geopolymer samples with maximum pore size

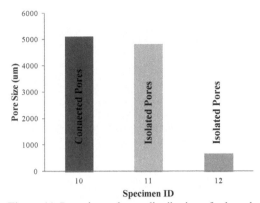

**Figure 11.** Pore size and pore distribution of selected geopolymer samples

The underlying principle of pores which is areas of stress concentration in concrete, and the increase in pore inter-connectivity results in decrease in its ability to withstand an applied load. The three dimensional volume of interest (VOI) view of GPC specimen shown in Figures 12-14, which exhibits the pore size distribution and the pore inter-connectivity of low, medium and high strength geopolymer concretes. GPC specimen number 10 (4.01 MPa) showed large pores with fully developed cracks, while high strength GPC (68.93 MPa) exhibited no signs of interconnectivity nor any micro cracks. Examination of geopolymer concrete using X-ray μCT-tomography highlights the importance of sizes and nature of the pore connectivity network among the GPC's prepared with class-C and Class-F fly ashes. Pore structure studies using advanced beamline technique such as X-ray μCT and small angle neutron scattering plays a vital role in understanding the overall mechanism of geopolymerization in the formulation of alkali activated concretes [8]. These studies are essential to develop an engineered geopolymer concrete which can withstand extreme mechanical and environmental adversities.

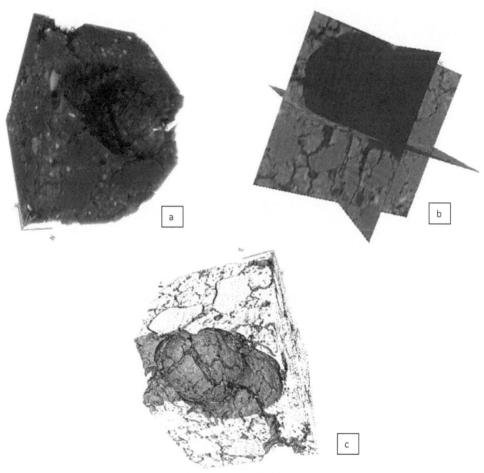

**Figure 12.** a) X-Ray μCT analysis showing the 3D view of GPC 10 which exhibited a strength of 4 MPa

**Figure 13.** X-Ray μCT analysis showing the 3D view of GPC 11 which showed a strength of 35 MPa

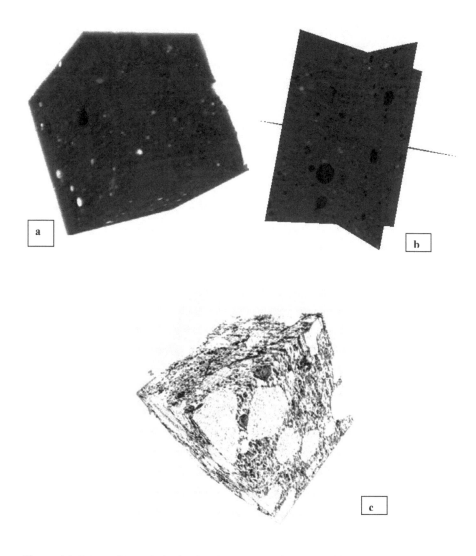

**Figure 14.** X-Ray µCT analysis showing the 3D view of GPC 12 which showed a strength of 69 MPa

CONCLUSIONS

This research lays a foundation for the characterization of fly ash in terms of its compressive strength which is one of the major engineering properties of any binder, and establishes it as a function of its CaO content, PSD, ratio of its amorphous to crystalline content and the ratio of activator solution to fly ash used to make the geopolymer mortar. Using the response surface method and thus the generated regression model, the response optimization characteristics which are necessary for an ideal fly ash with specific strength can be predicted. The surface and contour plots show that the increase in PSD and decrease in (S+A)/C leads to lower mechanical strength. This trend was mostly followed when the GPC's were subjected to experimental evaluation. In addition GPC's were subjected to chemical and microstructural studies which revealed the importance of certain strength increasing phases such as albite and hydrosodalite. X-Ray μCT analysis revealed the pore structure and inter-pore connectivity network for low (4 MPa), medium (~35 MPa) and high strength (69 MPa) geopolymer concrete. The high strength GPC exhibited no signs of cracking nor any inter pore connectivity network, therefore which makes it suitable for applications in harsh environments.

REFERENCES

[1] J.L. Provis, J.S.J. Van Deventer, Geopolymers: Structure, Processing, Properties and Industrial Applications, Woodhead Publishing Limited, 2009.
[2] P. Duxson, A. Fernández-Jiménez, J. Provis, G. Lukey, A. Palomo, J. van Deventer, Geopolymer technology: the current state of the art, Journal of Materials Science, 42 (2007) 2917-2933.
[3] J. Davidovits, Geopolymer Chemistry and Applications, Institut Géopolymère, 2011.
[4] E.I. Diaz-Loya, E.N. Allouche, S. Vaidya, Mechanical properties of fly-ash-based geopolymer concrete, ACI Materials Journal, 108 (2011) 300-306.
[5] P. Duxson, J.L. Provis, G.C. Lukey, S.W. Mallicoat, W.M. Kriven, J.S.J. van Deventer, Understanding the relationship between geopolymer composition, microstructure and mechanical properties, Colloids and Surfaces A: Physicochemical and Engineering Aspects, 269 (2005) 47-58.
[6] K. Kupwade-Patil, E. Allouche, S. Vaidya, E. Diaz-Loya, Corrosion analysis of reinforced geopolymer concretes, in: Concrete Solutions 2011, CRC Press, 2011.
[7] K. Kupwade-Patil, E. Allouche, Impact of Alkali Silica Reaction on Fly Ash-Based Geopolymer Concrete, Journal of Materials in Civil Engineering, 25 (2013) 131-139.
[8] J.-D. Gu, T.E. Ford, N.S. Berke, R. Mitchell, Biodeterioration of concrete by the fungus Fusarium, International Biodeterioration & Biodegradation, 41 (1998) 101-109.
[9] M. Sahmaran, M. Lachemi, V.C. Li, Assessing mechanical properties and microstructure of fire-damaged engineered cementitious composites, ACI Materials Journal, 107 (2010) 297-304.
[10] C. Shi, A.F. Jiménez, A. Palomo, New cements for the 21st century: The pursuit of an alternative to Portland cement, Cement and Concrete Research, 41 (2011) 750-763.
[11] I. Garcia-Lodeiro, A. Palomo, A. Fernández-Jiménez, D.E. Macphee, Compatibility studies between N-A-S-H and C-A-S-H gels. Study in the ternary diagram Na2O–CaO–Al2O3–SiO2–H2O, Cement and Concrete Research, 41 (2011) 923-931.
[12] M.D. Brown, C.A. Smith, J.G. Sellers, K.J. Folliard, J.E. Breen, Use of alternative materials to reduce shrinkage cracking in bridge decks, ACI Materials Journal, 104 (2007) 629-637.
[13] W. De Muynck, N. De Belie, W. Verstraete, Effectiveness of admixtures, surface treatments and antimicrobial compounds against biogenic sulfuric acid corrosion of concrete, Cement and Concrete Composites, 31 (2009) 163-170.
[14] J. Temuujin, A. van Riessen, R. Williams, Influence of calcium compounds on the mechanical properties of fly ash geopolymer pastes, Journal of Hazardous Materials, 167 (2009) 82-88.
[15] E.I. Diaz, E.N. Allouche, S. Eklund, Factors affecting the suitability of fly ash as source material for geopolymers, Fuel, 89 (2010) 992-996.
[16] W.D. Lindquist, D. Darwin, J. Browning, G.G. Miller, Effect of cracking on chloride content in concrete bridge decks, ACI Materials Journal, 103 (2006) 467-473.

# EVALUATION OF GRAPHITIC FOAM IN THERMAL ENERGY STORAGE APPLICATIONS

Peter G. Stansberry and Edwin Pancost
GrafTech International, Advanced Graphite Materials
Parma, OH 44130

## ABSTRACT

Enhanced performance in electronics, aerospace, and military systems require improved solutions in thermal management. Some of the solutions include developing new and novel materials that exhibit high thermal conductivity, dimensional stability at high temperature, high surface area, and low density. Graphite foam is an attractive candidate in thermal management systems because it can be manufactured with ligament thermal conductivities in excess of 1,500 W/m-K and bulk thermal conductivities greater than 150 W/m-K. These qualities, in combination with an open-pore structure and low density, offer distinct advantages over other types of materials. In order to understand the performance of graphite foam for thermal management a thermal energy storage device (TESD) was constructed using graphite foam cores and paraffinic phase change material (PCM). Heat transfer fluid at controlled temperature and flow rate was conveyed through a graphite foam core on the acquisition/rejection side of the TESD, which was interfaced with a graphite foam core saturated with PCM. The heating and cooling responses of the graphite foam infiltrated with PCM are reported.

## INTRODUCTION

Some thermal energy storage devices (TESD) use the latent heat of fusion of materials to absorb or release heat during a phase change. As a substance melts, for example, a considerable quantity of heat is required to transform it from a solid state to a liquid state. In a practical application of this phenomenon the ability to store thermal energy during transient peak loads, for instance as is found in certain electronics, is exploited. Also, since phase change occurs isothermally, a range of substances can be selected to perform depending on the particular temperature regime of interest. Although phase change materials (PCM) offer a number of advantages in thermal management systems, they generally exhibit low thermal conductivities such that response time to thermal stresses is slow and inconsistent.

Although metallic foams have been incorporated with PCM successfully in TESD and other heat management systems, their bulk thermal conductivities remain a function of the parent metal. For example, while the thermal conductivity of copper and aluminum is 400 W/m-K and 180 W/m-K, respectively, at 25% relative density their bulk thermal conductivity diminishes to only 45 W/m-K and 15 W/m-K[1]. On the other hand, graphitic foams have received considerable interest over the past decade or two because they can be manufactured with bulk thermal conductivities an order of magnitude higher than metal foams at comparable bulk densities, with controlled surface area and pore structure, and with dimensional stability at high temperatures and harsh environments.

The objective of the present study is to conduct a comparative study of the thermal response of graphitic foam in relation to thermal energy storage applications. The effects of metallic coatings and bonding to parting sheet on thermal performance are investigated.

METHOD

We produced billets of graphitic foam by heating mesophase pitch under pressure in a non-oxidizing environment. Under these conditions the pitch is molten and as temperature increases low molecular weight components volatilize and together with thermal cracking gases work to generate bubbles that tend to rise, orienting the molecules within cell walls and ligaments. With continued heat treatment, the viscosity of the pitch increases to the point that the foam is transformed into a rigid, infusible structure. Additional heating is required to drive off residual volatiles and to consolidate the foam, followed by graphitizing at temperatures above 2,500°C to develop crystallinity and to promote thermal conduction. Following graphitization the parent billet was machined into coupons ¼" thick by 7" long and 7" wide, as shown in Figure 1, which were used as conductive cores in the TESD. Table I provides some of the characteristics of the foam used in this study.

Figure 1. Machining of graphitic foam cores.

Table I. Characteristics of graphitic foam for TESD Evaluation.

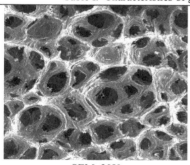

| | |
|---|---|
| SEM, 30X | Density = 0.13 g/cm³ |
| | Solid Volume Fraction = 6 % |
| | Bulk Thermal Conductivity = 14  W/m-K |

Some of the machined graphitic foam cores were coated with metallic copper using an electro-chemical process[2]. The deposition procedure created a layer of copper metal on the surfaces of the graphitic foam between 0.3 μm and 0.9 μm thick. Some of the coated graphitic foam cores were bonded to 0.012" thick aluminum parting sheet using a commercially available low-temperature solder in order to reduce thermal resistance. Coating with copper metal was chosen to promote joining the graphite foam to the aluminum parting.

A thermal energy storage test device was machined from polycarbonate. The design was of a 'two-pass" configuration in that the heat transfer fluid, in this case water, entered and existed the same side of the device, as shown in Figure 2. Both the acquisition and PCM side contain graphitic foam. Thermal couples were inserted one-half way into the thickness of the graphite foam on the PCM side of the unit. The placement of the thermocouples corresponded with the inlet, midpoint, and exit of the unit, as shown in Figure 2. A thermostated-water bath along with a variable-flow pump supplied heated fluid to the system under controlled conditions. All process variables such as pressure drop and temperature at various locations in the TESD were recorded using computerized data acquisition.

Octadecane was selected as the phase change material. It was infiltrated into the graphitic foam using a vacuum impregnation procedure[3]. Measured thermo-chemical properties of octadecane were: melting point 28.2 °C; heat of fusion 241.7 kJ/kg.

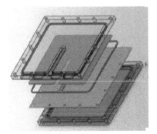

Figure 2. Design of TESD. Left, exploded view; Right, assembled unit without cores, showing placement of thermocouples.

RESULTS AND CONCLUSIONS

The response characteristics of the TESD were investigated under the conditions with water bath temperature at 36 °C at a flow rate through the unit at 3.5 liters per minute. Because of the highly porous nature of the graphitic foam the pressure drop between the entrance and exit of the heat transfer side was always less than 0.1 psi. The water bath was stabilized at the desired temperature before conducting each test, after which a valve was opened to convey the fluid through the TESD. Figure 3 shows the PCM temperature response in the case where the uncoated graphite cores were simply compressed against the aluminum parting sheet. Figure 4 shows the similar situation but with copper-coated foam. Initially, temperature rises rapidly and then remains unchanged as the melting of octadecane occurs. The period of melting is shorter on the inlet side while considerably longer on the exit side as heat is withdrawn from water as it moves through the unit.

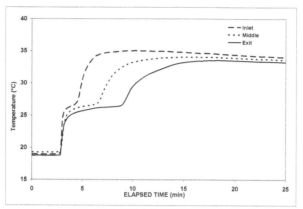

Figure 3. Thermal response of graphite foam cores compressed against aluminum parting sheet.

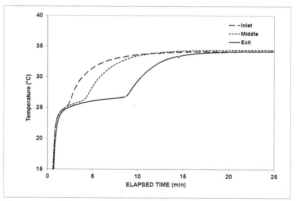

Figure 4. Thermal response of copper-coated graphite foam cores compressed against aluminum parting sheet.

Figure 5 displays the response of the copper-coated cores that were soldered to the aluminum parting sheet. In this situation thermal response and melting of octadecane are several times more rapid than the previous trials. This effect emphasizes the importance of minimizing thermal contact resistance in these applications.

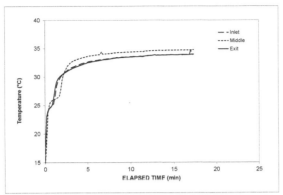

Figure 5. Thermal response of copper-coated graphite foam cores soldered to aluminum parting sheet.

ACKNOWLEDGEMENTS

The authors appreciate the support provided by the Air Force Research Laboratory (WPAFB, Dayton, Ohio) and help provided by the Ohio Aerospace Institute (Cleveland, Ohio).

REFERENCES

1. Klett, J. W., McMillan, N., D., Gallego, N. C., and Walls, C. A., The role of structure on the thermal properties of graphitic foams. *Journal of Materials Science* **39**, 3659-3676 (2004).

2. Lafdi, K., Almajali, M., and Huzayyin, O. Thermal properties of copper-coated carbon foam. *Carbon* **47**, 2620-2626 (2009).

3. Warzoha, R., Sanusi, O., McManus, B., and Fleischer, A. Evaluation of methods to fully saturate carbon foam with paraffin wax phase change material for energy storage. *13th IEEE ITHERM Conference*, 834-839 (2012).

# Thermal Management
# Materials and Technologies

# Q-STATE MONTE CARLO SIMULATIONS OF ANISOTROPIC GRAIN GROWTH IN SINGLE PHASE MATERIALS

J. B. Allen, C. F. Cornwell, B. D. Devine, and C. R. Welch
U.S. Army Engineer Research and Development Center, Vicksburg, MS, 39180-6199

ABSTRACT
    The Q-state Monte Carlo, Potts model is used to investigate 2D, anisotropic, grain growth in single-phase materials using hexagonal grain elements. While many factors can affect the microstructure anisotropy, this research focusses on the role played by grain boundary energy anisotropy. Specifically, for each computed grain orientation and surface normal, the corresponding surface energy is assigned through a mapping process using Wulff plots. Various Wulff plot geometries are considered, and their respective impact on the grain growth anisotropy is evaluated.

INTRODUCTION
    Increased control of material microstructure, leading to enhanced mechanical and physical properties continues to be an active area of research. Occupying a central role in this research effort has been the ability to control grain growth.    Numerous studies have demonstrated that a better understanding of grain growth processes is of fundamental importance in the fabrication of novel nanostructured materials. Several studies, for example, have shown that, as the average grain size of a material decreases, superior physical and mechanical properties can be achieved[1-3]. Other studies have emphasized the need to find ways of obtaining equiaxed, fine-grain microstructures without the occurrence of abnormal grain growth[4]. Issues relating to coarsening and shrinkage are of concern, as are the mechanisms which help to control the shape of grains.
    The vast majority of previous grain growth simulations have been concerned with the processes associated with normal grain growth, wherein the average grain size increases while the shape of the size distribution remains constant with time. While this idealized simplification may be useful for simulating purely homogeneous materials (those without any secondary constituents), or for theoretical purposes, the majority of materials fall outside this idealized category and consist of varying degrees of anisotropic grain distributions.    In fact, there is mounting evidence that suggests that certain material properties can be enhanced by deliberately introducing anisotropic grains within a fine-grained matrix.  Faber and Evans[5,6] for example, found that anisotropic grain microstructures can result in increased fracture toughness of a material.    Evidence for these findings came through observations of anisotropic grains responsible for crack deflection and bridging mechanisms in the crack wakes. Electrical and magnetic property enhancements have also been observed within anisotropic grain microstructures[7].
    Grain anisotropy is a result of numerous factors, including: differences in grain boundary energy and mobility, segregation of solutes on different boundaries[4], the presence of an amorphous liquid phase[8], interface growth velocity differences during phase transformation[9] and the anisotropy of the interfacial energy between two phases[10,11].    Unfortunately, there is little understanding about how these factors affect the evolution of an anisotropic microstructure. It is well known that the driving force for grain growth is the reduction in the total grain boundary energy, achievable by reducing the total grain boundary area of a system. Therefore, it is

hypothesized that one of the more important factors for controlling anisotropic grain growth is the grain boundary energy anisotropy.

The present work incorporates a two dimensional, Q-state Monte Carlo method that evaluates the effect of anisotropic grain boundaries on grain growth. Specifically, the anisotropic grain boundaries are simulated via the use of surface energies and binding energies; the former being assigned through a mapping process involving Wulff plots. For each computed grain orientation and surface normal, the corresponding surface energy will be assigned through a mapping process involving various polar Wulff plot configurations. For each configuration, the impact on the grain growth anisotropy will be evaluated. A practical application of this model would be to produce prescribed anisotropic polycrystalline ceramic structures through the purposeful orientation of the crystalline grains in the green form, and then use the model's predictive capabilities to engineer the sintering process to facilitate anisotropic grain growth.

THE MODEL

A two dimensional lattice of cells (100x100) representing the initial grain microstructure was created and initialized by randomly assigning an integer number $q_i$ between 1 and Q to each lattice site (i). Here Q is the total number of degenerate spins (grain orientations) within the system and was initialized with random integers between 1 and 60. Although in reality there may be a continuous range of possible orientations, previous research has shown that for sufficiently large values of Q (i.e., Q > 50) the results may be assumed independent of initial orientation number[12]. The cell resolution (100x100) was deemed sufficient based on the results of a benchmark, isotropic test case, which compared favorably (at this resolution) with analytical results based on a power law kinetics model[12-14] (see Figure 3). Further, hexagonal elements (see Fig. 1) were selected for this work rather than rectangular elements, in order to mitigate the negative effects of lattice pinning[15,16].

The interaction energy between two neighboring cells $E_\theta(q_1, q_2)$ can be expressed as:

$$E_\theta(q_1, q_2) = (1 - \delta_{q_1 q_2})\{J_\theta(q_1) + J_\theta(q_2) - J_b\} \tag{1}$$

Where $\delta_{q_1 q_2}$ is the Kronecker delta function ($\delta_{q_1 q_2}=1$ if $q_1 = q_2$; and $\delta_{q_1 q_2}= 0$ if $if q_1 \neq q_2$), $J_\theta(q_i)$ is the surface energy of a cell with state $q_i$ along a surface normal, defined by the direction $\theta$ (where $\theta= 0$, $\pi/3$, or $2\pi/3$), and $J_b$ is the grain boundary binding energy. Applying the Kronecker delta function, Equation 1 may be expressed more conveniently as:

$$E_\theta(q_1, q_2) = \begin{cases} 0 & if\ q_1 = q_2 \\ J_\theta(q_1) + J_\theta(q_2) - J_b & if\ q_1 \neq q_2 \end{cases} \tag{2}$$

The total system energy over all nearest-neighbors is thus given by the Hamiltonian:

$$E_{tot} = \frac{1}{2}\sum_{q_1}\sum_{q_2} E_\theta(q_1, q_2) \tag{3}$$

Grain growth is simulated using the method developed in several previous works[15,17,18]. To summarize, first a lattice site is chosen at random from the simulation space. Then a new state q is chosen at random from the Q possible states of the system. The change in energy ($\Delta E$) is computed using Eq. 3 and the probability (P) that the site will change orientation is then determined from the transition probability function:

$$P = \begin{cases} M \exp\left(-\frac{\Delta E}{k_B \tau}\right) & \Delta E > 0 \\ M & \Delta E \leq 0 \end{cases} \quad (4)$$

Where $\Delta E$ is the change in energy associated with the reorientation event, $k_B$ is the Boltzmann constant, M is the grain boundary mobility (M=1 was used in this work and refers to isothermal conditions), and $\tau$ is simulation temperature.

The method for determining the surface energies $J_\theta(q_i)$ where $\theta = 0, \pi/3$, or $2\pi/3$ closely follows that of Yang et al.[4]. For purposes of completeness we summarize the procedure here. The orientation angle ($\varphi$) for a given hexagonal cell is computed as $\varphi = \pi q/Q$, where q is any one of the Q integer orientations. This angle is then mapped to a Wulff plot (a polar plot described by the surface orientation angle and radii composed of surface energies), wherein $\varphi$ is used to locate the position of the surface orientation energy ($\gamma_1$) located at an angle $\varphi$ from the x-axis (see Figure 2). Due to the two-fold rotational symmetry of the Wulff plot, the surface energy of surface two ($\gamma_2$) is located along a surface orientation vector that is 90° from surface one, while $\gamma_3$ is equal to the minimum of $\gamma_1$ and $\gamma_2$. In this way, grains with orientation number from 1 to Q/3 have higher surface energies along the $\theta = 0$ direction than the other two directions; grains with orientation number from Q/3+1 to 2Q/3 have higher surface energies along the $\theta = \pi/3$ orientation; and grains with orientation number from 2Q/3 +1 to Q have higher surface energies along the $\theta = 2\pi/3$ direction.

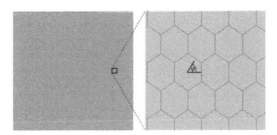

Figure 1. Hexagonal grain elements and definition of orientation angle ($\varphi$).

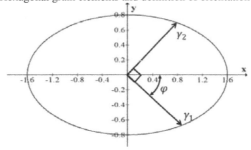

Figure 2. Elliptical (n=2) Wulff plot with surface energy determination.

In this study, various Wulff plot configurations were used, each based on the equation for a "super-ellipse" (or Lamé curve), which may be represented in rectangular coordinates as:

$$\left|\frac{x}{a}\right|^n + \left|\frac{y}{b}\right|^n = 1 \tag{5}$$

Where a and b are positive, real numbers corresponding to the semi-major and semi-minor axes, respectively, and n is a positive real number describing the shape of the curve. In polar coordinates $(r, \theta)$ Equation 5 may be written as:

$$r(\theta) = \left(\left|\frac{\cos(\theta)}{a}\right|^n + \left|\frac{\sin(\theta)}{b}\right|^n\right)^{-1/n} \tag{6}$$

Yang et al.[4] found that in order to develop highly anisotropic grain structures, the Wulff plot must have a maximum surface energy along one semi-major axis and a minimum surface energy along the other. Throughout this work we have selected constant semi-major and semi-minor axis values of a=1.6 and b=0.8, respectively.

The hexagonal grid was developed using the GAMBIT grid generation tool[19], and the anisotropic Monte-Carlo algorithm was developed within OpenFOAM[20]. For each of the four computer simulations, including one isotropic validation case (consisting of a circular Wulff plot), the microstructure was represented using an array of 100 ×100 hexagonal elements. The use of periodic boundary conditions was applied along all domain boundaries, the grain boundary binding energy was fixed at a constant value of 0.3, and the number of allowed initial grain orientations (Q) was 60. As customary, the simulation time was measured in Monte Carlo steps, wherein one Monte Carlo Step was equivalent to 100 x 100 reorientation attempts. A simulation temperature of $k_B\tau = 0$ was used throughout, and selected for its favorable ability to simulate grain growth phenomenon[15]. Note that the simulation temperature $(\tau)$ is not an actual physical temperature. Indeed setting $\tau = 0$ does not eliminate grain growth but merely eliminates thermal fluctuations and ensures the rejection of all steps for which $\Delta E > 0$. For post-processing, the average grain size was computed according to:

$$\langle R \rangle = \frac{\sum_i^N n_i R_i}{\sum_i^N n_i} \tag{7}$$

Where N is the total number of grains, and $n_i$ is the number of grains of size $R_i$.

RESULTS

For reference purposes, the model was first calibrated for isotropic grain boundary energies, that is, for surface energies which are assumed independent of grain orientation. In this case, the lengths of the semi-major and minor axes of the Wulff plot are identical, and a circle is produced with uniform grain boundary surface energy. The evolution of the average grain size <R> can be described by power-law kinetics[12-14]:

$$\langle R \rangle = \alpha t^{1/r} \tag{8}$$

Where $\alpha$ is a constant dependent on temperature, t is the time (in MCS units), and r is the grain growth exponent. For single phase systems, a grain growth exponent of 0.5 has been confirmed experimentally[12,21]. The slope of a straight fit through a plot of log $\langle R \rangle$ vs. log (t) by least squares fitting gives the inverse of the grain growth exponent (1/n). As shown in Figure 3, to a good approximation, the slope (0.49) is in agreement with the power law prediction.

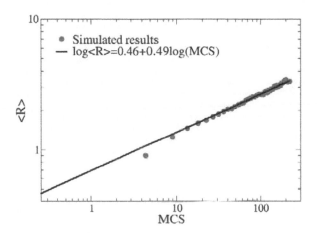

Figure 3. The slope of a straight fit through a plot of log $\langle R \rangle$ vs. log (t) by least squares fitting provides an estimate for the grain growth exponent (1/n). As shown, to a good approximation, the slope (0.49) is in agreement with the power law prediction.

The microstructure evolution corresponding to anisotropic grain growth (ellipse; n=2) is shown in Figure 4, and corresponds to 0, 100, 1000, and 4000 MCS. Unlike the isotropic grain growth model, the energy of a grain boundary is not constant, but a function of the orientations of grains on all sides of the grain boundary. Anisotropic grains result from the non-equilibrium effect of high energy grain boundaries which tend to migrate faster than corresponding low energy boundaries. Occasionally, neighboring grains will have equivalent (or near equivalent) surface energies, which results in some of the grains remaining isotropic.

As indicated, at time t=0 MCS (Figure 4 (a)), the grains appear fully isotropic (low aspect ratios), maintain a narrow size distribution, and cover the full range of orientation states. In this initial stage, the grain growth is primarily driven by grain boundary curvature. As shown, the initial grain states conform to a random distribution. At time t=100 MCS (Figure 4(b)), the isotropic growth tendency is maintained for all but a small subset of grains, which show early tendencies toward anisotropy, indicating elongated growth along directions corresponding to the grain surface normal (0°, 60°, and 120°). At time t=1000 MCS (Figure 4(c)), the number of isotropic grains have been reduced significantly. This is due to the higher grain boundary energies associated with the anisotropic grains, and the resultant increase in growth rate of these elongated grains. The consumption of the isotropic grains is a direct result of this process. Finally, at time t= 4000 MCS (Figure 4(d)), there are only a few isotropic grains remaining. The average grain size has more than quadrupled, and the number of original grain orientations (states) has been reduced significantly.

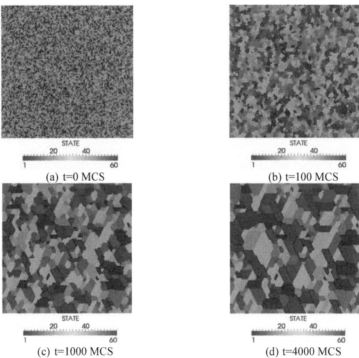

(a) t=0 MCS

(b) t=100 MCS

(c) t=1000 MCS

(d) t=4000 MCS

Figure 4. Anisotropic grain growth; 100x100 hexagonal elements; ellipsoid Wulff plot (a=1.6, b= 0.8, n=2); Q=60

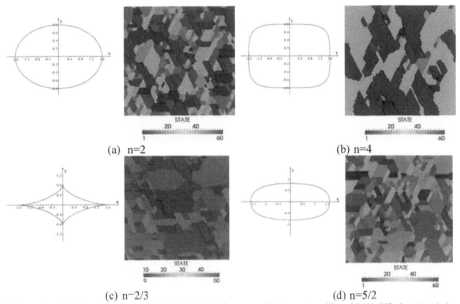

(a) n=2

(b) n=4

(c) n−2/3

(d) n=5/2

Figure 5. Anisotropic grain growth; 100x100 hexagonal elements; ellipsoid Wulff plots (a=1.6, b− 0.8) corresponding to n=2, n=4, n=2/3, and n=5/2; Q=60; t=4000 MCS

Figure 5 shows the anisotropic, microstructure results at time t= 4000 MCS. Also shown are the four different Wulff plot configurations corresponding to n=2, n=4, n=2/3, and n=5/2 (see Equation 6). As indicated, each plot shows grains having clear, directional growth preferences and rectangular dimensions with relatively high aspect ratios. Upon closer inspection, the n=2 and n=5/2 microstructures (Figure 5(a) and Figure 5(d)) appear quite similar in a variety of ways, including, grain number, average grain size, grain aspect ratio, and grain orientation distribution. The n=4, and n=2/3 plots (Figure 5(b) and Figure 5(c)), in contrast appear to contain a fair number of abnormal grains, easily distinguishable by the significant disparity in grain size between the largest and smallest grains. The n=4 plot (Figure 5(b)) is additionally unique, in that a much smaller subset of the original grain orientations remain than in any of the other three cases.

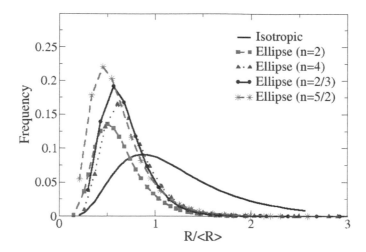

Figure 6. Normalized grain size distributions corresponding to the isotropic and elliptical Wulff plots (n=2, n=4, n=2/3, and n=5/2) at t=4000 MCS.

Figure 6 shows the frequency distribution of the normalized grain size (R/<R>) for each of the four Wulff plot configurations shown in Figure 5, as well as the isotropic case (corresponding to a circular Wulff plot). For each case, the simulations ran for a total of 4000 MCS. The normalized grain size was separated into 30 groups (bins) with values between 0 and 3, and the frequency was defined as the ratio of the number of grains within one group divided by the total number of grains.

As expected, each case is represented by a log-normal distribution with positive skew, and is easily distinguishable from its neighbor distributions by comparing values of mean ($\mu_{R/\langle R \rangle}$), standard deviation ($\sigma_{R/\langle R \rangle}$), and skewness (see Table 1). Here the degree of skewness (a measure of data symmetry) is represented as:

$$skewness = \frac{\sum_{i=1}^{N}((R/\langle R \rangle)_i - \mu_{R/\langle R \rangle})^3}{(N-1)\sigma_{R/\langle R \rangle}} \tag{9}$$

Where N is the total number of grains (for MCS=4000), and $(R/\langle R \rangle)_i$ is the normalized grain size for grain i.

As indicated, the isotropic case is represented by a log-normal distribution with a mean normalized grain radius of approximately one, a standard deviation 0.54, and a skewness of 0.84. In contrast, each of the elliptical distributions shows average grain radii significantly less than one and skewness values, in some cases greater than two. As shown, the n=2 and n=5/2 elliptical distributions have nearly equivalent mean radii of 0.66, and 0.67, respectively, but show significant disparity with respect to both the standard deviation and skewness values. The n=4 and n=2/3 elliptical distributions also have nearly equivalent mean radii (0.85), but unlike the n=2 and n=5/2 cases, show similar values for skewness. Table 1 also shows the total number of grains (N) that remained at time t=4000 MCS. As shown, the isotropic case resulted in the coarsest microstructure (N=176), while the elliptical distributions maintained grain numbers in correlation to their previous statistical similarities. That is, the n=2 and n=5/2 distributions averaged grain numbers close to 400, while the n=4 and n=2/3 distributions averaged approximately 200 grains.

Clearly these results serve to confirm the qualitative results observed in Figure 5. The abnormal grain growth seen for the n=4 and n=2/3 configurations are characterized by a disproportionately high number of unusually large grains which tend to consume smaller grains. It well understood that this naturally occurs in order to reduce the total grain boundary surface area and thus the total system energy. Less certain, is why this abnormal grain growth occurs for these two Wulff plot configurations and not for the n=2 and n=5/2 cases. Referring to Figure 5, one possible explanation may be due to the fact that the n=4 elliptical configuration shows maximum grain boundary surface energies lying not along a primary axis (as in the case of the n=2 and n=5/2 cases) but along intermediate x-y locations. Similarly, the n=2/3 configuration shows minimum grain boundary energies not along the y-axis but again along intermediate x-y locations. The excess of energy (in the n=4 case) or the lack of energy (in the n=2/3 case) along these intermediate locations may contribute to the occurrence of the observed abnormal grain growth.

Table 1. Grain size distribution statistics corresponding to the isotropic and elliptical Wulff plots (n=2, n=4, n=2/3, and n=5/2) at t=4000 MCS.

| Case | $\mu_{R/\langle R \rangle}$ | $\sigma_{R/\langle R \rangle}$ | Skewness | Total Grain Num. (N) |
|---|---|---|---|---|
| Isotropic (a=b=1) | 0.99 | 0.54 | 0.84 | 176 |
| Elliptical (a=1.6; b=0.8; n=2) | 0.66 | 0.32 | 1.30 | 412 |
| Elliptical (a=1.6; b=0.8; n=4) | 0.85 | 0.65 | 1.90 | 194 |
| Elliptical (a=1.6; b=0.8; n=2/3) | 0.85 | 0.58 | 2.01 | 209 |
| Elliptical (a=1.6; b=0.8; n=5/2) | 0.67 | 0.40 | 2.08 | 371 |

PRACTICAL APPLICATION

A practical application of the model is to allow prescribed anisotropic polycrystalline ceramic structures to be synthesized. As mentioned previously, such structures may provide increased fracture toughness[5,6] among other attributes. Key to the process would be to purposely orient crystalline grain structure in the green form. Such orientation might be achieved by depositing ions or iron particles on preferred faces of the grains, and then applying electric or magnetic fields combined with vibration to orient the grains along prescribed axes. The model's predictive capabilities could then be used to engineer the sintering process to cause anisotropic grain growth.

CONCLUSIONS

The Q-state Monte Carlo, Potts model was used to investigate 2D, anisotropic, grain growth in single-phase materials. While many factors can affect the microstructure anisotropy, this research focused on the role played by the grain boundary energy. Hexagonal grain elements were used in order to avoid the dependence of grain growth kinetics on the lattice geometry and various Wulff plot configurations were used in order to map surface energies to the lattice geometry. From this research, the following conclusions may be drawn:

(1) The Q-state Monte Carlo method provides an efficient and reliable means for the characterization of single phase grain growth studies.
(2) The shape of the Wulff plot, as governed by the values for the semi-major and semi-minor axes, as well as the exponential factor (n), plays a critical role in the resulting microstructure.
(3) Isotropic grains were shown to exhibit equiaxed growth due to equivalent surface energies of neighboring grains.
(4) Anisotropic grains resulted from the non-equilibrium effect of high energy grain boundaries which tended to migrate faster than corresponding low energy boundaries.
(5) Abnormal, anisotropic grain growth may be encountered for certain elliptical Wulff plot configurations (i.e., n=4, and n=5/2), and were characterized by a disproportionately high number of unusually large grains which tended to consume the smaller grains.
(6) Compared to isotropic grain growth, anisotropic grain growth may be characterized by lognormal grain size distributions with smaller average grain sizes and relatively large skewness values.
(7) The model provides a method to engineer the sintering process to enhance material properties by deliberately introducing anisotropic grain orientation within green form prior to sintering to achieve the desired anisotropic polycrystalline ceramic structures in the final product.

Future studies will be conducted in order to examine the role of various other factors affecting anisotropic grain growth, including the application of temperature gradients, the extension to three dimensions, and the addition of second phase materials.

ACKNOWLEDGMENTS

This study was funded through the U.S. Army Engineer Research and Development Center Directed Research Programs: "Nanoscale Studies of Polycrystalline Materials with Emphasis on Ceramics Synthesis" and "Effects of Grain Boundaries on Ceramic Properties" as part of the Advanced Material Initiative program.

REFERENCES

[1]S.C. Tjong, H. Chen, Nanocrystalline materials and coatings, *Mater. Sci. Eng.* R **45**, 1-88 (2004).
[2]Z.J. Liu, C.H. Zhang, Y.G. Shen, Y.-W. Mai, Monte Carlo Simulation of Nanocrystalline TiN/Amorphous SiNx Composite Films, *J. Appl. Phys.* **2**, 758-760 (2004).

[3]C. Lu, Y.-W. Mai, Y.G. Shen, Recent Advances on Understanding the Origin of Superhardness in Nanocomposite Coatings: A Critical Review, *J. Mater. Sci.* **41** (2006) 937-950.

[4]W.Yang, L. Chen, and G. Messing, Computer-Simulation of Anisotropic Grain-Growth, *Mat. Science and Engineering A,* **195**(1-2), 179-187 (1995).

[5] Faber, K. T., Evans, A. G., Drory, M. D. A Statistical Analysis of Crack Deflection as a Toughening Mechanism in Ceramic Materials, Fracture Mechanics of Ceramics **6**, 77-92 (1983).

[6]Faber, K. T., Evans, A. G. Crack Deflection Processes-I: Theory and II: Experiment, *Acta Metall.* **31**(4), 565-584 (1983).

[7] L.T. Bowen and EJ. Avella, Microstructure, Electrical Properties, and Failure Prediction in Low Clamping Voltage Zinc Oxide Varistors, J. Appl. Phys., **54**, 2764 (1983).

[8]W.A. Kaysser, M. Sprissler, C.A. Handwerker and J.E. Blendell, Effect of a Liquid Phase on the Morphology of Grain Growth in Alumina, *J. Am. Ceram. Soc.,* **70**, 339 (1987).

[9]O. Ito and E.R. Fuller, Jr., Computer Modeling of Anisotropic Grain Microstructure in Two Dimensions, *Acta Metall. Mater.,* **41**(1), 191-198 (1993).

[10]A.H. Heuer, G.A. Fryburg, L.U. Ogbuji, and T.E. Mitchell, $\beta$-$\alpha$ transformation in Polycrystalline SiC: I, Microstructural Aspects,"*J. Am. Ceram. Soc.,* **61**(9-10), *406-412* (1978).

[11]T.E. Mitchell, L.U. Ogbuji and A.H. Heuer, $\beta$-$\alpha$ transformation in Polycrystalline SiC: II. Interfacial Energetics, *J. Am. Ceram. Soc.,* **61**(9-10), 412-413 (1978).

[12]M.P. Anderson, D.J. Srolovitz, G.S. Grest, P.S. Sahni, Computer simulation of grain growth - I. Kinetics, Acta Metall. **32**, 783-791 (1984).

[13]M.A. Fortes, Grain Growth Kinetics: The Grain Growth Exponent, *Materials Science Forum, Trans Tech. Publications,* Switzerland, **94**, 319-324 (1992).

[14]S.K. Kurtz, and F.M.A. Carpay, Microstructure and normal grain growth in metals and ceramics. Part I. Theory, *J. Appl. Phys.* **51**, 5725 (1980).

[15]E.A. Holm, J.A. Glazier, D.J. Srolovitz, and G.S. Grest, Effects of Lattice Anisotropy and Temperature on Domain Growth in the Two-Dimensional Potts Model, *Phys. Rev. A,* **43**(6), 2662-2668 (1991).

[16]E.A. Holm, D.J. Srolovitz, and J.W. Cahn, Microstructural Evolution in Two-dimensional two-phase Polycrystals, *Acta Metall. Mater.,* **41**(4), 1919-1136 (1993).

[17]D.J. Srolovitz, G.S. Grest, M.P. Anderson, and A.D. Rollett, Computer Simulation of Recrystallization. II. Heterogeneous Nucleation and Growth, *Acta Metal.,* 36(8), 2115-2128 (1988).

[18]J. Wejchert, D. Weaire, J.P. Kermode, Monte Carlo Simulation of the Evolution of a Two-Dimensional Soap Froth, *Phil. Mag. B,* **53**, 15-24 (1986).

[19]*Gambit Users Guide,* Lebanon NH, 2007.

[20]OpenFoam, The Mews, Picketts Lodge, Surrey RH1 5RG, UK, 2006.

[21]H.V. Atkinson, *Acta Metall.* Overview no. 65: Theories of Normal Grain Growth in Pure Single Phase Systems, **36**, 469-491 (1988).

# Virtual Materials (Computational) Design and Ceramic Genome

# NUMERICAL CALCULATIONS OF DYNAMIC BEHAVIOR OF A ROTATING CERAMIC COMPOSITE WITH A SELF-HEALING FLUID

Benazzouk Louiza[1], Arquis Eric[1], Bertrand Nathalie[2], Cédric Descamps[3], Valat Marc[1]
[1] I2M/ TREFLE, 16 avenue Pey-Berland 33607 Pessac, France.
[2] LCTS, 3 Allee La Boetie, 33600 Pessac, France.
[3] SPS, Groupe SAFRAN, Les 5 Chemins, 33 187 Le Haillan Cedex.
lbenazzouk@ipb.fr

ABSTRACT
       Self-healing matrix composites are currently in full development. They are used for their remarkable physical and chemical properties. Yet, microscopic cracks can appear and spread on the material; they are aggravated by the conditions of use of the composite parts. A chemical reaction happens between the elements of the matrix and of the oxygen, allowing the formation of viscous glasses in a crack. By filling out the crack, these glasses protect the fiber and extend the life of the composite.
       To begin, we focus on the dynamic behavior of the glass drops formed in a crack, by studying their growth and the longitudinal and the radial coalescence. These cases of coalescence lead either to the formation of a liquid film or to the formation of a bridge. These phenomena are modeled using Thetis software, developed by I2M, to track the glass evolution in the crack when the composite material is submitted to intense external mechanical stresses such as rotation. Then, experimental studies are realized to study the wettability of various glasses on a mixed substrate, representative of the multilayer self-healing matrix.
       Finally, numerical simulations of these experiments are scheduled to compare the simulation results with those obtained experimentally.

INTRODUCTION
       The remarkable physicochemical properties of the Ceramic Matrix Composites (CMC) make them materials of choice for aerospace applications. Their introduction as structural elements of the hot parts of aircraft engines responds to very specific needs, such as higher combustion temperatures, while reducing polluting gas emissions (NOx, CO), in addition to significant weight savings, due to the low density of ceramic materials, thus leading to reduced fuel consumption[1].
       Particular attention is devoted to self-healing matrix composites, and more particularly to self-healing phenomena, which corresponds to the formation of a viscous glass due to the oxidation of matrix elements[2]. This viscous glass will gradually reseal the micro cracks thus slowing the diffusion of oxygen to the heart of the material and thus protecting carbon fibers from oxidation.

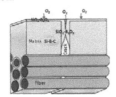 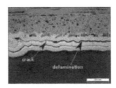

Figure1. Crack in a matrix composite a) schematic representation of a partially resealed crack, b) Photograph of a sample presenting cracks and delamination taken with an electron microscope

The (Figure 1a) shows a schematic representation of a cross-section view of CMC with a crack partially filled with viscous glass.

The (Figure 1b) is a photograph of a sample presenting cracks and delamination. The scale shown on the picture gives an idea of the crack width whose order of magnitude is about 1μm.

As a rough approximation, the crack may be viewed as the gap between two parallel planes. In presence of oxygen, oxidation of certain components of the composite surface feed the crack with viscous glass. Depending on the kinetics of formation of the glass and on the body force applied to the material, a liquid bridge may form. Evolution of the liquid bridge in the crack will be strongly influenced by centrifugal force (for rotating systems), and surface phenomena (capillary effects and wetting).

Some fluids properties and operational conditions are listed in Table 1.

Table 1. Operating conditions and viscous glass properties

| Temperature range (K) | rotation speed (rev/min) | oxides formed | densities ranges $(kg/m^3)$ | viscosities (Pa.s) | surface tensions ranges (N/m) |
|---|---|---|---|---|---|
| 923.15 | 100-600 | $B_2O_3$ | 2460 | 177.82 | 0.065 |
| 1273.15 | 100-600 | $SiO_2$ | 2650 | Quite solid | 0.276 |

Geometric and physical considerations show that flow in the crack will occur at very low Reynolds number. As body force and capillary effects will play a major role in flow in the crack, Bond number will be a key parameter for the problem.

As the lifetime of the composite material is strongly dependent on the plugging of the crack by the viscous glass formed, it is of interest to investigate numerically fluid configuration and flow in the crack under the action of external solicitations (e.g. centrifugal force for rotating systems).

However, multiphase flow involving surface phenomena and particularly moving three-phase contact line is not simple. Thetis® software developed by I2M is a suitable tool to model these phenomena. Model of wettability has been recently implemented in Thetis and its ability to precisely simulate those phenomena has to be investigating prior to examine the case of a crack.

In this work, we aim to study:

• Firstly, the simulation of surface phenomena (wetting, surface tension) that are not easily taken into account in numerical simulations. Implementation of a wettability model in Thetis allowed simulation of moving three-phase contact lines, subjected a prescribe static contact angle. Deposition of a drop on a horizontal surface led to a static configuration for which equilibrium contact angle could be estimated and compared to prescribed values.

• Secondly, the dynamic effects on contact angles, moving three-phase contact lines in an inclined plane surface, show a dynamic contact angle that may be very different from its static value. Experimental and theoretical results of Fujimoto [9] provide a convenient test case to study this phenomenon. The results of numerical simulations were compared too.

• Thirdly, the attention is focused on the dynamics on the formation of viscous glass produced by the oxidation of the matrix. The growth and coalescence of drops formed during the oxidation are exposed to the combined effects of rotation (i.e. volume force), surface tension and wetting. This can lead either to a viscous film on both sides of the crack, or to a liquid bridge which will reseal the crack.

- Finely, we will do an experimental study of the spreading of a glass drop on a surface of a bilayer materials (at the intersection between the $B_4C$ and SiC), the study will be conducted in a furnace, under an inert atmosphere and at a temperature of 550 °C. The aim is to remake these cases numerically and to compare the two results for different temperatures, atmospheres and glasses).

## MODELING AND COMPUTER SIMULATION FOR TWO-PHASE FLOW

Modeling and simulation of multiphase flows, whose characteristic interfacial length scale is larger than the smaller space step of the computational grid[4], is classically achieved with the 1-fluid model, also called model of Kataoka[3].For the model of wettability used thereafter, it is to clarify that the limits of the model are not well known, and that tests are needed to evaluate them. For that, studies of static contact angles and movement of viscous slug in a capillary are provided in this paper.

### The modeling approach

In this modeling framework, two-phase flow representation is restricted to non-miscible, incompressible and isothermal fluids. Moreover, a c onstant surface tension is assumed. In the 1-fluid model, the phase function C, which describes the interface evolutions over time through an advection equation, is assumed to behave like a Heaviside function. The interface between phases 0 and 1 is defined by the isosurface C = 0. 5.The surface tension force depending on the interface location is modeled by a C ontinuum Surface Force (CSF) approach based on the work of Brackbill[5].

### Smooth-VOF: smoothing controlled and improving of calculating the curvature

Different methods of smoothing of the VOF (Volume Of Fluid) function exist. La Faurie et al[6], for example, developed a Laplacian type filter, that is applied on the function from the VOF repeating these filtering process n times (typically 2 times) and using the neighboring cells at the interface to realize the smoothing. The approach of Popinet et al[7] is also interesting. They propose to go through a height function in order to recalculate the mean curvature via finite difference applied, to the derivative of the same height function.

All these methods present advantages and drawbacks. In Thetis software, the approach consists in improving the determination of interface curvature, and allowing imposing a precise wetting behavior (contact angle). The need to calculate a second order[8] derivation to approximate the local interface curvature and improve the calculation of the surface tension led us to develop a new approach. A smoothing function[9] from the VOF-PLIC, which we call Smooth VOF allows a more precise calculation of the local curvature. In this chapter, the SVOF function will be noted $C^S$.

The model is based on controlling the spread of the color function at the interface, in order to model the contact angles. The numerical calculation of the interfacial curvature required for simulating the effect of surface tension is difficult, with VOF method which is discontinuous on the scale of the mesh. So, we rebuilt a color function which describes the interface, based on geometric function resulting from VOF, representing its smoothing called $C^S$ (Smooth-VOF). Only the calculation of the curvature is affected by this new function but the properties of conservation of mass by the numerical algorithm VOF-PLIC[6] are preserved by maintaining the function outcome of VOF. The position of the interface is maintained in the new function SVOF by requiring that the position M, where C (M) = 0.5 is verified in the same cells as those in which $C^S$ (M) = 0.5.

Via an analogy with the diffusion equation for heat transfer, we configured the equation (1) for calculating SVOF.

$$-\nabla.D\nabla C^{S,n+l} + C^{S,n+l} + B_S (C^{S,n+l} - C_\infty^S) = C^{S,n} \tag{1}$$

This method is composed of the construction of a smooth function $C^S$ by solving a Helmholtz equation, with a source term corresponding to the VOF function of reference C. The relationship between contact angle and the value of the SVOF to the wall is shown by Guillaument[10].

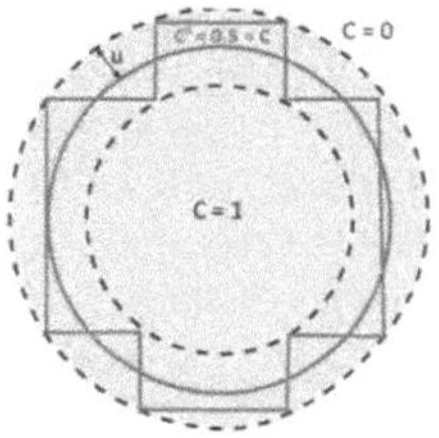

Figure 2. Smooth-VOF in green circle // VOF-PLIC slots in red

Modeling of wettability

To impose a contact angle, the idea is to create a numerical variable ($C_p$), which varies between 0 and 1, we can determine its value using the relationship that links it to the contact angle $C_p = f(\theta)$. For this, we need to apply a penalization for the Helmholtz equation previously cited, and that by introducing a variable (B) which is 0 in the whole domain of calculation except at the wall, where it is equal to infinity. So, $C_p = C^s$ this means that there is a virtual fluid without viscosity and density at the wall, where it will complete the first mesh $C_p = 0$ (non wetting state and therefore not filling the mesh) and $C_p = 1$ (completely wetting state, completely filled mesh) and using the correlation $C_p = f(\theta)$ we can determine the contact angle.

STATIC WETTING: DEPOSITION OF DROPS ON HORIZONTAL SURFACES

A first configuration to test the wettability model is the case of a drop deposited on a solid surface[11, 12]. The drop tends to adopt a shape conforming to a segment of a sphere (if gravity is negligible). At equilibrium, the liquid/gas interface joins the solid surface forming a contact angle, whose value depends on the nature of solid substrate as well as fluid phases.

The more the liquid wets the solid surface, the more the liquid will tend to spread on to the surface, showing an affinity with the solid surface, the lower the contact angle will be according to Young's equation. Numerically, the appears[13] contact angle as a property of the solid surface (wetting wall) and is fixed at a given value. Thus a drop of liquid initially hemispherical, placed beside a solid surface will deform and progressively reach its equilibrium shape. Numerical calculation of such a problem will provide evolution of contact angle[14], and geometric properties of the drop as a function of time till equilibrium is reached.

The problem presenting an axis of symmetry is two-dimensional. Solid surface is considered as perfectly smooth. The drop diameter is initially set at 1mm. The Bond number of the problem is $Bo = 10^{-2}$ where Bo is defined as $Bo = \rho g R^2/\sigma$. The bottom wall has a wetting property imposed by the model, and other boundary conditions will, in turn, be set to as impermeable walls. All information concerning the geometry and the boundary conditions are given on figure (3a).

All the parameters of the study were fixed, at the exception of the contact angle which was varied from 10° to 170°. All the simulations were performed with the same mesh.

The resolution of the Navier-Stokes equations was done using an augmented Lagrangian method, and the direct solver order to achieve residues of the order of $10^{-15}$. The time step was fixed at $10^{-4}$ s, and evolution of the drop were simulated during 50 s ($5.10^5$ iterations). The penalization of SVOF to ensure an imposition of the contact angle was formed on the first mesh line of the domain.

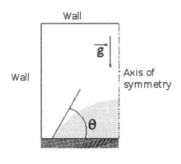

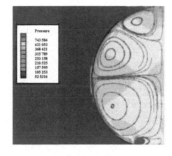

a) Initial condition

b) Drop in apparent equilibrium for large contact angle (170°)

Figure 3. Initial condition and apparent equilibrium of drops on a horizontal plane

Determination of the contact angle and pressure[15] were carried out over time. The values obtained at equilibrium ($\theta$ equilibrium) are presented in Table (2). We have also calculated the theoretical Laplace pressure ($\Delta P_{theoretic}$) inside the droplet at equilibrium, by measuring the radius of curvature at equilibrium and using Laplace's law ($\Delta P = 2\sigma / R$) and made the reverse via the pressure difference measured across the interface ($\Delta P_{measured}$), yielding radius of curvature of the interface via Laplace's equation.

Measurements and calculations show a good agreement when using the curvature determined by SVOF method. Similarly, the equilibrium contact angle compares well with the value imposed ($\theta$ imposed). For this simple case, the model of wettability provides satisfactory results regarding equilibrium contact angles.

However, it is to be mentioned that droplets placed on a non-wetted surface (e.g. Figure (3b), equilibrium contact angle of 170°), despite apparent equilibrium, presents internal flows, particularly near the contact line. It can be supposed that these flows result of the competition between gravitational effects and surface tension, but also include a numerical contribution due to the approximation of the curvature in the zone of the triple line.

Table 2. Comparison of theoretical and calculated drop properties at equilibrium for different contact angles

| Case n° | $\theta_{imposed}$ (°) | $R_{measured}$ (m) | $\Delta P_{measured}$ (Pa) | $R_{theoretic}$ via $\Delta P_{measured}$ (m) | $\Delta P_{theoretic}$ via $R_{measured}$ (Pa) | $\theta_{equilibrium}$ measured (°) |
|---|---|---|---|---|---|---|
| 1 | 10 | 0.00950 | 14.08 | 0,00923 | 13,68 | 13 |
| 2 | 50 | 0.00132 | 100.78 | 0,00128 | 98,48 | 50 |
| 3 | 70 | 0.00070 | 189.12 | 0,00068 | 185,71 | 69 |
| 4 | 90 | 0.00060 | 218.00 | 0,00059 | 216,30 | 92 |
| 5 | 110 | 0.00055 | 267.81 | 0,00048 | 260,00 | 113 |
| 6 | 130 | 0.00045 | 297.00 | 0,00043 | 288,88 | 134 |
| 7 | 170 | 0.00050 | 274.17 | 0,00047 | 260,00 | 167 |

## PHYSICAL VALIDATION IMPACT OF THE DROP ON AN INCLINED SURFACE

This part validates all models and methods by confronting them with a non-stationary real case. We studied the experiments carried out by Fujimoto et al[16]. These authors have made an experiment campaign concerning the impact of a liquid drop on a horizontal or inclined surface that has an important property of wettability. The main idea was and still is to study the contact angle hysteresis.

Fujimoto et al have created a device (detailed in the publication[16]), and we realized a numerical equivalent experiment. The chosen study concerns the impact of a drop of diameter $d_p = 0.53$ mm, with an initial velocity $v_0 = 4$ m.s$^{-1}$, on an inclined surface that respects the vertical axis of 30° ($\alpha = 30°$).

This initial configuration is reported on figure (4a). The definitions of the different statements are established on figure (4b). These variables are compared with the experiments, and the boundary conditions are given on figure (4c). We measured the elongation evolution –which is dimensionless out of respect to the initial diameter– as a function of time –also dimensionless, according to the ratio between initial diameter and initial velocity.

These elongation curves are a function of time and are reported on figure (5) where the experimental results of Fujimoto and our numerical results from different meshes are shown. We observe a v ery good agreement between the experiment and the simulation performed; numerical representations of the spreading phenomenon are given on figure (6). Figure (6a) shows a 2D view of the drop spreading where the streamlines are observable. Figure (6b) represents on the contrary a 3D view of the drop. These representations well agree with Fujimoto's experiments.

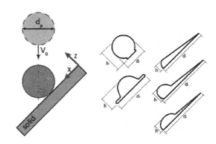

a) Initial situation     (b) Definition of measurements, height, diameter and spreading over time

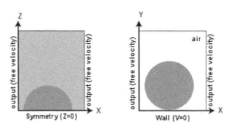

(c) Boundary conditions

Figure 4. Description of the initial situation for the case of Fujimoto

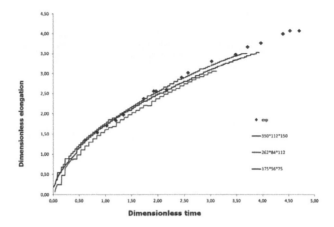

Figure 5. Dimensionless measure of the spreading of the drop for different mesh sizes

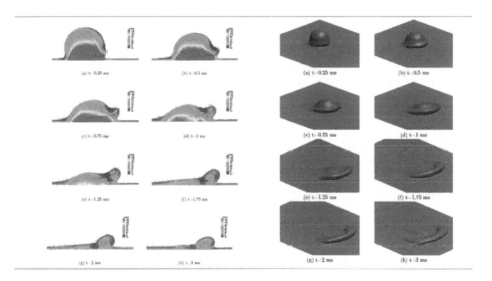

(a) 2D view of a drop for the Fujimoto case's     (b) 3D view of a drop for Fujimoto case's

Figure 6. Behavior description of a drop for the case of Fujimoto ($d_p$, $v_0$, $\alpha$) = (0,53 mm, 4 m/s, 30°).

COALESCENCE OF MICROMETER DROPLET:

The coalescence between two drops is a frequent phenomenon in nature or in industry. The previous works in that domain have demonstrated a difference between the coalescence of drops in a fluid and the coalescence of drops placed face-to-face on partly wetting walls. As for our case, the growth of the bridge formed between the drops evolves, following a power law[17] in the first times of coalescence. We study the propagation and migration of a liquid bridge placed between two plane and parallel surfaces.

The distance between the two plane surfaces was set at $2\,\mu m$ and the domain length was fixed at 100 microns. Initially, the domain is free of liquid (filled with gas) ; at the beginning of the calculation, a liquid is injected to two regions, set on two surfaces opposing one another, at a given flow rate. Two growing droplets are then formed (Figure 7).

A volume force is imposed and is parallel to the plane surfaces. The drops coalescence occurs when the droplets surfaces touch each other (if body force is not too intense). This problem is quite similar to the case of a liquid bridge placed in a rotating crack, as I said in the introduction.

The body force –represented by an acceleration noted $g^n$– is constant and equals gravity acceleration to the power n, where n varies between 5 and 7. To quantify the results, we use a modified Bond number which compares the effects of surface tension and the effects of body force.

$$\mathrm{Bo} = \frac{\rho g^n R^2}{\gamma} \tag{6}$$

Three behaviors were observed, according to the value of the Bond number :
- The Bond number equals 1 (when n=6) : after the formation of a liquid bridge, the bridge velocity is high and combined with a high injection flow rate (the Reynolds number equals $10^{-7}$), resulting in the movement of the liquid bridge which leaves a residual liquid film behind, on both sides of the crack (Figure 7b). As deposited films are partially fed by the liquid bridge, the bridge progressively reduces its thickness and finally breaks.
- The Bond number is inferior to 1 ( when n=5) : in this case, the interfacial effects are important ; the formed bridge continues to grow, due to liquid injection, and eventually completely fills the crack (Figure 7a).
- The Bond number is superior to 1 (when n=7) : the bridge adopts a rather special behavior, due to the competition that opposes the surface tension to the volume force. The instability of the interface (receding meniscus) provokes the entrapment of gas bubbles in a liquid bridge (Figure 7). These bubbles follow the bridge movement, which will eventually breaks : the protecting role of the liquid bridge is then not ensured.

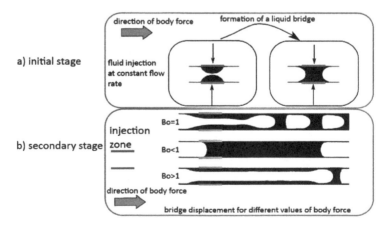

Figure 7. Characteristics of different flow configurations depending upon the value of the Bond number

PRESENTATION OF THE EXPERIMENTS

The sets of parameters are recapitulated in an experimental plan. The following Excel table shows the experiments to be achieved. It will allow us to quickly know the experimental conditions, results and comments.

Table 3. Summary of experiments to be carried

| Tests | Glass | Substrate | T (°C ) |
|-------|-------|-----------|---------|
| 1 | $B_2O_3$ | B4C | 550 |
| 2 | $B_2O_3$ | SiC | 550 |
| 3 | $B_2O_3$ | B4C | 550 |
| 4 | $B_2O_3$ | SiC | 550 |
| 5 | $B_2O_3$ | B4C/SiC | 550 |
| 6 | $B_2O_3$ | B4C/SiC | 750 |
| 7 | 25% $(SiO_2)$ + 75% $(B_2O_3)$ | B4C/SiC | 850 |
| 8 | 25% $(SiO_2)$ + 75% $(B_2O_3)$ | B4C/SiC | 850 |
| 9 | 25% $(SiO_2)$ + 75% $(B_2O_3)$ | SiC | 850 |
| 10 | 25% $(SiO_2)$ + 75% $(B_2O_3)$ | B4C | 850 |

The comparison of different results will allow us to study qualitatively and quantitatively the influence of various parameters on the phenomenon of wettability.

PREPARATION OF SAMPLES

Elaboration of glasses

Both used glasses were made from powders of boron oxide and silica oxide to 99.7% pure. The preparation is made in a platinum crucible doped on boron and gold (to prevent the glass from overflowing the crucible by capillary action) with a movable floor furnace, which can reach a temperature of 1500 °C. After calculating the masses of powder corresponding to the desired compositions, the amorphous structure of the glass is obtained by a heat treatment.

The $B_2O_3$ is very hygroscopic, once cooled and weighed; the glass is kept in a desiccator under a vacuum, to prevent the formation of boric acid due to the glass reaction with ambient humidity.

Preparation of substrates

The materials constituting the substrates are a pellet of $B_4C$, pellet of SiC (monolithic) and two pellets $B_4C$ /SiC bilayers. Monolithic pellets have each 20 mm diameter on 6 mm high, and the bilayers have 20 mm diameter on 12 mm high (6 mm of $B_4C$ and 6 mm of SiC). The major difficulty was to cut these cylinders to make several lamellae. The hardness of these materials is extremely high, which makes cutting very difficult.

The main objective was to maintain the intermolecular bonds at the interface of the multilayers. This objective could not be achieved. Indeed, the fragility of the interface between SiC and $B_4C$ did not allow the initially planned cut of the substrates; the solution is to use the entire half-cylinder for a single test.

EXPERIMENTAL RESULTS

Before starting the tests, it is necessary to calibrate the different components of the process, to ensure the reproducibility and the accuracy of the measures. Thus, the thermal calibrations of the oven and gas flow meters have been completed.

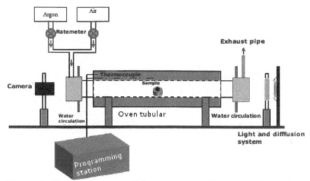

Figure 8. Characterization of the experimental process of wetting.

The beginning of the experiments was delayed by a series of problems; the only test to be realized was the experience number 5 (see table 3). Indeed, the wettability of glass ($B_2O_3$) on an interface between two materials (SiC/ $B_4C$) being still unknown, this test is a priority, to reproduce these results numerically (see table 4).

Table 4. Results obtained for the temperature, atmosphere, time and contact angle at equilibrium for a $B_2O_3$ glass on bilayer surface ($B_4C$ /SiC)

| N° photo | Room temperature (°C) | equilibrium time (min) | Atmosphere | $\theta°$ $B_4C$ | $\theta°$ SiC |
|---|---|---|---|---|---|
| **1** | **495** | / | **Argon** | **117** | **85** |
| 2 | 542 | / | Air | 107 | 84 |
| 3 | 550 | 0 | Air | 78 | 71 |
| **4** | **553** | **5** | **Air** | **65** | **64** |
| 5 | 552 | 10 | Air | 55 | 51 |
| 6 | 550 | 15 | Air | 49 | 47 |
| 7 | 550 | 20 | Air | 47 | 43 |
| **8** | **550** | **25** | **Air** | **43** | **40** |
| 9 | 550 | 30 | Air | 39 | 36 |
| 10 | 550 | 35 | Air | 36 | 35 |
| 11 | 550 | 40 | Air | 37 | 34 |
| 12 | 550 | 45 | Air | 32 | 31 |
| 13 | 550 | 50 | Air | 31 | 31 |
| 14 | 550 | 55 | Air | 30 | 30 |
| 15 | 550 | 60 | Air | 28 | 29 |
| 16 | 550 | 65 | Air | 30 | 30 |
| 17 | 550 | 70 | Air | 28 | 29 |
| 18 | 550 | 75 | Air | 28 | 26 |
| 19 | 550 | 80 | Air | 27 | 26 |
| 20 | 550 | 85 | Air | 28 | 26 |
| 21 | 550 | 90 | Air | 27 | 25 |
| **22** | **550** | **95** | **Air** | **28** | **25** |

Experiment 5 - photo 01 - B₂O₃ on (B₄C /SiC) - 495°C

Experiment 5 - photo 04 - B₂O₃ on (B₄C /SiC) - 553°C

Experiment 5 - photo 08 - B₂O₃ on (B₄C /SiC) - 550°C

Experiment 5 - photo 22 - B₂O₃ on (B₄C /SiC) – 550° - equilibrium

Table 5. Photograph of spreading of $B_2O_3$ on a bilayer material (SiC/ $B_4C$)

We noticed that under argon, the contact angle of $B_2O_3$ on the $B_4C$ is higher than the one on the SiC, and deforms the glass drop (see photograph 1). But under air, the contact angle of $B_2O_3$ on the $B_4C$ decreases abruptly, and five minutes after reaching the experimental temperature, the contact angles on both materials are similar. The introduction of oxygen results in a sudden drop spreading, which means that interfacial reactions exist, and that they favor the wettability of $B_2O_3$ on the bilayer material.

Over time, we notice that the two contact angles are relatively close. However, we find that for a time of 95 minutes, the spreading of $B_2O_3$ on the SiC is slightly better than on the $B_4C$. Unfortunately, it is impossible to compare this result with other experimental cases, because at the moment, they are still currently being tested.

6. CONCLUSION

One of the aims of the study of the spreading and the wettability of glass in a crack is essentially to compare the experimental results with the numerical results. In order to do so, we will create and validate a model that allows us to take into account the phenomenon of wetting, and more specifically to follow the behavior and the displacement of the triple line in a liquid / solid / vapor system. Therefore, it was necessary to test this model through different cases of deposition of drops on horizontal and inclined surfaces.

Thereafter, the attention was focused on the formation and the displacement of a liquid bridge in a capillary crack, due to the growth and the coalescence of the two drops. The originality of this case consisted in the submission of the liquid bridge to a body force (a constant body force, assimilated to a centrifugal effect). That force allowed us to obtain different configurations of the liquid bridge, according to the value of the Bond number. This case is of practical interest for the self-healing matrix composites. It has to be pursued, in order to provide a range of the Bond number, in which the liquid bridge may preserve the material from oxidation.

Eventually, we plan to make a numerical study of the experimental case in progress, involving the simulation of the spreading of a glass drop on a bilayer surface (B4C/SiC), in order to see the behavior of the drop under different temperatures, atmospheres, types of glasses and types of substrates.

REFERENCES AND CITATIONS

[1]Thebault, J. Résistance à la corrosion des composites thermostructuraux. J .Phys IV France, 831. (1993).

[2]Lara-Curzio E. Oxydation induced stress-rupture of fiber bundles. Journal of Engineering Materials and Technology, 120, 105. (1998).

[3]Dimitrakopoulos & Higdon. Local instant formulation of two-phase flow. Int. Journal of Multiphase Flows, 12(5) 745-758. (1986).

[4]Scardovelli, R., Zaleski, S. Direct numerical simulation of free-surface and interfacial flow. Ann. Review Fluid Mech, 31, 567-603. (1999).

[5]J.U. Brackbill, D.B. Kothe, and C. Zemach. A continuum method for modeling surface tension. Journal of Computational Physics, 100(2) : 335–354. (1992).

[6]Lafaurie, B., Nardone, C., Scardovelli, R., Zaleski, S., Zanetti, G. Modeling merging and fragmentation in multiphase flows with surfer. Journal of Computational Physics, 113, 134–147. (1994).

[7]Popinet, S. An accurate adaptive solver for surface-tension-driven interfacial flows. Journal of Computational Physics, 228 , 5838–5866. (2009).

[8]Youngs, D.L., & Morton, K.W., & Baines, M.J. Time-dependant multilateral flow with large fluid distortion Numerical Method for Fluid Dynamics. Academic Press, New York. (1982).

[9]Pianet, G., Vincent, S., Leboi, J., Caltagirone, J.P., Anderhuber M. Simulating compressible gas bubbles with a smooth volume tracking 1-fluid method. Journal of Multiphase Flow, (36) , 273–283. (2010).

[10]Guillaument & a l. Modélisation 1-Fluide de l'angle de contact et de la ligne triple – Application à l'impact d'une goutte d'émulsion sur une paroi Proceedings in ICMF. (2010).

[11]Basu, S., Nandakumap, K., Masliyahi, J.H. A Study of Oil Displacement on Model Surfaces Department of Chemical Engineering, University of Alberta, Edmonton, Alberta, Canada T6G 2G6. (1996).

[12]Tanner, H. The spreading of silicone oil drops on horizontal surfaces. J.Phys. D : Appl. Phys, 12, 1473–1484. (1979).

[13]Bretherton, F. The motion of long bubbles in tubes. J. Fluid Mech, 10, 166. (1961).

[14]De Gennes, P. Wetting : statics and dynamics. Rev. Mod. Phys. 57 (3), 827–863.

[15]Quéré, D. (1991). Sur la vitesse minimale d'étalement en mouillage partiel. C.R.Acad. Sc. Paris, Série II, 313, 313–318. (1985).

[16]Fujimotot, H., Shiotani, Y., Tong, A.Y., Hama, T., Takuda, H. Three dimensional numerical analysis of the deformation behavior of droplets impinging onto a solid substrate. International journal of Multiphase Flow, 33 : 317-332, (2007).

[17]Ristenpart, W., McCalla, P., Roy, R., & Stone, H. Coalescence of spreading drop on a wettable substrate. Physical Review Letters, 97. (2006).

EXPLICIT MODELLING OF CRACK INITIATION AND PROPAGATION IN THE
MICROSTRUCTURE OF A CERAMIC MATERIAL GENERATED WITH VORONOI
TESSELLATION

S Falco[a*], N A Yahya[b], R I Todd[b], N Petrinic[a]
[a] Department of Engineering Science, University of Oxford, Parks Road, Oxford, OX1 3PJ, UK
[b] Department of Materials, University of Oxford, Parks Road, Oxford, OX1 3PH, UK

*Corresponding author: simone.falco@eng.ox.ac.uk

ABSTRACT
    Explicit modelling of crack propagation in the microstructure of a ceramic material generated with
Voronoi tessellation. Ceramic materials are currently used for both vehicle and personnel armour since they
can be very effective in stopping ballistic projectiles by breaking and/or eroding them. However, such
armour is generally fairly heavy and does not have multihit capability, mainly due to excessive
fragmentation during impact. The development of new ceramics for armour is further hindered by the
limited understanding of the mechanisms involved in their success and therefore what the characteristics of
ideal ceramics would be. This work studies the initiation of cracks in ceramic microstructures, as in many
armour ceramics (e.g alumina) the main cause of failure is inter-granular crack propagation. To simulate
this phenomenon the mesh of a polycrystalline aggregate is first created and then crack propagation along
grain boundaries is modelled. A method to generate three-dimensional meshes of polycrystalline aggregates
based on Hardcore Voronoi tessellation of random points distribution is presented. The mesh created is
enriched by introducing contacts along the grain boundaries to explicitly simulate the propagation of inter-
granular cracks. Finally, the numerical results are compared with micro-cantilever beam tests

INTRODUCTION
    The numerical simulation of the behaviour of polycrystalline materials at grain level is crucial to
understand the behaviour of the material at the macroscopic level.
In particular for ceramic materials, the development of new materials with improved performance is
hindered by the lack of understanding of the mechanisms involved in their success, and hence what the
characteristics of the ideal ceramic for a given application should be. Therefore the investigation of the
deformation and cracking mechanisms of the microstructure is fundamental to determine both the
macroscopic response of the material (i.e. stress-strain response) and the localised phenomena (i.e.
initialization and propagation of cracks), in order to improve the design of ceramic material.
In the past decades several computational methods, especially finite element method (FEM) have been
developed to simulate the behaviour of polycrystalline materials. The topology of the polycrystalline
aggregate can either be experimentally measured or virtually generated.
    Several experimental techniques have been used to measure the real microstructure topology of a
material. They have provided important information (e.g. the distribution of grain size and the number of
first neighbours) but generally they are fairly complex and expensive.
Depending on the material chemistry, processing and microstructural length of interest, a three-dimensional
structure can be rendered using numerous methods, either non-destructive or based on sectioning (e.g.
acoustic microscopy, laser ultrasonics [1], X-ray, focused ion beam FIB, electron backscatter diffraction
EBSD [2]). These methods – besides being quite expensive – produce a huge amount of data, which
requires complex post-processing that can lead to non-unique topologies due to grain overlapping [3].

In the past 20 years many authors numerically investigated the micro-mechanics in polycrystalline materials. Initially using 2D virtual structures, then – with the increase of computational power – building 3D polycrystalline models. Different methods were used to discretize the space with regular shapes (e.g cubes, dodecaedra, truncated octaedra [4-7]), but they are not representative of the variability of grain size and shape observed in real polycrystalline materials. Among random generated morphologies Voronoi tessellations have the advantages of being defined analytically and having grains with straight edges and planar faces [8].

In particular Poisson-Voronoi, Harcore-Voronoi and Laguerre-Voronoi tessellation algorithms are widely accepted to generate models of real microstructures of metallurgic and ceramic materials [9-12].

In the present article the algorithm developed to generate finite element (FE) models of polycrystalline structures based on Hardcore-Voronoi tessellation is presented. The method defines and meshes the grain separately, and then the structure is assembled. This approach allows assigning different material properties to each grain, and introducing contacts along the grain boundaries – to simulate the propagation of inter-granular cracks.

GENERATION OF POLYCRYSTALS

Voronoi Tessellation

Given a finite set of $N$ points (hereinafter called nuclei) – arbitrarily positioned in space – the Voronoi tessellation decomposes the entire space into $N$ convex polyhedra (i.e. one per each nucleus). Each convex polyhedron (hereinafter called cell) generated from the generic nucleus $P$ contains all the points closer to $P$ than to any other nuclei. All the cells are convex polyhedra bounded by polygonal surfaces (hereinafter called *faces*).

The Voronoi tessellation can be interpreted in terms of homogeneous growth process of metallurgic or ceramic grains under the following assumptions [11]:

- All crystalline nuclei have the same weight and appear at the same time and remain fixed in the same location during the growth process.
- The growth is isotropic, uniform and has a constant rate.
- Grain growth in a direction stops when two grain boundaries contact each other, and there are no voids in between the grains.

Mathematically the Voronoi tessellation in a 3D space can be expressed as:

$$\{R_P\} = \{x \in X \mid d(P_i, x) \leq d(P_k, x), k = 1,2, \dots, N : k \neq i\}$$

Where $P$ are the nuclei, $R_P$ is the set of point associated with the nucleus $P_i$ (i.e. points composing the Voronoi cell) and $x$ any point in the space $X$.

However, unconstrained nucleus seeds generate tessellation with highly distorted or excessively irregular cells. The mesh of those cells leads either to a high refinement of the mesh or to using distorted elements. In both cases the numerical analysis is badly affected.

Within the different techniques that have been used to generate tessellation with more regular cell shapes, the Hardcore Voronoi represents a balanced choice between effectiveness and simplicity [13]. The Hardcore assumption imposes a minimum distance $\rho$ in between the nuclei. If a nucleus falls inside the sphere of radius $\rho$, it is moved to another position in the unit cell. The Hardcore assumption produces more regular grains and drastically reduces the number of small entities (Fig 1).

Mathematically the Hardcore Voronoi is defined as:

$$|P_i - P_j| > \rho \ \forall i,j = 1,2,\dots,N : i \neq j$$

A physical explanation for using the Hardocere Voronoi method is that the size of the nuclei before the growth process is finite, and different nuclei may not intersect each other.

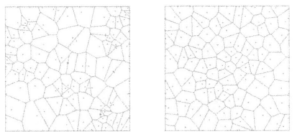

Figure 1: Samples of Voronoi (left) and Hardcore Voronoi (right) tessellations

In this paper, unless otherwise specified, all the space decompositions are performed using the Harcore Voronoi method.

Generation of the polycrystalline microstructure

   Contrary to most of the techniques developed to generate aggregate microstructures from Voronoi tessellations, in the method presented in this paper the domain in which the nuclei are seeded does not determine the final shape of the structure. The peculiarity of the procedure proposed is that the final shape of the aggregate structure is not defined before the cell generation.

The nuclei domain is defined as a box (hereinafter called unit cell) that has the only constraint of being big enough to contain the final shape of the structure. The nuclei are positioned randomly – following the Hardcore assumption – in the unit cell. Then one nucleus is added in correspondence of each unit cell vertex. The addition of the nuclei on the vertices of the unit cell ensures that the cells generated from the internal nuclei are bounded (i.e. all the vertices of the cell have finite coordinates). The cells generated from the external nuclei, instead, are all unbounded, and to bind them the nucleus itself is added to the list of the vertices of the respective cell.

   The commercial software Matlab is used to compute the Voronoi tessellation. The output of the "voronoin" command consists of the list of the vertices for each cell and the list of the coordinates of the vertices. The classic "convhulln" command computes the Delaunay triangulation (i.e. defines the cell faces as triangles), hiding the real connectivity of the planar faces.

The tool "minConvexHull" (part of the "geom3d" library by Legland [14]), instead, provides the connectivity of the perimeter of the polygonal faces by merging the coplanar Delaunay triangles, hence is the one used to define the cell faces.

   To let the cells grow unconfined (but bounded) leads to a tessellation that decomposes the whole unit cell and part of the space outside it. Therefore a tessellation of any solid shape inscribed in the unit cell can be cut out from the unconfined cells, as a sample is cut from a bigger medium.

The cutting method developed in this paper follows the concept of a real cut along an arbitrarily oriented plane. Since there is no theoretical limit to the number, the position and the orientation of the cutting planes, the approach presented in this paper allows for the generation of any convex solid shape.
The cutting procedure for a sample cell intersecting the cutting plane is:

- Detect the vertices of the cell beyond the cutting plane.
- Determine the internal vertices connected to the each external vertex.
- Project the external point on the cutting plane along the connecting edge.
- Replace each external vertex with its projection on the faces, respecting the connectivity order.
- Erase the external vertices.
- Create a new face with the projected points.
- Erase the potential degenerated faces and grains.

The procedure is repeated for all the cells and all the cutting planes.
The result is complete decomposition of the desired solid shape in Voronoi cells, topologically equivalent to a section of a real polycrystalline microstructure.

The cutting operations may generate duplicate entities (vertices and faces). Their removal is necessary to avoid errors in the further mesh generation steps, but this can be very demanding in terms of computational time. Moreover the cutting procedure may lead to small edges (i.e. vertices very close to each other), which, if not removed, would imply a mesh with either distorted elements or high refinement. The same strategy is adopted to solve both the issues. It relies on the indexing of the vertices to detect all the vertices of a cell that lay closer to each other than a numerical precision value $\rho_P$. The points are then merged into one point and the indices shifted. Eventually all the duplicated faces in a cell are eliminated.

With the proposed procedure any desired solid shape can be decomposed with Hardcore Voronoi cells (Fig. 2). The output is a virtual model of a polycrystalline structure, ready for meshing, since it is clear of duplicate entities and has an imposed minimum edge length that will lead the choice of the element size.

Figure 2: Examples of solid shapes

MODELLING OF THE POLYCRYSTALLINE MICROSTRUCTURE
In this section the procedure to create a finite element (FE) model of the microstructure is described.
First the meshing technique is illustrated, and then the material model and the contacts are defined.
It is important to specify that the FE models generated are virtually exportable in any FE analyser. In particular – at present –complete and ready-to-analyse input files for both "Abaqus FEA" and "LS-DYNA" can be automatically generated.

Mesh of the grains
The approach proposed in this paper for the generation of the FE model is to mesh each cell separately using the same meshing options, and then to homogenize the 2D discretization on the coincident faces.

The meshing approach is based on a widely utilized bottom-up procedure, consisting of meshing vertices, edges, faces and cells (i.e. 0D, 1D, 2D and 3D entities) successively. The n-D mesh is generation is constrained by the (n-1)-D mesh, and the element lengths are interpolated from it.

The 0D meshing consists of attributing to each vertex a characteristic length. Afterwards each edge is meshed into 1D elements, with lengths derived from the characteristic length of the two vertices. To each new node is assigned a characteristic length that determines the length of the 2D elements, and then each face is meshed into triangular elements.

The same procedure is applied for the 3D meshing; each cell is discretized into tetrahedral elements. According with the quality of the mesh (and the number of nodes) the tetrahedral elements can either be classic (4 nodes) or quadratic (10 nodes). The meshing libraries from Gmsh package are used to perform all the meshing steps of the bottom-up procedure.

Material model

In the present work, each grain is modelled as a three dimensional linearly elastic anisotropic domain with arbitrary orientation.

The constitutive relation for the grain material – using the compact Voigt notation – is:

$$\sigma_i = C_{ij}\varepsilon_j$$

In particular, all the simulations in the next sections examine the behaviour of $\alpha$-Alumina, $Al_2O_3$ (hereinafter referred to as Alumina).

Alumina is a hard brittle material having hexagonal-rhombohedral structure, and its properties depend on the crystallographic orientation [15]. Alumina crystal has a trigonal structure of class ($3\bar{m}$), and its stiffness matrix – after reducing the number of independent constants $C_{ij}$ with symmetry considerations – is:

$$\begin{Bmatrix} \sigma_{xx} \\ \sigma_{yy} \\ \sigma_{zz} \\ \sigma_{yz} \\ \sigma_{zx} \\ \sigma_{xy} \end{Bmatrix} = \begin{bmatrix} C_{11} & C_{12} & C_{13} & C_{14} & 0 & 0 \\ C_{12} & C_{11} & C_{13} & -C_{14} & 0 & 0 \\ C_{13} & C_{13} & C_{33} & 0 & 0 & 0 \\ C_{14} & -C_{14} & 0 & C_{44} & 0 & 0 \\ 0 & 0 & 0 & 0 & C_{44} & C_{14} \\ 0 & 0 & 0 & 0 & C_{14} & C_{66} \end{bmatrix} \times \begin{Bmatrix} \varepsilon_{xx} \\ \varepsilon_{yy} \\ \varepsilon_{zz} \\ \varepsilon_{yz} \\ \varepsilon_{zx} \\ \varepsilon_{xy} \end{Bmatrix}$$

Where $C_{66} = 0.5 \cdot (C_{11} - C_{12})$.

The crystal has only six independent elastic constants, and – despite its three-fold symmetry – is not orthotropic. Many authors have experimentally measured the elastic constants for Alumina crystals, and their results are quite consistent [16-19]. For this work, the elastic constants measured by Hovis et al. [19] have been used:

| | GPa | | $10^{-11}Pa^{-1}$ |
|---|---|---|---|
| $C_{11}$ | 497 | $S_{11}$ | 2.35 |
| $C_{12}$ | 163 | $S_{12}$ | -0.70 |
| $C_{13}$ | 116 | $S_{13}$ | -0.38 |
| $C_{14}$ | 22 | $S_{14}$ | -0.48 |
| $C_{33}$ | 501 | $S_{33}$ | 2.16 |
| $C_{44}$ | 147 | $S_{44}$ | 6.93 |

Table 1: Non-zero terms of the stiffness ($C_{ij}$) and compliance ($S_{ij}$) matrices for $\alpha$-$Al_2O_3$

Finally a local coordinate system is assigned to each grain by defining the angles of the local axes with respect to the global ones.

## Assembly of the grains

Assembling the meshed grains as they are after the meshing, usually, implies that the node distribution on a face does not overlap perfectly with the one on the coincident face.

The mismatch occurs even when meshing two identical faces with the same meshing properties, because of numerical discrepancies in creating the 2D discretization.

The perfect match is necessary to avoid errors in the definition of the grain boundary interface within grains, whether it is implemented with contact or with cohesive elements.

The first step to homogenize the meshes on the coincident faces is to create a list of the "twin" faces and verify that they have the same number of 2D elements.

Any difference in the number of elements on coincident faces or an excessive distance between correspondent nodes (i.e. distorted elements after the homogenization) indicates the presence of errors in the geometry definition, hence is carefully monitored and avoided.

Figure 3: Meshed grain and cubic structure obtained assembling 50 grains

## Contact definition

Since alumina fails mainly because of inter-granular crack propagation (i.e. the grain interface is definitely weaker than the grain itself) the interface must be defined only between the element across the grain boundaries and not inside the grains.

The algorithm developed imposes that the meshes on two coincident surfaces are identical. This approach leaves the freedom of defining any desired interface between the grains. In this paper the interface is simulated using "Tiebreak surface-to-surface contacts" [20] within the correspondent elements, because of their simplicity and versatility. The "Tiebreak contact" binds together two sets of triangular segments (i.e. the coincident element faces across the grain boundary). The segments are perfectly bonded until the tension ($\sigma_N$) and the shear ($\sigma_T$) stresses across the contact satisfy the inequality:

$$\left(\frac{\sigma_N}{S_N}\right)^2 + \left(\frac{\sigma_T}{S_T}\right)^2 \geq 1$$

Where $S_N$ and $S_T$ are, respectively, the tensile and the shear failure stresses.
After the failure the contact behaves as a surface-to-surface contact, without any interface tension.

## EVALUATION OF THE MACROSCOPIC ELASTIC CONSTANTS

In this section the FE models of polycrystalline aggregates are used to estimate the overall Young's modulus of the isotropic material.

Given a polycrystalline structure, consisting of $N$ grains and subjected to a uniaxial traction – since the material is not supposed to develop micro-cracks – the stress and strain volume average can be used to extract the apparent Young's Modulus as follows:

$$E = \frac{\bar{\sigma}_1}{\bar{\varepsilon}_1}$$

Numerical models
For a unit cell with $N$ crystals, the assumption of macroscopic isotropic behaviour is restrictive if the number of grains is not large enough. Therefore a scatter is expected in the predicted elastic modulus, which should decrease with the increasing of the number of the grains to, ideally, converge to the exact value for an infinite number of grains.
To explore the effect of the number of grains on the prediction of the elastic properties of the polycrystalline aggregate, three different cubic structures were created, formed by 25, 50 and 100 grains

| $n_{gr}$ | $n_{struct}$ | $l\ [\mu m]$ | $V\ [\mu m^3]$ | $\overline{V_{gr}}\ [\mu m^3]$ | $d_{gr}\ [\mu m]$ | $\rho\ [\mu m]$ | $\rho_P\ [\mu m]$ |
|---|---|---|---|---|---|---|---|
| 25 | 4 | 4.00 | 64.00 | 0.64 | 1.07 | 0.5 | 0.05 |
| 50 | 4 | 3.17 | 31.86 | 0.64 | 1.07 | 0.5 | 0.05 |
| 100 | 2 | 2.52 | 16.00 | 0.64 | 1.07 | 0.5 | 0.05 |

Table 2: characteristic dimensions of the three structures

Table 2 summarises the dimensions of the three structures. The structure size increases with the number of grains in order to maintain the average grain size constant.
For all the structures the maximum element length is imposed to be equal to the minimum edge length $\rho_P$, to minimize the mesh distortion.
The two parameters affecting the elastic modulus of a polycrystalline aggregate – for a given number of grains – are the geometry of the grains and their orientation.
In order to study the influence of both the parameters 10 different structures ($n_{struct}$ in Table 2), with different geometries, have been generated. Finally 5 random coordinate system distributions were assigned to each structure, giving a total of 20 different models for 25 and 50 grains structures and 10 different models for 100 grains structures (50 different models in total).
In all the simulations, on one face of the cube a displacement perpendicular to the face is applied, whilst the opposite face is constrained only in the displacement direction.

Results
The results of the simulations are summarised in Fig 4. The graph shows the values of the Young's modulus predicted against the number of grains of the structure. The small asterisks represent the result of a single simulation, whilst the big asterisks are the average of the results for a single structure (i.e. grain geometry). The results referring to the same structure – but with different grain orientation distribution – share the same marker colour. Finally the dashed line represents the Young's modulus of a structure with the direction 1 of all the grains aligned with the load direction.
The average results fall very close to the reference value of the Young's modulus measured experimentally (398 GPa), and well within the range of values found in literature. The dashed line moreover matches perfectly the theoretical value of the Young modulus for the given grain alignment.

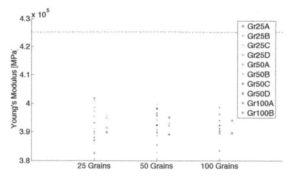

Figure 4: Results

The number of grains has a massive influence on the scatter of the results. Fig. 5 shows the histograms of the predicted Young's modulus for a given number of grains. For the structures with 25 grains the results are quite uniformly distributed over the range, whilst increasing the number of grains the predicted values focus in a narrower range.

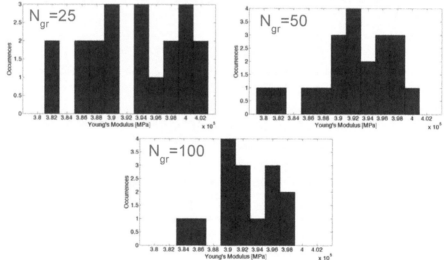

Figure 5: Occurrences of the values of the Young's modulus for the 3 set of structures

It is important to underline that the three histograms have the same range of values on the horizontal axis.

SIMULATION OF THE FAILURE OF A MICRO-CANTILEVER BEAM

This section studies the capability of the FE models generated to simulate the failure of a polycrystalline material. In particular a micro-cantilever beam test was simulated, from the unloaded condition to the complete failure of the beam.

The experiment consists of an Alumina micro-beam, with a triangular cross section – produced using focused ion beam (FIB) machining –deformed to fracture. A nanoindenter applies the deformation linearly and records the reacting force (Fig 6). The top pictures show the intact beam before the test, whilst the bottom pictures show the beam after the fracture.

In this paper only one of the several tests performed is presented. A more accurate description of the experiment and of the results of all the tests carried out will be found in [21].

Figure 6: Micro-cantilever beam before (top) and after (bottom) the failure

As is known – and perhaps confirmed by the bottom right picture in Fig. 5 – cracks in alumina propagate preferably along the grain boundaries that are notably weaker then the grain itself. Therefore the "Tiebreak surface-to-surface contacts" are defined only across the grain boundaries.

Four different FE models – with the same polycrystalline structure – are generated to analyse the effect of the contact strength on the crack initiation and propagation. In particular – due to the nature of the test – only the effect of the variation of the normal stress across the contact is taken into account. The shear stress is neglected by imposing $S_T \gg S_N$ in the constitutive equation of the tie-break contact.

The polycrystalline structure consists of a total of 13 grains, similar in size and shape to the ones that belong to the real structure (Fig. 7)

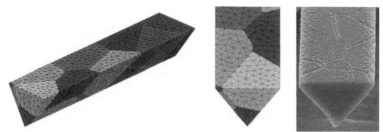

Figure 7: FE model of the micro-cantilever beam and comparison with the real structure

All the nodes on the back face of the beam are fully constrained, and a linear increasing force is applied on the node corresponding to the loading point in the experimental setup. Finally – to eliminate the variability linked to the anisotropy of the grains – the material model adopted for the whole model is isotropic elastic, using the values of the Young's modulus and the Poisson ratio measured experimentally (i.e. respectively $E = 398GPa$ and $v = 0.22$).

As expected, also the cracking surfaces (i.e. grain faces) of the model and of the experimental beam have evident visual similarity (Fig. 8).

Figure 8: Failure of the FE model and comparison of the virtual and the real cracking surfaces

Finally, the effect of the variation of the parameter $S_N$ on the behaviour of the beam are summarised in a graph, and compared with the experimental results (Fig. 9).

The first important observation is that the presence of the tie-break contacts does not affect the behaviour of the model before the failure condition is satisfied (i.e. the model without the contacts on the grain boundaries behave identically to the others before the failure).

Moreover both the simulations and the experiment exhibit a sudden failure (i.e. the crack very quickly propagates through the whole beam), and the failure load can be tuned by varying the value of $S_N$.

To find the exact value of $S_N$ that reproduces the exact failure point of a particular experimental test with a particular numerical model does not give any information about the real behaviour of the grain boundaries. The difference in the position and the inclination of the grain boundaries – both in real experiments and numerical simulations – result in completely different results. Hence the only sensible values to compare are the statistical ones. Moreover there are also other factors affecting the behaviour of the beam (e.g. presence of defects, anisotropy of the grains) that the simulations presented do not take into account.

Finally it must be taken into account a possible difference in the dimensions, because the dust generated from the machining deposits on the beam surfaces, making the beam appear bigger than it is.

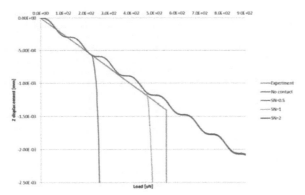

Figure 9: Results of the simulations varying the contact strength

A more complete comparison of statistical experimental and numerical results will be the topic of a separate work.

CONCLUSIONS

A new method to build numerical models of 3D realistic polycrystalline microstructure has been developed. The approach presented shapes the polycrystalline structure in any solid form by cutting the unconfined aggregate along an arbitrary number of planes. Since the planes do not have any constraint on position and inclination in the space, theoretically any convex solid can be obtained, and the initial unit cell does not influence the final shape of the structure.

The grains are meshed separately with a bottom-top meshing approach, and then the grains are assembled together, in order to allow the definition of interfaces on the grain boundaries.

The FE models are used to estimate the overall Young's modulus of the isotropic material. A total of 50 FE models of polycrystalline structures are generated – categorised into three groups of structures composed of different numbers of grains (25, 50 and 100). The average values of all the three groups have a very good agreement with the experimental measurements, but the histograms of the predicted Young's modulus shows that a higher number of grains corresponds to a less scattered distribution of the results.

Finally a micro-cantilever beam test is simulated, from the unloaded condition to the complete failure of the beam. The algorithm developed is used to generate the FE model of a polycrystalline beam with a triangular cross section. "Tie-break contacts" are defined across the grain boundaries to allow for inter-granular crack propagation. The effect of the variation of the strength of the contact is explored, obtaining very promising results.

Summarising, the models generated by the algorithm described in this paper showed the capability of simulating the macroscopic behaviour of a polycrystalline medium, from the elastic deformation to the complete failure. All the simulations have been compared with experimental evidence, showing an excellent agreement.

ACKNOWLEDGEMENTS

This research was carried out as part of the Understanding and Improving Ceramic Armour (UNICAM) Project. We gratefully acknowledge funding from EPSRC and the Ministry of Defence, UK.

REFERENCES

[1] Burke SK, McK.Cousland S, Scala CM (Nondestructive characterization of advanced composite materials) 1994: Mater Forum 18, 85-109

[2] Bhandari Y, Sarkar S, Groeber M, Uchic MD, Dimiduk DM, Ghosh S (3D polycrystalline microstructure reconstruction from FIB generated serial sections for FE analysis) 2007: Computational Materials Science 41, 222-235

[3] Benedetti I, Aliabadi MH (A three-dimensional grain boundary formulation for microstructural modelling of polycrystalline materials) 2013: Computational Material Science 67, 249-269

[4] Sarma GB, Dawson PR (Effects of interactions among crystals on the inhomogeneous deformations of polycrystals) 1996: Acta Mater. 44, 1937-1953

[5] Raabe D, Zhao Z, Mao W (On the dependence of the in-grain subdivision and deformation texture of aluminium grain interaction) 2002: Acta Mater. 50, 4379-4394

[6] Zhao Z, Kuchnicki S, Radovitzky R, Cuitiño A (Influence of in-grain mesh resolution on the prediction of deformation textures in fcc polycrystals by crystal plasticity FEM) 2007: Acta Mater. 55, 2361-2373

[7] Ritz H, Dawson PR (Sensitivity to grain discretization of the simulated crystal stress distributions in FCC polycrystals) 2009: Model. Simul. Mater. Sci. Engrg. 17, 1-21

[8] Wakai F, Enomoto N, Ogawa H (Three-dimensional microstructural evolution in ideal grain growth – General statistics) 2000: Acta Mater. 48, 1297-1311

[9] Barbe F, Decker L, Jeulin D, Cailletaud G (Intergranular and intragranular behavior of polycrystalline aggregates Part. 1: F.E. Model) 2001: Int J Plast 17, 513-536

[10] Barbe F, Forest S, Cailletaud G (Intergranular and intragranular behavior of polycrystalline aggregates. Part 2: Results) 2001 Int J Plast 17, 537-563

[11] Kumar S, Kurtz S (Simulation of material microstructure using a 3d Voronoi tesselation: Calculation of effective thermal expansion coefficient of polycrystalline materials) 1994: Acta Metallurgica et Materialia 42(12), 3917-3927

[12] Kumar S, Kurtz S, Agarwala V (Micro-stress distribution within polycrystalline aggregate) 1996: Acta Mech 114, 203-216

[13] Fritzen F, Bohlke T, Schnack E (Periodic three-dimensional mesh generation for crystalline aggregates based on Voronoi tessellations) 2009: Comput Mech 43, 701-713

[14] Legland D (Graphics Library geom3d) 2009: http://www.mathworks.com/matlabcentral/fileexchange/24484-geom3d

[15] Vodenitcharova T, Zhang LC, Zarudi I, Yin Y, Domyo H, Ho T, Sato M (The effect of anisotropy on the deformation and fracture of sapphire wafers subjected to thermal shocks) 2007: J. Mater. Process. Technol. 194(1-3), 52-62

[16] Mayer WG, Parker PM (Method for the determination of elastic constants of trigonal crystal systems) 1961: Acta Crystallogr. 14, 725

[17] Winey JM, Gupta YM (r-Axis sound speed and elastic properties of sapphire single crystals) 2001: J. Appl. Phys. 90 (6), 3109

[18] Gladden JR, So JH, Maynard JD, Saxe PW, Page YL (Reconciliation of ab initio theory and experimental elastic properties of Al2O3) 2004: J. Appl. Phys. Lett. 85 (3), 392

[19] Hovis DB, Reddy A, Heuer AH (X-ray elastic constants for alpha-Al2O3) 2006: Appl. Phys. Lett. 88, 131910

[20] (LS-Dyna Keyword User's Material. 2007): Livermore Software Technology Corporation

[21] Yahya NA, Norton A, Todd RI (in preparation)

# KINETIC MONTE CARLO SIMULATION OF CATION DIFFUSION IN LOW-K CERAMICS

Brian Good
Materials and Structures Division, NASA Glenn Research Center
Cleveland, OH, USA

## ABSTRACT

Low thermal conductivity (low-K) ceramic materials are of interest to the aerospace community for use as the thermal barrier component of coating systems for turbine engine components. In particular, zirconia-based materials exhibit both low thermal conductivity and structural stability at high temperature, making them suitable for such applications. Because creep is one of the potential failure modes, and because diffusion is a mechanism by which creep takes place, we have performed computer simulations of cation diffusion in a variety of zirconia-based low-K materials. The kinetic Monte Carlo simulation method is an alternative to the more widely known molecular dynamics (MD) method. It is designed to study "infrequent-event" processes, such as diffusion, for which MD simulation can be highly inefficient. We describe the results of kinetic Monte Carlo computer simulations of cation diffusion in several zirconia-based materials, specifically, zirconia doped with Y, Gd , Nb and Yb. Diffusion paths are identified, and migration energy barriers are obtained from density functional calculations and from the literature. We present results on the temperature dependence of the diffusivity, and on the effects of the presence of oxygen vacancies in cation diffusion barrier complexes as well.

## INTRODUCTION

The high melting point of zirconia makes the material useful for high-temperature applications. However, zirconia undergoes two phase transitions between room temperature and the melting point; it exists in a monoclinic structure below about 1100°C, a tetragonal structure between 1100°C and 2300°C, and a cubic structure between 2300°C and the melting point.[1] While this would seem to render the material unsuitable for such applications, the addition of substitutional aliovalent cation dopants such as $Y^{3+}$ can stabilize the tetragonal and cubic phases. The high-temperature cubic phase can be fully stabilized, retaining the cubic structure from room temperature to the melting point. Fully stabilized zirconia exists in a cubic fluorite structure in which the cations are located on a face-centered cubic sublattice, with the oxygen ions located on a simple cubic sublattice whose lattice constant is one-half that of the cation sublattice.

Yttria-stabilized zirconia (YSZ) exhibits both thermal stability and low thermal conductivity, making it suitable for use as a component of coating systems for aerospace applications, for example, the thermal protection component of a coating system for turbine engine components. It exhibits oxygen and cation diffusive ionic conduction, both of which are generally considered to occur via the hopping of ions to nearest-neighbor vacancies on the appropriate sublattice. However, the detailed mechanisms of the two cases are somewhat different, resulting in very different dependences of the diffusivity on the cation dopant concentration.

The high ionic oxygen conductivity makes YSZ-based materials of interest for use as oxygen sensors, or as solid electrolytes for fuel cells. However, in the case of YSZ the oxygen diffusivity does not increase monotonically with $Y^{3+}$ concentration; it increases at low concentrations, but reaches a maximum between 8 and 15 mol % $Y_2O_3$, and decreases at higher concentrations.[2]

When oxygen ions move diffusively on the oxygen sublattice, they pass through a two-cation barrier complex, the composition of which determines the migration energy barrier. Substitutional doping by cations whose ionic charge is smaller than that of $Zr^{4+}$ results in the formation of additional oxygen vacancies beyond the intrinsic concentration, as needed to maintain electrical neutrality. On the other hand, increasing the cation dopant concentration also increases the fraction of diffusion barrier complexes that contain one or two dopant ions, and these energy barriers will be different from the energies of barriers consisting of two $Zr^{4+}$ cations; for barriers containing one or two $Y^{3+}$ cations, the barrier energies are larger than that of a two- $Zr^{4+}$ barrier.

The behavior of the oxygen diffusivity as a function of dopant concentration may be understood as a competition between the two effects described above. As the Y concentration increases, the number of oxygen vacancies available as target sites for diffusive hops increases, increasing the diffusivity. However, increasing the Y concentration also increases the number of higher-energy ZrY and YY complexes, which tends to reduce the diffusivity.[3,4]

Cation bulk diffusion in these materials is several orders of magnitude slower than oxygen diffusion.[5] Bulk diffusion takes place via diffusive hopping of cations among vacancies on the cation sublattice, with the cations passing through a barrier complex consisting of two oxygen atoms. In contrast to oxygen diffusion, $Y^{3+}$ doping has no strong direct effect on the concentration of vacancy sites on the cation sublattice, and the number of target vacancy sites remains at the intrinsic value. However, because cation doping increases the concentration of vacancies on the oxygen sublattice, there will be a change in the fraction of barrier complexes containing fewer than two oxygen atoms, which may have an indirect effect on the cation diffusivity.

Cation diffusion is a mechanism by which creep takes place, which can affect the mechanical integrity and phase stability of the material, and the utility of alternate dopants will be limited if their inclusion substantially increases cation diffusion. Zhu and Miller investigated the thermal conductivity of YSZ with additional dopants, with the goal of developing reduced-k materials.[6] They found that, compared with a YSZ 4.55 mol% $Y_2O_3$ baseline concentration, both YSZ-Nd-Yb and YSZ-Gd-Yb showed a reduced thermal conductivity at a temperature of 1316C, in some cases showing a reduction in thermal conductivity by more than a factor of two. The best YSZ-Nd-Yb samples showed a thermal conductivity slightly lower than that of the best YSZ-Gd-Yb. The thermal conductivity was found to increase with thermal cycling time. The increase was largest for YSZ, with smaller increases for YSZ-Nd-Yb and YSZ-Gd-Yb. Other aspects of the thermal cycling results were ambiguous. For plasma-sprayed coatings, YSZ-Nd-Yb showed a lower cycles-to-failure than YSZ-Gd-Yb. However, for electron-beam physical-vapor-deposited coatings, the trend was reversed. In both cases, there was considerable scatter in the data

In view of the range of potential applications, considerable experimental and theoretical effort has gone into understanding the microstructure and phase behavior of YSZ. Of interest here, computer simulations using a variety of techniques have been performed. The tetragonal-to-cubic phase transition in YSZ has been investigated by Schelling et al. using molecular dynamics simulation with a Coulomb+Buckingham potential.[7] They correctly predicted the experimentally observed stabilization due to the doping of zirconia with yttria. Fabris et al. have carried out similar work using empirical self-consistent tight-binding molecular dynamics in conjunction with a Landau approach, to describe the thermodynamics of the transition.[8] Fevre et al. have investigated the microstructure of YSZ experimentally via neutron scattering and theoretically via Monte Carlo and molecular dynamics simulation, also using a Coulomb+Buckingham potential.[9] Fevre et al. have also investigated the thermal conductivity of YSZ using nonequilibrum molecular dynamics, finding that the conductivity increases with yttria concentration for yttria-rich compositions and for temperatures below 800K, but decreases with concentration at high temperatures for all concentrations.[10]

A number of computational studies of oxygen diffusion in YSZ have been performed using a variety of techniques. Kahn et al. performed molecular dynamics simulations of oxygen diffusion in YSZ doped with a variety of dopants.[11] In particular, they considered the detailed bonding of oxygen vacancies to various dopant cations, including rare earths. They found that for $Y^{3+}$, $La^{3+}$, $Nd^{3+}$ and $Gd^{3+}$, the binding of a dopant ion to a nearest neighbor oxygen vacancy was not favored energetically, though binding to a next nearest neighbor vacancy was favored. Other MD studies of oxygen diffusion have been performed by Perumal[12] and Shimojo.[13] Okazaki et al. investigated the effect of the ordering of Y dopants in YSZ, and showed that the oxygen diffusivity is enhanced compared to YSZ containing randomly distributed Y dopants.[14] Krishnamurthy et al. have performed kinetic Monte Carlo (kMC) simulations of oxygen diffusion in YSZ[3] and lanthanide-doped YSZ[4], and produced diffusivities in reasonable agreement with experiment.

Because cation diffusion in YSZ is orders of magnitude slower than oxygen diffusion, it is less amenable to study via molecular dynamics simulation, especially at lower temperatures. Some theoretical work does exist, however. Kilo et al. have analyzed Zr diffusion from creep data, dislocation loop shrinkage data, and Zr tracer diffusion data to identify the defects responsible for cation diffusion, and show that diffusion involving single cation vacancies as the most likely mechanism.[15] However, they note that the measurement of activation energies remains problematic; the relative ordering of the energies for Zr and Y diffusion are not consistent among various studies.

Kilo et al. performed simulations of cation diffusion in YSZ (as well as doped lanthanum gallates), using NPT molecular dynamics and a Buckingham+Coulomb potential.[16] They considered the hopping of cations via vacancy sites, introduced in the form of Schottky defects, at a mole fraction of 0.004, for yttria mole fractions of 0.11, 0.19 and 0.31. They found that cation diffusion is controlled by cation vacancies; interstitial diffusion is not significant. They also found that the diffusion coefficients for Y and Zr are significantly different, with Y diffusion 3-5 times faster than Zr diffusion. They reported calculated enthalpies, for YSZ with a yttria concentration of 11 mol %, of 4.8eV (Y) and 4.7eV (Zr). These vary in both magnitude and ordering from experimental results from Kilo et al.[15], who reported 4.2eV (Y) and 4.6eV (Zr). Their molecular dynamics results also showed that cation diffusivities were independent of $Y_2O_3$

concentration, or slightly increasing with increasing $Y_2O_3$ concentration, in contrast with the results of Fevre.[9,10]

Kilo et al. performed tracer diffusion studies of Y, Ca and Zr in yttria- and calcia-stabilized zirconia, and found a correlation of cation diffusivity with the ionic radius of the diffusing ion.[5] Zr and Y bulk diffusivities were maximized for a stabilizer content of 10-11 mol %. Activation enthalpies were found to be 4.2eV for Y and 4.5eV for Zr. In addition, the experimentally measured prefactor was of the same order for the two materials, 0.041 for $Zr^{4+}$, and 0.024 for $Y^{3+}$, suggesting that the vibrational frequencies for the two ions are not very different. However, for other cations, the prefactors differed from these by up the three orders of magnitude. The yttrium diffusivity decreased with increasing yttrium concentration, in contrast to the behavior observed in molecular dynamics simulations.[16]

Kilo et al. described experimental studies of lanthanide diffusion in YSZ and CSZ.[5] They use empirical potentials to calculate the energy landscape and to identify the lowest-energy diffusion path. They found that the path is nonlinear, with the saddle point located a significant distance from the midpoint of the linear path connecting a hopping cation and its target nearest neighbor vacancy site. They also reported activation enthalpies for a variety of lanthanides in YSZ and CSZ, along with Arrhenius prefactors.

In order to understand potential trade-offs between reducing thermal conductivity and degrading the mechanical performance of these materials, we have investigated the effect on cation diffusivity of the inclusion of substitutional Yb, Nb, and Gd in YSZ via kinetic Monte Carlo simulation. We discuss the dependence of cation diffusivity on doping species concentration, and on the presence of oxygen vacancies in the barrier complexes.

KINETIC MONTE CARLO METHOD

Kinetic Monte Carlo simulation differs from the more widely used Metropolis Monte Carlo method in that it is explicitly aimed at capturing the important features of "infrequent event" systems. While the first applications of the method date back to the 1960s, the method has enjoyed considerable recent popularity.[18-20]

Infrequent event systems are systems which evolve from state to state, but for which the system resides in a given state for a relatively long time, with infrequent and relatively rapid transitions between states. Diffusion is such a phenomenon; the state of the system is characterized by the location of all of its atoms, which typically reside in local energy minima located at lattice or interstitial points. Each atom undergoes thermal vibrations, and, provided that the temperature is not too high, will only rarely escape from its local minimum and move to an adjacent vacant lattice or interstitial site. Further, because of the relatively large time an atom remains in the same minimum, it undergoes a large number of vibrations, and is effectively randomized, retaining no information about its location prior to its most recent hop.

The kMC method complements the more widely used molecular dynamics (MD) method. Provided that a system's dynamics can be accurately represented using classical or quantum-approximate potentials, MD simulations can produce a detailed trajectory for each particle in the simulation. However, accurately representing atomic vibrations in such simulations requires that the numerical integration of the equations of motion be carried out using a time step on the order of femtoseconds. This places a limit on the size of systems that may be effectively studied using MD, and on simulation duration as well.

This restriction means that MD can be very inefficient when used to study infrequent event systems, notably the diffusive hopping systems of interest here. In an MD simulation of a system containing only a small fraction of vacancies, a computational cell of reasonable size will experience a relatively small number of diffusive hops, with most of the computational resources spent computing the trajectories of atoms between the infrequent hops. By contrast, the kinetic Monte Carlo method allows one to concentrate on the events of interest, and to effectively consider only the average behavior of the system between such events, while giving up information on the detailed trajectories of all atoms in the simulation.

A diffusive hop typically takes place on a time scale much slower that the typical period of atomic vibration, so that the system effectively loses any memory of the details of the hop, i.e. which of the vacancy's neighbor atoms was involved in the most recent hop. Each such hop may therefore be considered to be an independent event. The probability per unit time that a vacancy will undergo a hop is constant, with the survival probability decreasing exponentially. The probability distribution $p(t)$ of the time of first escape is given by $p(t) = k_{tot} \exp(-k_{tot}t)$, and the average time of first escape $\tau$ is given by $\tau = \int_0^\infty tp(t)dt = 1/k_{tot}$. Because all hopping events are independent, the effective total rate constant $k_{tot}$ is just the sum of rate constants $k_{AB}$ for all possible paths, with each rate constant determined by the height of the migration energy barrier in the direction of the hop:

$$k_{tot} = \sum_B k_{AB}$$

When the migration barrier energies are known, the hopping rates may be computed from $v_{AB} = v^0 \exp(-E_{AB}/k_B T)$ in which $v_{AB}$ and $E_{AB}$ are the hopping rate and migration barrier energy for a hop between oxygen or cation sites A and B respectively, and $v^0$ is the frequency factor. $v^0$ is typically assigned a value between $10^{12}$ and $10^{13}$ for these materials; given that the measured diffusivities (both oxygen and cation ) from different studies can differ substantially, we assume a baseline value of $10^{13}$ with the understanding that the values of the diffusivities presented here involve considerable uncertainty. A species-specific correction to the frequency factor will be described later. For each possible hop, the hopping probability can be computed from the hopping rate, with $P_{AB} = v_{AB}/\Gamma$, where $\Gamma$ is the sum of hopping rates for all possible hops in the computational cell. A catalog of all possible hops, and the corresponding hopping rates and probabilities, is created.

During the kMC process, one of the possible events (that is, a hop defined by the hopping ion and the target vacancy site) is chosen probabilistically from the catalog and executed. Hopping rates for all possible hops involving the new vacancy location are computed and added to the catalog, while rates involving the vacancy's previous location are deleted, and the sum of the hopping probabilities is updated. Finally, the simulation clock is advanced by a stochastically chosen time step $\Delta t = -\ln(R)/\Gamma$ where R is a random number greater than zero and less than or equal to unity.

When the simulation has run long enough to accumulate statistically useful information, the mean square displacement, averaged over all vacancies, is computed. The vacancy diffusivity $D_v$ is obtained from the Einstein relation $\langle R^2 \rangle = 6 D_v t$, and the ionic diffusivity $D_i$ is obtained by balancing the number of vacancy and ionic hops:

$$D_i = \frac{C_v}{1 - C_v} D_v$$

where $C_v$ is the concentration of vacancies on the appropriate sublattice.

ENERGY BARRIERS

The most energetically favorable hopping path for cation diffusion is in the [110] direction, from a cation site to a nearest neighbor cation vacancy site, as confirmed by DFT calculations, and by the MD simulations of Kilo.[17]

Cation diffusivity is sensitive to the choice of migration barrier energy. However, the values of those quantities have not been definitively established. Solmon et al. report hopping enthalpies of 4.8-4.95eV in the range of 1300-1700°C.[21] Gomez-Garcia et al. report enthalpies of 5.5-6.0eV above 1500C.[22] Chien and Heuer[23] report a value of 5.3eV at 1100-1300°C, and Dimos and Kohlstedt[24] find a value of 5.85eV at 1400-1600°C. Kilo et al. find values of 4.4-4.8eV at 1125-1460°C.[15] Mackrodt et al. have calculated Y and Zr migration energies of 2-7eV for YSZ.[25]

A set of energy barriers has been calculated using density functional theory (DFT) as implemented in the Quantum Espresso code.[26] These barriers are in reasonable agreement with experiment, with the exception of the Zr barrier energy, which is about 35 percent too large. Kilo[17] presents activation enthalpies and Arrhenius prefactors $D_0$ for a variety of lanthanide dopants in YSZ, and we have used these data in our kMC simulations. These authors also reference barrier energies for Zr and Y, but more recent work by these authors suggests different barrier energies for Zr and Y, and we have used these newer values in our work. The barrier energies and prefactors are shown in Table 1.

Table 1. Migration barrier energies and prefactors ($D_0$).

| Species | Energy, eV | $\ln(D_0)$ |
|---------|-----------|------------|
| Zr | 4.6 | -2.3 |
| Y | 4.2 | -3.7 |
| Yb | 4.9 | 0.01 |
| Nd | 4.7 | -0.04 |
| Gd | 5.4 | 4.1 |

RESULTS AND DISCUSSION

Kilo et al. find that although the migration enthalpies are similar for a number of lanthanides, the diffusivities are different, which may be due to the substantially different Arrhenius prefactors. In order to incorporate these prefactors in the kMC methodology, we note the hopping probabilities are given by $\nu^0 = e^{(-E_a/k_bT)}$ and the inverse of the sum of the probabilities for all hopping vacancies in the simulation provides a time scale.

The baseline value for the vibrational frequency $\nu^0$ is taken to be $10^{13}$, as is commonly used in kMC simulations. This value has given reasonable agreement with experiment in the case of oxygen diffusion in YSZ.[3] Further, because neither the barrier energies nor the prefactors are very different between Y and Zr, assuming the same value for $\nu^0$ is also reasonable for cation diffusion simulations of $ZrO_2$. However, while the lanthanide dopants considered here do have similar barrier energies, the prefactors vary by several orders of magnitude, and the assumption of a constant values for $\nu^0$ for these materials is more problematic. We therefore assume that the values of $\nu^0$ for all dopants, including Y, differ by the same factor as the ratios of the prefactors. All kMC runs were performed with these corrections to the vibrational frequencies.

Chien and Heuer obtained estimates of the concentrations of zirconium vacancies and cation-anion vacancy complexes.[23] Over a temperature range of 1100-1300°C, they estimated concentrations up to $10^{-8}$. Kilo et al. extrapolated this data to temperatures above 3000°K and obtained concentrations up to $10^{-3}$.[16] In this work we use a fixed concentration of $3\times10^{-3}$, not very different from the value of $4\times10^{-3}$ used in the MD simulations of Kilo.

For most of the simulations described below it was assumed that the barrier complexes were fully populated, with no vacancies in the two-atom complex. In some cases, however, oxygen vacancies were assigned to the barrier complex sites stochastically, consistent with the concentration of cation dopants, as discussed below.

Arrhenius plots of kMC results for $ZrO_2$ and YSZ are presented in Figure 1, for a temperature range of 1000K to 3000K. The plots are close to identical, but there is a slight variation in the computed activation energies. This can be understood by noting that the migration barrier energy for $Y^{3+}$ is smaller than that of $Zr^{4+}$, so that increasing the concentration of yttria is expected to reduce the total activation energy and increase the diffusivity. The activation energy for $ZrO_2$ and YSZ with 4.55, 10 and 15 mol% yttria are 4.60eV, 4.504eV, 4.422eV and 4.364eV respectively, with the value for $ZrO_2$ reflecting the assumed activation energy for pure Zr diffusion.

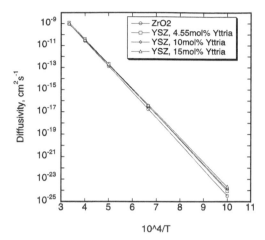

Figure 1. Temperature dependence of cation diffusivity in $ZrO_2$ and YSZ.

Figure 2 shows experimental results for YSZ, and some representative kMC results for YSZ 4.55 mol % $Y_2O_3$, and the same material additionally doped with 11.5 mol % $Yb_2O_3$ or 22.5 mol % $Gd_2O_3$. The kMC results are in good agreement with the experimental results of Kilo[16] and Chien[23], but agree less well with the results of Solmon.[21]

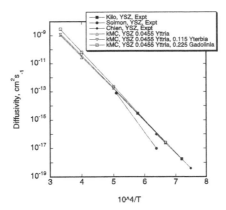

Figure 2. Comparison of experimental cation diffusivities with results from kinetic Monte Carlo simulations.

Arrhenius plots are shown in Figure 3 for the baseline YSZ/4.55 mol % $Y_2O_3$, and for the basline YSZ doped with $Gd_2O_3$, $Nd_2O_3$ or $Yb_2O_3$, at total dopant concentrations (including $Y_2O_3$) of 6 and 16 mol %. The greatest variation among the diffusivities is at the highest temperature, where the YSZ-Gd diffusivity is the largest, but by less than an order of magnitude.

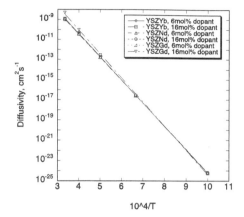

Figure 3. Diffusivities for YSZ doped with various aliovalent cations.

The concentration dependence of the diffusivity is shown in more detail in Figures 4a-b, where the diffusivities are plotted against dopant concentration, At 1500K there is no systematic concentration dependence, while at 3000K the diffusivity of YSZ-Gd diverges from that of the other two dopants, with the difference at 3000K being somewhat less than an order of magnitude.

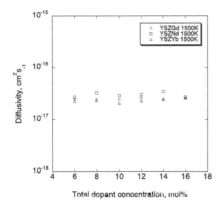

Figure 4a. Concentration dependence of diffusivities of doped YSZ. 1500°K.

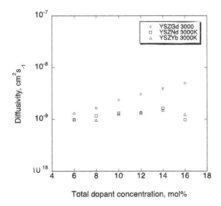

Figure 4b. Concentration dependence of diffusivities of doped YSZ. 3000°K.

The cation diffusivities of YSZ-Nd and YSZ-Yb are lower than that of YSZ-Gd at high temperature, suggesting that a YSZ-based material containing either of those dopants might be less susceptible to mechanical degradation than a material containing Gd, for degradation mechanisms, such as creep, that depend on the cation diffusivity. This conclusion is consistent with thermal cycling results on electron-beam physical-vapor-deposited coatings, but is not consistent with results from plasma-sprayed coatings.[6] The reason for this discrepancy is not yet clear, though there are pronounced differences between the microstructures in the two cases.

Finally, the effects of including oxygen vacancies in the barrier complexes are considered. We note that the concentration of oxygen vacancies depends on the concentration and charge state of the dopants. All the dopants considered here have a charge of +3, (compared with the Zr ions' charge of +4) so that a single oxygen vacancy is created for every pair of substitutional dopant cations. The concentration of these induced oxygen vacancies is much larger than the intrinsic concentration at the temperatures of interest here. The concentration of the dopant species therefore determines the fraction of barrier pairs on the oxygen sub lattice that include one or two oxygen vacancies.

DFT calculations of relaxed barrier energies for barriers containing a single oxygen were performed for Zr and Y, and the reductions in the barrier energies compared with the previously-calculated energies for fully-populated barriers. The ratios of the single- and double-atom barriers were used to scale the experimental energies shown in Table 1, with the result that the Zr and Y single-atom barrier energies were reduced to 1.776 and 1.717eV, respectively.

Results are shown in Figures5a-b. The plots were taken from kMC runs at 1500°K and 3000°K, for YSZ/4.55 mol % $Y_2O_3$, with and without barrier vacancies. At both temperatures, the diffusivity is larger when vacancies are included in the barriers. At 1500°K, both diffusivities increase with increasing yttria concentration. The diffusivity for the vacancy-containing material shows a much larger increase with concentration, with the difference between the two reaching a bit less than an order of magnitude at a yttria concentration of 20 mol %. At 3000°K, the diffusivity for YSZ without barrier vacancies is approximately independent of yttria concentration, while the vacancy-containing material increases by more than an order of magnitude.

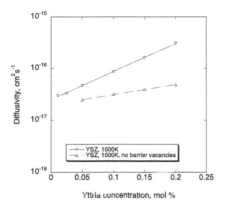

Figure 5a. Effect of vacancy-containing diffusion barriers on diffusivity of YSZ. 1500°K.

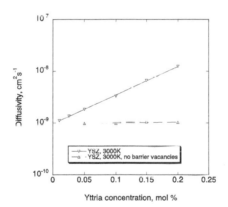

Figure 5b. Effect of vacancy-containing diffusion barriers on diffusivity of YSZ. 3000°K.

CONCLUSION

We have performed kinetic Monte Carlo computer simulations of cation diffusion in $ZrO_2$, undoped YSZ and YSZ doped with Gd, Nd and Yb. The kMC results for diffusivities and activation energies are in reasonable agreement with available experimental data. The activation energyy of YSZ is smaller than that of $ZrO_2$, and decreases with increasing yttria concentration. The diffusivity does not vary a great deal among the various dopants considered, with the largest diffusivity occurring in YSZ-Gd having a dopant concentration of 16 mol%, the largest dopant concentration investigated. Only YSZ-Gd shows a significant increase in diffusivity with dopant concentration.

The presence of an oxygen vacancy in the barrier complex reduces the migration barrier energy substantially, with the results that the diffusivity of YSZ is increased by as much as an order of magnitude at high yttria concentration. We expect similar behavior from the other dopants considered here.

The cation diffusivity of YSZ-Gd is somewhat larger than those of YSZ-Nd and YSZ-Yb. While this may indicate that YSZ-Gd is more likely to exhibit degradation via creep, the experimental evidence is ambiguous, and a more detailed understanding of the microstructures produced by different coating application methods is needed.

REFERENCES

[1]P. Aldebert and J. P. Traverse, J. Am Ceram. Soc. 68 [1], 34-40 (1985).
[2]R. E. W. Casselton, Phys. Status Solidi A, 2, 571-585 (1970).
[3]R. Krishnamurthy, Y.-G. Yoon, D. J. Srolovitz and R. Car, J. Am. Ceram. Soc. 87 1821-1830 (2004).
[4]R. Krishnamurthy, D. J. Srolovitz, K. N. Kudin and R. Car, J. Am. Ceram. Soc. 88 [8],2143-2151 (2005).
[5]M. Kilo, M.A. Taylor, Ch. Argirusis and G. Borchardt, B. Lesage, S. Weber, S. Scherrer, H. Scherrer, M. Schroeder and M. Martin, J. Appl. Phys. 94, 7547 (2003).
[6]D. Zhu and R. A. Miller, "Thermal Conductivity and Sintering Behavior of Advanced Thermal Barrier Coatings," NASA/TM-2002-211481
[7]P. K. Schelling, S. R. Phillpot and D Wolf, J. Am Ceram. Soc. 84 [7], 1609-1619 (2001).
[8]S. Fabris, A. T. Paxton and M. W. Finnis, Phys. Rev. B 63, 094101 (2001).
[9]M Fevre, A. Finel and R. Caudron, Phys. Rev. B 72, 104117 (2005).
[10]M Fevre, A. Finel, R. Caudron and R. Mevrel, Phys. Rev. B 72, 104118 (2005).
[11]M. S. Kahn, M. S. Islam and D. R. Bates, J. Mater. Chem. 8 [10], 2299-2307 1998.
[12]T. P. Perumal, V. Sridhar, K. P. N. Murthy, K. S. Easwarakumar and S. Ramasamy, Comp. Mat. Sci. 38, 865-872 (2007).
[13]F. Shimojo, T. Okabe, F. Tachibana, M. Kobayashi and H. Okazaki, J. Phys. Soc. Jpn 61, 2848-2857 (1992), and F. Shimojo and H. Okazaki, J. Phys. Soc. Jpn 61, 4106-4118 (1992).
[14]H. Okazaki, H. Suzuki and K. Ihata, Phys. Let. A 188, 291-295 (1994).
[15]M. Kilo, G. Borchardt, C. Lesage, O. Kaitsov, S. Weber and S. Scherer, J. Eur. Ceram. Soc., Faraday Trans. 5, 2069 (2000).
[16]M. Kilo, M. A. Taylor, C. Argirusis, G. Borchardt, R. A. Jackson, O Schulz, M. Martin and M. Weller, Solid State Ionics 175, 823-827 (2004).

[17]M. Kilo, M.A. Taylor, C. Argirusis, G. Borchardt, S. Weber, H. Scherrer and R.A. Jackson, J. Chem. Phys. 121, 5482 (2004).

[18]W. M. Young and E. W. Elcock, Proc. of the Phys. Soc. 89, 735 (1966).

[19]A. B. Bortz and M. H. Kalos and J. L. Lebowitz, J. of Comput. Physics 17, 10 (1975).

[20]A. F. Voter, Kinetic Monte Carlo, in Radiation Effects in Solids, Proceedings of the NATO Advanced Study Institute on Radiation Effects in Solids, K. E. Sickafus, E. A. Kotomin, B. P. Uberuaga, eds., 1-23, Springer, Dordrecht, The Netherlands, 2007.

21H. Solmon, C. Monty, M. Filial, G. Petot-Ervas and C. Petot, Solid State Phenom. 41, 103 (1995).

[22]D. Gomez-Garcia, J. Martines-Fernandez, A. Dominguez-Rodriguez and J. Castaing, J. Am. Ceram. Soc. 80, 1668-1672 (1997).

[23]F. R. Chien and A. H. Heuer, Philos. Mag. A 73, 681-697 (1996).

[24]D. Dimos and D. L. Kohlstedt, J. Am. Ceram. Soc. 70, 277 (1987).

[25]W. C. Mackrodt and P. M. Woodrow, J. Am. Ceram. Soc. 68, 277 (1986).

[26]P. Giannozzi, S. Baroni, N. Bonini, M. Calandra, R. Car, C. Cavazzoni, D. Ceresoli, G. L. Chiarotti, M. Cococcioni, I. Dabo, A. Dal Corso, S. Fabris, G. Fratesi, S. de Gironcoli, R. Gebauer, U. Gerstmann, C. Gougoussis, A. Kokalj, M. Lazzeri, L. Martin-Samos, N. Marzari, F. Mauri, R. Mazzarello, S. Paolini, A. Pasquarello, L. Paulatto, C. Sbraccia, S. Scandolo, G. Sclauzero, A. P. Seitsonen, A. Smogunov, P. Umari, R. M. Wentzcovitch, J.Phys.:Condens.Matter, 21, 395502 (2009)

# EFFECTIVE THERMOELASTIC PROPERTIES OF C/C COMPOSITES CALCULATED USING 3D UNIT CELL PRESENTATION OF THE MICROSTRUCTURE

Galyna Stasiuk[1], Romana Piat[1], Vinit V Deshpande[1,2], and Puneet Mahajan[2]

[1]KIT, Institute of Engineering Mechanics, 76131 Karlsruhe, Germany
[2]Indian Institute of Technology Delhi, Department of Applied Mechanics, Hauz khas, New Delhi 110 016, India

ABSTRACT

Both the micro constituents of C/C composites, carbon fibers and pyrolytic carbon (PyC) are orthotropic and, for this reason, effective properties of these composites are strongly dependent on the fibers arrangement in the preform. For a better understanding of the relationship between fiber orientations and the corresponding elastic properties, 3D unit cell models for different fiber distributions have been created.

These cells are numerically generated and reproduce 3D fibers arrangements in the real preform. For this purpose statistical studies of a carbon felt and 2D preforms were provided and corresponding distribution functions were identified.

The numerical procedure for microstructure generation consists in the creation of a set of non-intersecting cylinders. It is considered that the fibers are straight and there is no "packing problem" due to their low volume fraction (less than 25%). The distribution of the cylinders is random and it statistically corresponds to the distribution of the fibers studied by micro computer tomography (μCT).

Then, a growth algorithm is used for reproduction of the PyC coating on the fiber interfaces. After that, the generated geometry of the microstructure is meshed and utilized for calculation of the effective mechanical properties by the FE method. The calculated elastic and thermal properties are close to those obtained by using homogenization procedures.

## INTRODUCTION

C/C composites exhibit high elastic modulus, low thermal expansion, low weight and high chemical and corrosion resistance in a wide temperature range. Thus they are very important in modern automobile, aeronautic and space industry and despite their high cost they become more and more important and replace metals which are utilized in this branch.

Fabrication of the C/C composites by chemical vapour infiltration (CVI) of carbon fiber preforms leads to a very complicated structure of the composite on the micro and mesolevel which provides high stiffness of the material[1, 2]. The carbon fibers of the preform are embedded in a matrix of PyC, which has a cylindrically layered structure, and it is strongly influenced by the CVI parameters[3]. Experimental studies of the PyC-matrix on the micrometer scale were provided using nanoidentation[4, 5] and the cylindrically orthotropic elastic constants of PyC were obtained. Information concerning mechanical properties of carbon fiber were provided in[6,7] and show a high anisotropic Young's modulus of carbon fibers.

The anisotropy of the constituents as well as the presence of pores makes it very difficult to estimate the elastic properties. Due to the high anisotropic properties of the fibers the structure of the preform plays an important role in the overall response of the composite.

The microstructure of the C/C composite and its properties were discussed in[8, 9]. Using the computer tomography these results were extended and the distribution of the pores and fiber in composite were obtained as discrete orientation distribution functions [10].

The real pores in the composite are irregular. By using the principal component analysis the pores obtained by computer tomography were approximated by ellipsoids[11] which allowed classifying pores and also in this case accordingly to the obtained results to simplify calculations without accuracy loss.

The aim of our studies is to compare properties of the composite material obtained by different methods, namely the finite element analysis (FEA), the Mori-Tanaka method and the composite cylinder assemblage (CCA) model.

The FEA is widely used in modelling of composites and studies of their thermal and mechanical properties. It was successfully used in studying porosity[11] and in modeling of woven composites[12]. However there are some difficulties during developing a good model of the composite due to anisotropy of its constituents and complex structures.

The homogenization scheme for determination of the poroelastic properties of the materials with transversely isotropic constituents (geomaterials or rock-like composites with arbitrary-oriented ellipsoidal inhomogenities and pores) is proposed in[13]. These and other[14, 12] studies are based on the well-known Mori-Tanaka method[15].

For composites with aligned cylindrical inclusions CCA is developed and it was discussed in [16]. This model does not include the presence of pores in the composite.

STUDIES ON COMPOSITE ARCHITECTURE

C/C composites considered in this paper are made of carbon fiber preform densified by a CVI process (Figure 1a). During the infiltration using an ethylene precursor, at a high temperature and pressure the PyC deposits on the preform (more about it in[17, 18]) and builds the matrix of the composite. The carbon fiber preform plays the role of a skeleton of the composite.

Carbon fiber and PyC have different mechanical properties based on different arrangements of chemical bonds between atoms. The carbon fibers are extremely anisotropic, they are very strong in the longitudinal direction and soft in the cross section direction, and the PyC is also anisotropic, but the stiffness in the direction of the deposition and in the perpendicular plane differ only by a factor two and they are of the same order as the stiffness of the carbon fiber in the cross section.

The studied carbon fiber preform is made of layers of the unidirectional carbon fiber and the felt which is needle-punched. The direction of the unidirectional layers changes in the plane perpendicular to punching (direction 3) and it corresponds to the angles 0/0/90/90° (Figure 1b). Here "unidirectional" does not mean the perfect unidirectional orientation of the fibers but only the orientation of most of fibers, i.e. the preferred orientation of the fibers in the layer.

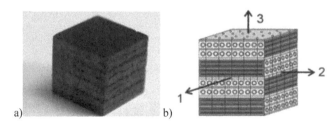

Figure 1. (a) C/C composite; (b) Schematic presentation of the C/C composite.

The studies on the fiber distribution were provided using micro computer tomography (μCT). After image processing the 3D discrete orientation distribution function in the terms of the Euler angles is obtained[9, 10]. The distribution function is built on a sphere, which surface is divided into

N hexagons of equal areas. Then the number of the fibers oriented in the direction perpendicular to each hexagon is calculated and normalized (Figure 2a).

During the infiltration process micropores are formed depending on CVI parameters. Porosity is one of the important characteristics of the material. The elastic parameters of the composite are strongly dependent on the volume fraction, the shape and the orientation of the pores. The volume fractions of the pores were obtained experimentally by the water impregnation method, and by metallographic and µCT studies[9].

The morphological studies of the microporosity were provided using µCT. During this procedure the pore surfaces were identified, labeled and the shape of each pore was stored as a cloud of points. In order to provide statistical studies and to simplify calculation procedures the pores were fitted to oblate and prolate spheroids of the same volume and preserving the orientation of the pore using principal component analysis (Figure 2b) [10].

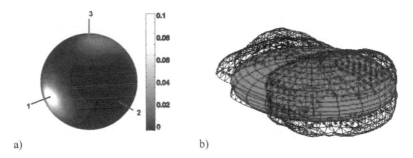

a)                              b)

Figure 2. (a) Experimental obtained fiber distribution function; (b) Approximation of a prolate pore by an ellipsoid.

This procedure was provided only for pores situated inside the unidirectional layers. Most of the pores in the felt layers are difficult to approximate by ellipsoids, because their shape is much different from a convex one (see studies and modeling of the irregular porosity in ref[10]).

The provided studies give us understanding of the composite architecture (layers, pores, volume fractions of the constituents, orientation and placing) which are used in the homogenization procedure.

MATERIAL MODELING

The homogenization procedure for calculation of the effective properties of the material which consists of fibers, pores and PyC matrix was performed using either the Mori-Tanaka method, FEA or CCA.

3D UNIT CELL PRESENTATION OF THE MICROSTRUCTURE

The modeling was provided using ABAQUS software. In this environment the n fibers as non-intersectable cylinders were generated as randomly distributed in a cube with orientation vectors following a normal distribution $N(a,\sigma)$ along the direction of the axis, where the parameters of the distribution are taken from the experimentally obtained distribution function: $a$ is the mean value (the UD direction) and $\sigma$ is the mean deviation of the axis from the UD direction in terms of the body angle. Then cylinders of greater radii (PyC coating) were created using the coordinates of the fibers and merged with fibers in the unit structure (Figure 3).

Figure 3. Model of a unit cell structure.

In the next step the surfaces of the unit cell were subjected to 6 kinds of loading: 3 axial and 3 shear loadings:

$$
\begin{aligned}
&X^- :u_x = 0;\ \ X^+ :u_x = \varepsilon^{(0)} \cdot a;\ \ X^-, X^+, Y^-, Y^+, Z^-, Z^+ :u_y = 0;\ u_z = 0;\\
&Y^- :u_y = 0;\ \ Y^+ :u_y = \varepsilon^{(0)} \cdot a;\ \ X^-, X^+, Y^-, Y^+, Z^-, Z^+ :u_x = 0;\ u_z = 0;\\
&Z^- :u_z = 0;\ \ Z^+ :u_z = \varepsilon^{(0)} \cdot a;\ \ X^-, X^+, Y^-, Y^+, Z^-, Z^+ :u_y = 0;\ u_x = 0;\\
&X^- :u_y = 0;\ \ X^+ :u_y = \varepsilon^{(0)} \cdot a;\ \ X^-, X^+, Y^-, Y^+, Z^-, Z^+ :u_x = 0;\ u_z = 0;\\
&Y^- :u_z = 0;\ \ Y^+ :u_z = \varepsilon^{(0)} \cdot a;\ \ X^-, X^+, Y^-, Y^+, Z^-, Z^+ :u_y = 0;\ u_x = 0;\\
&Z^- :u_x = 0;\ \ Z^+ :u_x = \varepsilon^{(0)} \cdot a;\ \ X^-, X^+, Y^-, Y^+, Z^-, Z^+ :u_y = 0;\ u_z = 0;
\end{aligned}
\tag{1}
$$

The values of the strains and the stresses were obtained by FEM and averaged by the volume of the cell. Then they were used in the tensor equation $\sigma = C : \varepsilon$ (Hookes law), which forms 36 linearly independent equations, the 36 unknowns being the components of the stiffness tensor[11].

MORI-TANAKA METHOD

The homogenization procedure consists of two steps: homogenization of the fiber and PyC matrix and then embedding pores into the obtained material.

In the first step the Mori-Tanaka scheme for n-phase composite was utilized for fibers oriented according to a discretized distribution function. Each set of fibers in a given direction was assumed to be a separate inclusion. The average stiffness tensor is given by

$$
\mathbf{C}^* = \mathbf{C}^M + \sum_{i=1}^{n_f} f_i^* \left( \mathbf{C}_i^F - \mathbf{C}^M \right) : \mathbf{B}_i : \left( \mathbf{I} + \sum_{i=1}^{n_f} f_i^* \left( \mathbf{B}_i - \mathbf{I} \right) \right)^{-1},
\tag{2}
$$

$$
\mathbf{B}_i = \left( \mathbf{I} + \mathbf{P}_i^M : \left( \mathbf{C}_i^F - \mathbf{C}^M \right) \right)^{-1},\quad \mathbf{P}_i^M = \mathbf{s}_i^{EshC} : \left( \mathbf{C}^M \right)^{-1},
$$

$C^M, C_i^F$ are the stiffness of the PyC matrix and fibers in the i-th orientation, $n_f$ – the number of discretized directions of the fiber, $f_i^* = f_i / (1 - v_{pores})$ - the volume fractions of the fibers in the i-th orientation in the composite without pores ($f_i$ are the volume fractions of the fibers in the composite in the i-th orientation, $\sum_{i=1}^{n_f} f_i = v_{fiber}$), $s_i^{EshC}$ is the Eshelby tensor for the fiber in the i-th orientation in PyC matrix.

In this case the Eshelby tensor can be calculated with explicit expressions[15] for cylindrical inclusion in an isotropic matrix and then rotated in the i-th direction[19].

Rotation for the stiffness matrix $C_{ijkl}$ and the Eshelby tensor $s^{EshC}$ to i-th orientation can be performed with the transformation $Q_{ijkl} = a_{ip} a_{jq} a_{kr} a_{js} Q_{pqrs}$ accordingly to the rotation matrix $a_{ij}(\theta, \varphi)$ where $\theta$ and $\varphi$ are the Euler rotational angles[14].

To obtain the stiffness tensor of the PyC matrix the stiffness tensor of PyC $C_{pqrs}^{PyC}$ is averaged in all possible directions:

$$C_{ijkl}^M = \frac{1}{4\pi} \int_0^{2\pi}\int_0^{\pi} a_{ip} a_{jq} a_{kr} a_{js} C_{pqrs}^{PyC} \sin\theta \, d\theta \, d\varphi. \tag{3}$$

In the second step the pores are embedded into the material obtained in the first step using the Mori-Tanaka method. The pores were interpolated by ellipsoids.

$$C^{eff} = C^* - \left(\sum_{i=1}^{n_p} f_i^P C^* : A_i^P\right) : \left(I + \sum_{i=1}^{n_p} f_i^P \left(A_i^P - I\right)\right)^{-1}, \tag{4}$$

$A_i^P = \left(I - s_i^{EshS}\right)^{-1}$, $C^*$ is the stiffness of the homogenized material in the first step, $s_i^{EshS}$ is the Eshelby tensor for a spheroid in the i-th orientation, $n_p$ – the number of pores, $f_i^P$ is the volume fraction of the pores (approximated by the spheroid) in the i-th orientation

The calculations performed using the two-step homogenization procedure include calculations of the Eshelby tensor for fibers in isotropic matrix $C^M$ and then calculation of the Eshelby tensor for pores in the anisotropic matrix $C^*$. The expressions for the Eshelby tensor for regular inclusions in the isotropic matrix can be obtained analytically[15], but for regular (ellipsoidal) pores in the anisotropic matrix the Eshelby tensor can be obtained only numerically[19].

The coefficient of the thermal expansion for the composite with unidirectional fiber according to the Mori-Tanaka model can be expressed in the form[20]:

$$\alpha^* = \alpha^M + \left[(1 - v_{fiber})H^{-1} + (\alpha^F - \alpha^M)\right]^{-1}, \tag{5}$$

$$H = (S^F - S^M) : C^M : T : S^M : (S^F - S^M)^{-1} : (\alpha^F - \alpha^M), \quad T = \left[I + s^{EshC} : S^M : (C^F - C^M)\right]^{-1},$$

Where $S^F = (C^F)^{-1}$ and $S^M = (C^M)^{-1}$ are compliances and $\alpha^M$, $\alpha_i^F$ are CTEs of the PyC matrix and fibers.

## CCA MODEL

CCA is a micromechanical approach to determine the effective properties of CVI densified carbon-carbon composites, which was proposed by Hashin[21] and used for unidirectional composites by Tsukrov[16]

A composite cylinder consists of a fibre surrounded by concentric layers of a PyC matrix. According to the model, the volume of the composite material can be filled with composite cylinders with diminishing radii. If these cylinders are chosen to be geometrically similar, then the entire construction is a CCA with the effective properties of any member composite cylinder. Assuming that the assemblage is transversely isotropic, the cylinder is subjected to three loading cases: uniaxial elongation, transverse hydrostatic loading and axial shear. The elasticity solutions to these loading cases give the effective elastic properties of the composite. The loading case of transverse shear is solved by Generalised self consistent scheme since the composite cylinder is not CCA-admissible under transverse shear[16, 21].

## NUMERICAL RESULTS

Numerical results are provided for UD layer of the sample produced of 2D preform infiltrated during 60 hours by pressure of 20kPa at constant temperature of 1095°. Mechanical and thermal properties of the carbon fiber and PyC[6, 16, 22] are given in Table I. Properties of PyC are averaged in the plane perpendicular to fiber accordingly to (3) (integration was provided only over $\theta$).

Table I. Mechanical properties of carbon fiber and PyC at ambient temperature.

|  | $E_1 = E_2$ [GPa] | $E_3$ [GPa] | $G_{12} = G_{32}$ [GPa] | $G_{13}$ [GPa] | $G_{23}$ [GPa] | $\alpha_1 = \alpha_2$ [$10^{-6}$/K] | $\alpha_3$ [$10^{-6}$/K] |
|---|---|---|---|---|---|---|---|
| Fiber | 14.7 | 204 | 22.2 | 22.2 | 5 | 8.9 | -0.67 |
| PyC$^{22}$ | 14.58 | 27.01 | 5.9 | 5.9 | 4.9 |  |  |
| PyC$^{16}$ | 15.11 | 15.11 | 5.9 | 5.9 | 5.9 | 3.3 | 3.3 |

Figure 4. FEM-mesh generated on the unidirectional unit cell.

The volume fractions of the fiber and pores were obtained experimentally. The volume fraction of the fiber in the UD layer is $v_{fiber} = 0.23$ and the volume fraction of pores is $v_{pores} = 0.08$. The fibers are distributed randomly and parallel to direction 3.

The results presented in the Table II were obtained for the composite without pores by the Mori-Tanaka method (the first step (2)), CCA model and for those with porosity by the Mori-Tanaka method (the second step (4)) and by FEA (Figure 4).

The thermal expansion coefficients (Table III) are obtained by CCA model and the Mori-Tanaka method (5) for different properties of the PyC and compared with ref[23, 24].

Table II. Calculated Young's modulus for composite with PyC[22]

| | Without porosity | | With porosity | |
|---|---|---|---|---|
| | $E_1, E_2$ [GPa] | $E_3$ [GPa] | $E_1, E_2$ [GPa] | $E_3$ [GPa] |
| Mori-Tanaka | 15.194 | 70.371 | 13.996 | 58.373 |
| CCA | 14.796 | 75.315 | – | – |
| FEA | – | – | 23.969 | 59.843 |

Table III. Thermal expansion coefficient of the material without pores

| | PyC[16] | |
|---|---|---|
| | $\alpha_1, \alpha_2$ [$10^{-6}$/K] | $\alpha_3$ [$10^{-6}$/K] |
| Mori-Tanaka | 5.32 | 0.14 |
| CCA | 5.34 | 0.05 |
| Ref [23, 24] | 5.2 | 0.05 |

CONCLUSION

Mechanical and thermal properties of C/C composite were obtained by the Mori-Tanaka, CCA and FEA methods for unidirectional composite. The results show good coincidence in the case of unidirectional fiber and can be generalized in the future work for calculation of the thermo-elastic properties of the C/C composite with non-aligned fiber.

ACKNOWLEDGEMENTS

The financial support of the DFG (DFG PI 785/3-2, PI 785/1-2)) is gratefully acknowledged. The authors thank S. Dietrich for the µCT studies.

REFERENCES
[1]W. Benzinger, KJ. Hüttinger, *Carbon*, **37**, 941-946 (1999).
[2]W. Benzinger, KJ. Hüttinger, *Carbon*, **37**, 1311-1322 (1999).
[3]B. Reznik, D. Gerthsen, KJ. Hüttinger, *Carbon*, **39**, 215-229 (2001).
[4]T. Gross et al, *Carbon*, **49**(6) (2011).
[5]R. Piat et al, *Carbon*, **41**( 9), 1858-1862 (2003).
[6]LH. Peebles, Carbon fibers: formation, structure, and properties, *CRC Press*, Boca Raton, 1994.
[7]G. Wagoner, R. Bacon, Elastic constants and thermal expansion coefficients of various carbon fibres. *In Extended abstracts of 19th biennial carbon conference*, 00296-297, 1989.
Peebles Carbon fibers: formation, structure, and properties, *CRC Press*, Boca Raton, 1994.
[8]WG. Zhang, KJ. Hüttinger, *Carbon*, **41**, 2325-2337 (2003).
[9]J-M. Gebert, Intrinsische Risse und Poren in Kohlenstoff Verbundwerkstoffen, In Werkstoffwissenschaft und Werkstofftechnik NR.068, *Shaker*, Aachen, 2011.
[10]S. Dietrich et al, *Composites Science and Technology*, **72**(15), 1892-1900 (2012).
[11]B. Drach et al, *Int. J. of Solids and Structures*, **48**, 2447-2457 (2011).

[12]J. Skoček, J. Zeman, M. Šejnoha, *Modelling and Simulations in Materials Science and Engineering*, **16**(8), 085002 (2008)

[13]A. Giraud et al, *International Journal of Solids and Structures*, **44**, 3756-3772 (2011).

[14]J. Schjodt-Thomsen,R. Pyrz *Mechanics of Materials*, **33**, 531-544 (2001).

[15]S. Nemat-Nasser, M. Hori, *North-Holland Elsevier*, Amsterdam, 1993

[16]I. Tsukrov, B. Drach, T. Gross. *Int Journal of Engineering Science*, **58**, 129–143 (2012).

[17]A. Li et al *Proceed Combustion Institute*, **33**, 1843–1850 (2001).

[18]A. Li, O. Deutschmann, *Chem Eng Sci*, **62**(18-20), 4976-4982 (2007).

[19]A.C. Gavazzi, D.C. Lagoudas. *Computational mechanics*, **7**, 1319 (1990),

[20]I. Sevostianov, *Mechanics of Materials*, **45**, 20-33 (2011).

[21]Z. Hashin, *Mechanics of Materials*, **8**, 293–308 (1990).

[22]J.-M. Gebert et al, *Carbon*, **48**(12), 3647-3650 (2010).

[23]Z. Hashin, *Journal of Applied Mechanics*, 50, 481-505 (1983).

[24]R.A. Schapery, *Journal of Composite Materials*, 2, 380 (1968).

# INELASTIC DESIGN OF MMCS WITH LAMELLAR MICROSTRUCTURE

Yuriy Sinchuk, Romana Piat
Institute of Engineering Mechanics, KIT
Karlsruhe, Germany

## ABSTRACT

The purposes of this paper are the optimal design and modeling of inelastic behavior of MMCs with lamellar microstructure. The material under consideration consists of two phases: ceramic ($Al_2O_3$) and aluminium alloy (Al12Si). For modeling of the elasto-plastic behavior of the metal the J2 flow theory was used. The brittle damage model was used for modeling of the ceramics failure. A single domain of this material with preferred orientation of lamellae was modeled using FE analysis and homogenization procedure. Results of the modeling were verified by comparing with experimental data and good correspondence between obtained results was observed. The optimization problem for the determination of the sample microstructure with minimal compliance was formulated. The design variables of the posed problem are local orientation of the lamellar domains and the local volume fractions of ceramic in the domain. Solution of the optimization problem is carried out for a prescribed global volume fraction of the ceramic in the whole specimen. Resulting optimal microstructures were obtained for different geometries of the specimen and for different loading cases. The obtained results show that the optimal microstructure design obtained using inelastic model is partially different from the elastic one.

## INTRODUCTION

Metal/ceramic composites (MMCs) offer several advantages over monolithic metals and their alloys – these include high specific stiffness and strength, better creep, fatigue and wear resistance, and good thermal properties[1, 2]. A further advancement has recently been achieved by the availability of open porous ceramic preforms fabricated by freeze-casting[3, 4]. The description of the freeze-casting process is given by Deville[5] and Wegst et al.[6]. Ceramic bodies obtained by freeze-casting of water suspensions exhibit a hierarchical lamellar structure composed of individual domains[7] within which the ceramic lamellae are oriented parallel to each other. The domain size and the lamellae spacing are functions of the freeze-casting parameters.

Several research studies have been carried out in the last few years to investigate the mechanical properties of MMCs fabricated by infiltrating Al alloy melt in freeze-cast ceramic preforms. Roy and Wanner[8] and Ziegler et al.[9, 10] studied the elastic anisotropy of single- and polydomain samples. The results showed that the stiffness of the composites is highest along the freezing direction. In the plane orthogonal to the freezing direction, stiffness depends strongly on the lamellae orientation – being highest (and similar to the stiffness when parallel to the freezing direction) along lamella orientation and lowest when perpendicular to lamella orientation. Elastic-plastic flow behavior and damage evolution under external uniaxial compression were studied by Roy et al.[11]. These results showed that along the freezing direction and along the lamella orientation in the plane that is normal to the freezing direction, the composites display a ceramic-controlled behavior with high compressive strength and limited or no ductility. Roy et al.[11] also studied the mechanism of internal load transfer in one single-domain sample. The plastic deformation starts in the aluminum solid phase and, correspondingly, more load is transferred to the stiffer alumina phase. In further studies, Launey et al.[12] showed that the 3-point bending strength of the composite is similar to alumina with fracture toughness about 40 MPa√m.

For better understanding and prediction of the mechanical behavior of these materials and also their targeted development, a material modeling and, based on it, a microstructure optimization should be provided. Microstructure modeling using homogenization or FE methods is most suitable for these studies. A review of homogenization methods in the context of the behavior of inelastic composites has been presented by Kanoute et al.[13]. Classical methods, such as asymptotic homogenization and mean-field approaches, and some recent developments in multiscale methods, such as $FE^2$ and transformation field analysis, are described. The internal microstructure of the composites has a highly significant impact on their mechanical properties and its understanding is essential for application of these materials. Analytical and semi-analytical methods take into account only the content of the micro constituents and make some approximations on their geometry. The results of the application of these methods for modeling elastic behavior of the given type of material are presented in[10, 14].

In this work, a classical secant homogenization approach[15, 16] is used for modelling inelastic behaviour of the microstructure. Some results of application of the secant method for MMC model with plastic and elastic phases are presented in [17]. Here the nonlinear homogenization method uses plastic and damage models for the simulation of inelastic behaviour of the metal and ceramic phases respectively.

The homogenization model has been used for microstructure optimization. A solution of the elastic optimal compliance problem[18], gradient-based optimization and the SIMP approach are used. Usual works of topology optimization only take the distribution of material volume fraction as the design variables. In ref. 14, the solution of the optimal design problem for elastic layered MCCs has been worked out using both the ceramic content and the material orientation as design variables. The topology optimization problem with volume fraction as a design variable for inelastic Voigt-Reuss homogenization model is considered in[19]. Here we are going to apply the same methodology as ref[19] for optimization with ceramic content and material orientation as design variables for the secant homogenization microstructure.

INELASTIC MATERIAL HOMOGENIZATION: SECANT METHOD

As the applied load increases, the elastic behaviour of the ceramic and metallic phases changes to inelastic. Plastic deformation of the metallic phase and brittle failure of the ceramic phase are the main reasons for this change.

Let $f_c, \sigma_c, \varepsilon_c$ and $f_m, \sigma_m, \varepsilon_m$ denote the content, stress, and strain fields of the ceramics and metal phases. The classical secant method based on following relations:

$$f_c + f_m = 1,$$
$$\boldsymbol{\sigma} = f_c \boldsymbol{\sigma}_c + f_m \boldsymbol{\sigma}_m,$$
$$\mathbf{C}_{\text{sec}} = f_c \mathbf{C}_c^{\text{sec}} : \mathbf{A}_c + f_m \mathbf{C}_m^{\text{sec}} : \mathbf{A}_m, \tag{1}$$

where $\boldsymbol{\sigma}$ is a total stress tensor, $\mathbf{C}_c^{\text{sec}}$ and $\mathbf{C}_m^{\text{sec}}$ are secant stiffness tensors of the ceramic and metal phases, respectively; $\mathbf{A}_c$ and $\mathbf{A}_m$ are the corresponding strain concentration tensors. Tensor $\mathbf{C}_c^{\text{sec}}$ is provided by the damage model. The secant tensor for the plastic metal phase[16], $\mathbf{C}_m^{\text{sec}}$ can be found in the form:

$$\mathbf{C}_m^{\text{sec}} = 2K_m \mathbf{I}^h + 2\mu_m \mathbf{I}^d,$$
$$\mu_m = \frac{\sigma_m^{eq}}{3\varepsilon_m^{eq}} = \frac{1}{2}\sqrt{\frac{\mathbf{s}_m : \mathbf{s}_m}{\mathbf{e}_m : \mathbf{e}_m}},$$
$$I_{ijkl}^h = \frac{1}{3}\delta_{ij}\delta_{kl}, \quad I_{ijkl}^d = \frac{1}{2}\left(\delta_{ik}\delta_{jl} + \delta_{il}\delta_{jk} - \frac{2}{3}\delta_{ij}\delta_{kl}\right). \tag{2}$$

where $\delta_{ij}$ is the Kronecker delta, and $K_m$ and $\mu_m$ are the secant bulk and shear moduli of the metal phase, respectively. In the last relation, $\mathbf{s}_m$ and $\mathbf{e}_m$ are the deviatoric parts of the tensors $\boldsymbol{\sigma}_m$ and $\boldsymbol{\varepsilon}_m$, respectively. For the calculation of the stress and strain fields into each micro constituent, the following system of nonlinear equations is solved iteratively [16]:

$$
\begin{aligned}
\boldsymbol{\varepsilon}_m &= \mathbf{A}_m : \boldsymbol{\varepsilon}, \\
\boldsymbol{\varepsilon}_c &= f_c^{-1}(\boldsymbol{\varepsilon} - f_m \boldsymbol{\varepsilon}_m), \\
\mathbf{C}_c^{\text{sec}} &= \mathbf{C}_c^{\text{sec}}(\boldsymbol{\varepsilon}_c), \quad \mathbf{C}_m^{\text{sec}} = \mathbf{C}_m^{\text{sec}}(\boldsymbol{\varepsilon}_m), \\
\mathbf{C}_{\text{sec}} &= \mathbf{C}_{\text{sec}}(\mathbf{C}_c^{\text{sec}}, \mathbf{C}_m^{\text{sec}}, f_c), \\
\mathbf{A}_m &= f_m^{-1}(\mathbf{C}_{\text{sec}} - \mathbf{C}_c):(\mathbf{C}_m^{\text{sec}} - \mathbf{C}_c)^{-1},
\end{aligned}
\tag{3}
$$

where $\boldsymbol{\varepsilon}$ is total strain tensor.

The components of the effective elastic stiffness tensor for the layered material are calculated analytically [18, 16] into the form:

$$
\begin{aligned}
(\mathbf{C}_{\text{sec}})_{ij} &= \sum_{k=1,2} f_k \left( C_{ij}^{(k)} - \frac{C_{i2}^{(k)} C_{j2}^{(k)}}{C_{22}^{(k)}} + \frac{C_{i2}^{(k)}\left(f_1 C_{2j}^{(1)}/C_{22}^{(1)} + f_2 C_{2j}^{(2)}/C_{22}^{(2)}\right)}{C_{22}^{(k)}\left(f_1/C_{22}^{(1)} + f_2/C_{22}^{(2)}\right)} \right), \quad i,j=1,2,3, \\
(\mathbf{C}_{\text{sec}})_{44} &= \frac{C_{44}^{(1)} C_{44}^{(2)}}{f_2 C_{44}^{(1)} + f_1 C_{44}^{(2)}}, \quad (\mathbf{C}_{\text{sec}})_{i4} = (\mathbf{C}_{\text{sec}})_{4i} = 0, \mathbf{C}^{(1)} = \mathbf{C}_c^{\text{sec}}, \mathbf{C}^{(2)} = \mathbf{C}_m^{\text{sec}}, f_1 = f_c, f_2 = f_m.
\end{aligned}
\tag{4}
$$

From the solution (3) the overall secant stiffness $\mathbf{C}_{\text{sec}}$ is obtained, and the total stress tensor can be updated according to $\boldsymbol{\sigma} = \mathbf{C}_{\text{sec}} : \boldsymbol{\varepsilon}$.

INELASTIC MATERIAL HOMOGENIZATION: TANGENT OPERATOR

The following expression can be obtained for the calculation of the tangent stiffness by the secant homogenization method:

$$
\mathbf{C}_{tg} = \frac{\partial \boldsymbol{\sigma}}{\partial \boldsymbol{\varepsilon}} = f_c \frac{\partial \boldsymbol{\sigma}_c}{\partial \boldsymbol{\varepsilon}_c} \frac{\partial \boldsymbol{\varepsilon}_c}{\partial \boldsymbol{\varepsilon}} + f_m \frac{\partial \boldsymbol{\sigma}_m}{\partial \boldsymbol{\varepsilon}_m} \frac{\partial \boldsymbol{\varepsilon}_m}{\partial \boldsymbol{\varepsilon}} = \mathbf{C}_c^{tg}\left(\mathbf{I} - f_m \frac{\partial \boldsymbol{\varepsilon}_m}{\partial \boldsymbol{\varepsilon}}\right) + f_m \mathbf{C}_m^{tg} \frac{\partial \boldsymbol{\varepsilon}_m}{\partial \boldsymbol{\varepsilon}} = \mathbf{C}_c^{tg} + f_m \left[\mathbf{C}_m^{tg} - \mathbf{C}_c^{tg}\right] \cdot \frac{\partial \boldsymbol{\varepsilon}_m}{\partial \boldsymbol{\varepsilon}},
$$

in which $\mathbf{C}_c^{tg}$ and $\mathbf{C}_m^{tg}$ are the algorithmic stiffness tangent operators provided by the corresponding plasticity (metal) and damage (ceramics) models. Viewed from a different side

$$
\mathbf{C}_{tg} = \frac{\partial(\mathbf{C}_{\text{sec}} : \boldsymbol{\varepsilon})}{\partial \boldsymbol{\varepsilon}} = \mathbf{C}_{\text{sec}} + f_c^{-1}\left(\frac{\partial \mathbf{C}_{\text{sec}}}{\partial \boldsymbol{\varepsilon}_c} : \boldsymbol{\varepsilon}\right) + f_m \left[f_m^{-1}\left(\frac{\partial \mathbf{C}_{\text{sec}}}{\partial \boldsymbol{\varepsilon}_m} : \boldsymbol{\varepsilon}\right) - f_c^{-1}\left(\frac{\partial \mathbf{C}_{\text{sec}}}{\partial \boldsymbol{\varepsilon}_c} : \boldsymbol{\varepsilon}\right)\right] \cdot \frac{\partial \boldsymbol{\varepsilon}_m}{\partial \boldsymbol{\varepsilon}},
$$

derivative $\dfrac{\partial \boldsymbol{\varepsilon}_m}{\partial \boldsymbol{\varepsilon}}$ is obtained from the last two relations in the following

$$
\mathbf{C}_c^{tg} + f_m\left[\mathbf{C}_m^{tg} - \mathbf{C}_c^{tg}\right] \cdot \frac{\partial \boldsymbol{\varepsilon}_m}{\partial \boldsymbol{\varepsilon}} = \mathbf{C}_{\text{sec}} + f_c^{-1}\left(\frac{\partial \mathbf{C}_{\text{sec}}}{\partial \boldsymbol{\varepsilon}_c} : \boldsymbol{\varepsilon}\right) + f_m \left[f_m^{-1}\left(\frac{\partial \mathbf{C}_{\text{sec}}}{\partial \boldsymbol{\varepsilon}_m} : \boldsymbol{\varepsilon}\right) - f_c^{-1}\left(\frac{\partial \mathbf{C}_{\text{sec}}}{\partial \boldsymbol{\varepsilon}_c} : \boldsymbol{\varepsilon}\right)\right] \cdot \frac{\partial \boldsymbol{\varepsilon}_m}{\partial \boldsymbol{\varepsilon}}
$$

and

$$\frac{\partial \varepsilon_m}{\partial \varepsilon} = f_m^{-1} [\mathbf{L}_m - \mathbf{L}_c]^{-1} : [\mathbf{C}_{\text{sec}} - \mathbf{L}_c],$$

$$\mathbf{L}_m = \mathbf{C}_m^{tg} - f_m^{-1} \left( \frac{\partial \mathbf{C}_{\text{sec}}}{\partial \varepsilon_m} : \varepsilon \right), \quad \mathbf{L}_c = \mathbf{C}_c^{tg} - f_c^{-1} \left( \frac{\partial \mathbf{C}_{\text{sec}}}{\partial \varepsilon_c} : \varepsilon \right). \tag{5}$$

## DAMAGE OF THE CERAMIC PHASE

For modeling of the damage in the ceramic phase we utilize a model initially proposed by Lapczyk and Hurtado[20] with Hashin's initiation criteria for damage initiation. For our composites, we suppose that a crack can propagate in directions transverse to the lamella plane only and that there is no debonding and no damage in the metal phase. Only the monotonic loading problem is considered. In each material point, the degradation of the ceramic phase because of tension or compression is described by scalar damage variable $d \in [0,1]$. Account these assumptions, the damaged stiffness matrix are simplified to the following form:

$$\mathbf{C}_c^{\text{sec}} = [\mathbf{C}_c^{\text{sec}}] = \frac{1}{D} \begin{bmatrix} (1-d)E_c & (1-d)v_c E_c & 0 \\ (1-d)v_c E_c & E_c & 0 \\ 0 & 0 & D(1-d)G_c \end{bmatrix},$$

$$D = 1 - (1-d)v_c^2, \tag{6}$$

where $E_c, G_c, v_c$ are the corresponding Young and shear moduli and the Poisson ratio of the undamaged ceramics. Hashin's damage initiation criteria have been reduced to following:

$$F^T = \left( \frac{\hat{\sigma}_{11}}{X^T} \right)^2 = 1, \quad \hat{\sigma}_{11} \geq 0,$$

$$F^C = \left( \frac{\hat{\sigma}_{11}}{X^C} \right)^2 = 1, \quad \hat{\sigma}_{11} < 0, \tag{7}$$

where $\hat{\sigma}_{11} = \frac{1}{1-d} \sigma_{11}$ is the effective stress in the direction of lamella orientation, $\sigma_{11}$ is the true stress component, and $X^T, X^C$ – are the average tensile and compressive strengths for the ceramic phase. If $d = 0$, $\hat{\sigma} = \sigma^0 = \mathbf{C}_c^0 : \varepsilon_c$, in which $\mathbf{C}_c^0$ is the stiffness tensor of the undamaged ceramics. The damage variable $d$ is defined by[20]:

$$d = \frac{\delta^f (\delta - \delta^0)}{\delta(\delta^f - \delta^0)}; \quad \delta^0 \leq \delta \leq \delta^f, \tag{8}$$

where the so called equivalent displacement $\delta$ can be obtained using the following relations:

$$\delta = l_c \langle \varepsilon_{11} \rangle, \quad \delta^0 = \frac{\delta}{\sqrt{F^T}}, \quad \delta^f = \frac{2G^T}{X^T}, \quad \hat{\sigma}_{11} \geq 0,$$

$$\delta = l_c \langle -\varepsilon_{11} \rangle, \quad \delta^0 = \frac{\delta}{\sqrt{F^C}}, \quad \delta^f = \frac{2G^C}{X^C}, \quad \hat{\sigma}_{11} < 0, \tag{9}$$

In (9) $\delta^0$ and $\delta^f$ are the equivalent displacements for the damage initiation and completely damaged states, $l_c$ is the characteristic length of a finite element, and $\langle \varepsilon_{11} \rangle = 0.5(\varepsilon_{11} + |\varepsilon_{11}|)$. Introducing the parameter $l_c$ is necessary to overcome the mesh dependency that appears due to the strain localization effect. However, in general, this approach does not guarantee protection from the mesh dependency of the result, and solution control by mesh refinement can be required.

For the sake of stabilizing of the computation process, an artificial viscosity in the evaluation of the damage variable is involved[20] in the form:

$$\dot{d}^v = (d - d^v)/\eta, \tag{10}$$

in which $\eta$ is the viscosity coefficient. The equation (10) is discretized by the backward Euler scheme for the pseudo-time step $\Delta t_{j+1} = t_{j+1} - t_j$ in the following way:

$$d_{j+1}^v = \frac{\Delta t_{j+1}}{\eta + \Delta t_{j+1}} d_{j+1} + \frac{\eta}{\eta + \Delta t_{j+1}} d_j^v.$$

Notethat for sensitivity analysis on each time increment we have $\dfrac{\partial d^v}{\partial d} = \dfrac{\Delta t}{\eta + \Delta t}$.

Due to the presence of artificial viscosity, the sensitivity analysis of the damage process will be much more complicated. So this regularization will not be considered ($\eta \equiv 0$) in the calculated analytical derivative, and, accordingly, excluded from the damage model during optimization process.

CALCULATION OF THE DAMAGE VARIABLE

Regarding the damage initiation increment, after calculation of $\boldsymbol{\sigma}_c = \mathbf{C}_c^{sec} : \boldsymbol{\varepsilon}_c$, it is possible to update $\widehat{\sigma}_{11}$, recalculate $\delta^0$, and to store it for the next increments. In this way, $\delta^0$ will be updated for each FEM-Newton's iteration of this increment. We have found a way to calculate $d$ analytically and obtain a value close to the one provided by recalculating $\delta^0$ for each time increment. The damage variable can be calculated analytically from the following set of equations:

$$\delta^0 = \frac{\delta\, X^I (1-d)}{|\sigma_{11}|}, \quad I \in \{T, C\},$$
$$d = d(\delta^0), \tag{11}$$
$$\boldsymbol{\sigma}_c = \mathbf{C}_c^{sec}(d) : \boldsymbol{\varepsilon}_c,$$

in which index $I$ indicates tension or compression loading.

After some algebraic transformations, and with $P = \dfrac{2 v_c^2 G^I}{1 - v_c^2}$ the system (11) can be rewritten in the form of a 2$^{nd}$ order equation with unknown $\delta^0$:

$$\delta^0 |\sigma_{11}^0| + P \frac{\delta}{v_c^2 \delta^0} = P + |\sigma_{11}^0| \delta^f + X^I \delta. \tag{12}$$

Between the two roots of the equation (2) that were found, it is necessary to select the one that belongs to the interval $(0, \delta^f)$. After elimination of $\delta^0$ and $\sigma_{11}$ from the system (11), the equation for $d$ can be rewritten in the form:

$$d \frac{P\delta}{\delta^f} + \frac{\delta^f |\sigma_{11}^0| - 2G^I}{d} = P + |\sigma_{11}^0| \delta^f - X^I \delta. \tag{13}$$

The solution of this equation, which belongs to the interval $(0,1)$, is the desired solution. This method requires more computational efforts but it is stable with respect to time increment changes.

CALCULATION OF THE TANGENT OPERATOR FOR THE DAMAGE MODEL
The constitutive tangent operator is calculated in form:

$$\mathbf{C}_c^{tg} = \frac{\partial \boldsymbol{\sigma}_c}{\partial \boldsymbol{\varepsilon}_c} = \mathbf{C}_c^{sec} + \left( \frac{\partial \mathbf{C}_c^{sec}}{\partial d^v} \right) : \left( \frac{\partial d^v}{\partial d} \frac{\partial d}{\partial \boldsymbol{\varepsilon}_c} \right) \tag{14}$$

The derivative $\dfrac{\partial d}{\partial \boldsymbol{\varepsilon}_c}$ can be calculated using the implicit differentiation rule in following:

$$\frac{\partial d}{\partial \boldsymbol{\varepsilon}_c} = -\left( d^2 \frac{\partial Q_2}{\partial \boldsymbol{\varepsilon}_c} + d \frac{\partial Q_1}{\partial \boldsymbol{\varepsilon}_c} + \frac{\partial Q_0}{\partial \boldsymbol{\varepsilon}_c} \right) \Big/ (2dQ_2 + Q_1)$$

$$Q_0 = \delta^f |\sigma_{11}^0| - 2G^I, \quad Q_1 = X^I \delta - P - |\sigma_{11}^0| \delta^f, \quad Q_2 = \frac{P\delta}{\delta^f}$$

$$\frac{\partial Q_0}{\partial \boldsymbol{\varepsilon}_c} = \delta^f \frac{\partial |\sigma_{11}^0|}{\partial \boldsymbol{\varepsilon}_c}, \quad \frac{\partial Q_1}{\partial \boldsymbol{\varepsilon}_c} = X^I \frac{\partial \delta}{\partial \boldsymbol{\varepsilon}_c} - \frac{\partial Q_0}{\partial \boldsymbol{\varepsilon}_c}, \quad \frac{\partial Q_2}{\partial \boldsymbol{\varepsilon}_c} = \frac{P}{\delta^f} \frac{\partial \delta}{\partial \boldsymbol{\varepsilon}_c}.$$

with $\dfrac{\partial |\sigma_{11}^0|}{\partial \varepsilon_{ij}} = \mathrm{sign}(\sigma_{11}^0) \left( C_c^0 \right)_{11ij}$ and

$$\frac{\partial \delta}{\partial \varepsilon_{ij}} = \begin{cases} l_c, & \text{if } i = j = 1, \ \hat{\sigma}_{11} \geq 0 \text{ and } \varepsilon_{11} > 0, \\ -l_c, & \text{if } i = j = 1, \ \hat{\sigma}_{11} < 0 \text{ and } \varepsilon_{11} < 0, \\ 0, & \text{in other cases.} \end{cases}$$

For the procedure with $\delta^0 = \text{const}$ after damage onset, from (8), we have

$$\frac{\partial d}{\partial \boldsymbol{\varepsilon}_c} = \frac{\delta^f \delta^0}{(\delta^f - \delta^0) \delta^2} \frac{\partial \delta}{\partial \boldsymbol{\varepsilon}_c}. \tag{15}$$

## NUMERICAL MODELING OF THE INELASTIC MATERIAL BEHAVIOR OF THE SINGLE DOMAIN

In our numerical studies, we have considered a single-domain microstructure (size 1.25x1.25 mm) loaded by compression at the top and supported at the bottom. The effective stress-strain behaviors of the differently-orientated domains with $f_c = 0.35$ were calculated using the previously described homogenization method and the ABAQUS FE model. The material behavior of the domain with elastic ceramic and plastic metal behavior was calculated using the secant method with $\mathbf{C}_c^{sec} \equiv \mathbf{C}_c^0$. Parameters of the metal measured by Ziegler[10] (Al alloy stress-strain response for unidirectional loading) were used as input for plasticity modeling. The resulting stress-strain behavior of the domain is presented in Figure 1a. For the domain with plastic metal behavior and damage of the ceramics following parameters for damage model were utilized: $X^T = 180\,\text{MPa}$, $X^C = 1370\,\text{MPa}$, $G^T = 70\,\text{N/m}$, $G^C = 150\,\text{N/m}$, $\eta = 0.01$. The result of the calculations is presented in Figure 1b. The ABAQUS FE modeling of the domain was provided using the damage model with Hashin's initiation criteria. The stress-strain responses for the FE model were averaged over all integration points of the domain; and the obtained result is presented in Figure 1c.

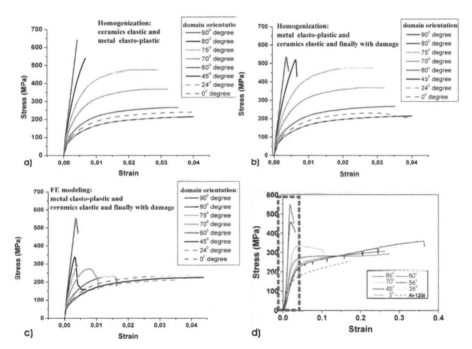

Figure 1. Stress-strain behavior of the single domain for different lamella orientations: calculated by homogenization for a) elastic ceramics and plastic metal behavior; b) for plastic metal and damage ceramics behavior; c) FEM modeling for plastic metal and damage ceramics behavior; d) Measured compressive stress–strain response for different domain orientations[18].

An analysis of the obtained stress-strain behavior shows that for the homogenization including damage, only 90°, 80° and 0° degree models demonstrate significant influence of the damage, whereas, by FE modeling, the influence of the damage can be clearly observed for all orientations between 60° and 90° (compare Fig.1b and c). These differences in damage response for domain orientations between 60° and 90° may be caused by the strain localization effect induced by the influence of the boundary condition of FE modeling. Also the experimentally-obtained stress-strain behavior for different lamella orientations[21] was utilized (see Figure 1d). The experimentally measured strains correspond to large deformations, and the calculated stress-strain behavior corresponds only to the part of the graph Fig. 1d that is highlighted by a red dashed rectangular zone, and for this reason, matching the numerical and experimental results is difficult. Nevertheless, the common nature of the behavior can be observed.

INELASTIC MATERIAL OPTIMIZATION: DIRECT DIFFERENTIAL METHOD

The proposed inelastic material behavior model can be used for microstructure optimization. The minimum compliance problem that was solved for elastic material model [14] will be solved for inelastic behavior. The design variables $\mathbf{b}$ are the local ceramic content and lamella orientation for given boundary conditions and ceramic content in the entire specimen. As in [14], we utilize a 2D formulation and suppose that the specimen occupies a domain $\Omega$ and is loaded by force (e.g., three-point bending test, see Fig. 2a). We assume that the overall response of the microstructure can be calculated by the model described above, which is path-independent for $\eta=0$ in the damage model. This assumption is significantly simplifies the analytical derivative calculations. The direct differentiation method [19] is used for sensitivity analysis. The objective function of the minimization problem is the global strain energy:

$$F_E(t) = \int_0^t \int_\Omega \boldsymbol{\sigma} : \dot{\boldsymbol{\varepsilon}} \, d\Omega \, d\tau, \tag{16}$$

in which $\tau$ is a pseudo-time variable. Using the results of the boundary value problem from the FEM analysis, the strain energy for a the given time increment $n+1$ can be presented in the form:

$$(\Delta F_E)_{n+1} = \frac{1}{2} \left( \mathbf{f}_n^{ext} + \mathbf{f}_{n+1}^{ext} \right) \cdot \Delta \mathbf{u}_{n+1} = \mathbf{f}_{n+1/2}^{ext} \cdot \Delta \mathbf{u}_{n+1}, \tag{17}$$

in which $\Delta \mathbf{u}_{n+1}$ is the displacement increment and $\mathbf{f}_{n+1}^{ext}$ is the increment of loading.
The derivative of the incremental energy with respect to $\mathbf{b}$ has the form

$$\frac{d(\Delta F_E)_{n+1}}{d\mathbf{b}} = \mathbf{f}_{n+1/2}^{ext} \cdot \frac{d\Delta \mathbf{u}_{n+1}}{d\mathbf{b}}, \tag{18}$$

note that the external forces $\mathbf{f}_{n+1/2}^{ext}$ are independent from $\mathbf{b}$. According to the direct differentiation technology presented in [19] for path-independent models that describe the force-loaded problem, the derivative is expressed with respect to the design variable in the following matrix form:

$$\frac{d\Delta \mathbf{u}_{n+1}}{d\mathbf{b}} = -\mathbf{K}_{n+1}^{-1} \int_\Omega \mathbf{B}^T \frac{\partial \Delta \boldsymbol{\sigma}_{n+1}}{\partial \mathbf{b}} \, d\Omega, \tag{19}$$

in which $K_{n+1}$ is the tangent matrix obtained from the last iteration of the Newton method of nonlinear FEM analysis, and B is the matrix of nodal derivatives of the FEM interpolation basis functions.

It is worth noting that, for the given problem, $K_{n+1}$ is not symmetric because the tangent operator of the damage model is nonsymmetric. Now, the problem is reduced to the analytical calculation of the derivatives $\Delta\boldsymbol{\sigma}_n$ with respect to $b_i$.

INELASTIC MATERIAL OPTIMIZATION: SENSITIVITY ANALYSIS.

Consider an analysis for the local ceramic content design variable. The component $b_i$ of the unknown design variables vector $\mathbf{b}$ corresponds to the ceramic content in the element $i$ (with $i = 1,...,2N_{elem}$, in which $N_{elem}$ is the number of the finite elements in the model). The gradient of stress increment is calculated in a form that is convenient for secant-based homogenization: $\dfrac{\partial\Delta\boldsymbol{\sigma}_n}{\partial b_i} = \dfrac{\partial\boldsymbol{\sigma}_n}{\partial b_i} - \dfrac{\partial\boldsymbol{\sigma}_{n-1}}{\partial b_i}$. It is easy to show that, for monotonic load with equal force increment $\Delta\mathbf{f}^{ext} = \Delta\mathbf{f}_{n+1}^{ext} = \mathbf{f}_{n+1}^{ext} - \mathbf{f}_n^{ext}$, $n = \overline{1, N-1}$, ( $N$ – number of increments), (18) can be written in the form

$$\frac{dF_E(t_N)}{d\mathbf{b}} = \Delta\mathbf{f}^{ext} \cdot \left( \left( N - \frac{1}{2} \right) \frac{d\mathbf{u}_N}{d\mathbf{b}} - \sum_{n=1}^{N-1} \frac{d\mathbf{u}_n}{d\mathbf{b}} \right). \tag{20}$$

So, for each design variable $b_i$, $i = 1,...,2N_{elem}$, on the given time increment, it is necessary to solve the system of linear equations

$$K_n \frac{d\mathbf{u}_n}{db_i} = -\int_\Omega B^T \frac{\partial\boldsymbol{\sigma}_n}{\partial b_i} d\Omega. \tag{21}$$

For simplicity, we will omit indexes in following: $\dfrac{\partial\boldsymbol{\sigma}_n}{\partial b_i} = \dfrac{\partial\boldsymbol{\sigma}}{\partial f_c} = -\dfrac{\partial\boldsymbol{\sigma}}{\partial f_m}$, so $\dfrac{\partial\boldsymbol{\sigma}}{\partial f_m}$ refers to the stress derivative with respect to the metal content for any finite element $i$ and for any time increment $n$. The following set of equations is derived from the secant method (1-3):

$$\frac{\partial\boldsymbol{\sigma}}{\partial f_m} = \frac{\partial\mathbf{C}^{sec}}{\partial f_m} : \boldsymbol{\varepsilon} + \left( \frac{\partial\mathbf{C}^{sec}}{\partial\boldsymbol{\varepsilon}_c} : \boldsymbol{\varepsilon} \right) : \frac{\partial\boldsymbol{\varepsilon}_c}{\partial f_m} + \left( \frac{\partial\mathbf{C}^{sec}}{\partial\mathbf{c}_m} : \boldsymbol{\varepsilon} \right) : \frac{\partial\boldsymbol{\varepsilon}_m}{\partial f_m},$$

$$\frac{\partial\boldsymbol{\sigma}}{\partial f_m} = -\boldsymbol{\sigma}_c + \boldsymbol{\sigma}_m + f_c\mathbf{C}_c^{tg} : \frac{\partial\boldsymbol{\varepsilon}_c}{\partial f_m} + f_m\mathbf{C}_m^{tg} : \frac{\partial\boldsymbol{\varepsilon}_m}{\partial f_m}, \tag{22}$$

$$f_c \frac{\partial\boldsymbol{\varepsilon}_c}{\partial f_m} = \boldsymbol{\varepsilon}_c - \boldsymbol{\varepsilon}_m - f_m \frac{\partial\boldsymbol{\varepsilon}_m}{\partial f_m}.$$

From (22) follows $\dfrac{\partial \varepsilon_m}{\partial f_m} = f_m^{-1}\left[\mathbf{L}_m - \mathbf{L}_c\right]^{-1} : \left[\sigma_c - \sigma_m + \dfrac{\partial \mathbf{C}^{sec}}{\partial f_m} : \varepsilon - \mathbf{L}_c : (\varepsilon_c - \varepsilon_m)\right]$, with $\mathbf{L}_m, \mathbf{L}_c$ from (5).

The obtained derivative $\dfrac{\partial \varepsilon_m}{\partial f_m}$ is used for the calculation of $\dfrac{\partial \sigma}{\partial f_m}$.

Now, let us consider an analysis in with lamella orientation is the design variable. We need to calculate the stress gradient for the design variable lamella orientation $\dfrac{\partial \sigma_n}{\partial b_i}$, $i = N_{elem} + 1,...,2N_{elem}$. The stress tensor derivative $\dfrac{\partial \sigma_n}{\partial b_i}$ is written in the form $\dfrac{\partial \sigma_{glb}}{\partial \theta}$, that refers to the differentiation of the stress vector with respect to design variable lamella orientation $\theta$ for any finite element $i$ and for any time increment $n$ in the global coordinate system. Taking coordinate transformations into account, we obtain the following relation in the matrix form:

$$\dfrac{\partial \sigma_{glb}}{\partial \theta} = \dfrac{\partial Q'}{\partial \theta}\sigma_{mat} + Q'\dfrac{\partial \sigma_{mat}}{\partial \theta} = \dfrac{\partial Q'}{\partial \theta}C_{mat}^{sec}\varepsilon_{mat} + Q'\dfrac{\partial \sigma_{mat}}{\partial \varepsilon_{mat}}\dfrac{\partial \varepsilon_{mat}}{\partial \theta} =$$

$$= \left[\dfrac{\partial Q'}{\partial \theta}C_{mat}^{sec}Q + Q'C_{mat}^{tg}\dfrac{\partial Q}{\partial \theta}\right]\varepsilon_{glb} = C^{\theta}\varepsilon_{glb}. \tag{23}$$

The indexes *glb* and *mat* correspond to global and material coordinates; Q is the coordinate transformation matrix (between the global and the material coordinates) and $Q'$ is its transpose. For the linear elastic case $C_{mat}^{sec} = C_{mat}^{tg} = C^0$, and $C^{\theta}$ is reduced to the matrix used in the strain-based method [22].

## INELASTIC MATERIAL OPTIMIZATION: NUMERICAL RESULTS.

The solution of the optimal material orientation problem presents some difficulties. First of all, there are a lot of local minimuma of the objective function.

Another difficulty is related to the application of the gradient-based methods for the optimization problem with damage of the ceramics. The reason for this is that the first derivative of the strain energy (objective function) with respect to the design variable has a discontinuity due to the damage. To solve this problem, the optimization iteration process can be stopped near the point of discontinuity (because of the small step size), but the derivative of the objective function does not tend to zero near this point. Here we proposed to use the solution of the *principal stress directions* optimization method obtained for inelastic problem for the calculation of the initial orientation for starting the nonlinear optimization process using gradient-based optimization methods.

As a demonstration of the proposed methods, the minimum compliance problem for three-point bending (see Figure 2a) was solved. Only half of the specimen (size 28 mm x12 mm) was considered for reasons of symmetry. The load is applied into three nodes (490kN at each node, Figure 2a) as concentrated forces. The initial ceramic content and lamella orientation in all elements are 0.5 and 90°, respectively.

To solve the optimization problem, the interior-point method[23] from MATLAB Optimization Toolbox was used. This method transforms the constrained optimization problem into a sequence of equality-constrained problems which are easier to solve than the original inequality-constrained problem. The elastic solution obtained by the principal stress direction method was used as a starting point for the estimation of the optimal lamella orientation for inelastic material behavior. The optimal solutions estimated by different models of the ceramic and metal phases are presented in Figure 2. It can be observed that, in comparison to the elastic solution in regions with higher strains (see Figure 2c), the ceramic content rises due to the plasticity of the metal (see a short overview of the studies on the internal load transfer in composites in[17,24]).

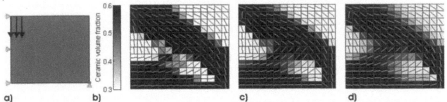

a)     b)     c)     d)

Figure 2. Microstructure optimization for three-point bending: a) geometry and boundary conditions; optimal microstructure for different behaviors of the phases: b) elastic of both phases; c) elastic for the ceramics and plastic for the metal; d) damage for the ceramics and plastic for the metal.

Table I. Strain energy in initial and optimized microstructures for different models of the phases and different design variables.

| In [J] | Elastic | Plastic | Damage |
|---|---|---|---|
| Initial strain energy | 106.83 | 272.31 | 272.56 |
| Principal stress direction (only one design variable: lamella orientation) | 75.63 (70,8%) | 90.93 (33,4%) | 99.22 (36,4%) |
| Full gradient-based optimization (two design variables) | 68.8 (64,4%) | 79.57 (29,2%) | 82.04 (30,1%) |

After the onset of damage, this phenomenon is enhanced: the damaged ceramics is weaker and, as a consequence, in the regions with higher strains and in its neighborhood, the ceramic content continues to rise. The decrease of the compliance energy by optimization is presented in Table 1. The column "damage" refers to the plastic behavior of the metal and damageable elastic behavior of the ceramics. A comparison of the optimal strain energy calculated for different numbers of the design variables shows the same tendency as for pure elastic material behavior. Lamella orientation has major influence on the resulting energy. From a practical point of view, the optimal design of the studied material can be based only on the optimization of the lamella orientation, using the principal stress direction method.

SUMMARY

A model for predicting the inelastic behavior of MCCs has bean formulated using the secant homogenization procedure, that includes plasticity and damage simulation of the micro phases behavior. Tangent stiffness operators are derived for secant homogenization method and for damage model. A method of analytical computation of the damage variable is proposed. A sensitivity analysis of the two-phase inelastic secant homogenized properties is performed with respect to two design variables: ceramic content and lamella orientation. An optimization procedure based on these sensitivities has been utilized for the optimal design problem.

The obtained results show that both FE and nonlinear homogenization (taking into account the J2 plasticity of the metal phase) are accurate and applicable for prediction of the inelastic behavior of the single-domain microstructure. The results of the microstructure optimization with inelastic material behavior show that the design variable lamella orientation has a more significant impact on the goal function than the ceramic content. The optimal lamellae orientations for elastic and inelastic models are close to each other.

ACKNOWLEDGEMENTS
The financial support of the DFG (DFG PI 785/3-2, PI 785/1-2)) is gratefully acknowledged.

REFERENCES
[1]N. Chawla, K.K. Chawla, Metal Matrix Composites, *Springer*, New York, 2006.

[2]A. Mortensen, J. Llorca, Metal Matrix Composites, *Annu. Rev. Mater. Res.*, **40**, 243-270 (2010).

[3]S. Deville et al, *Science*, **311**, 515-518 (2006).

[4]A. Mattern, Interpenetrating Metal-Ceramic Composites with Isotropic and Anisotropic $Al_2O_3$ Reinforcement (in German), Doctoral thesis, University of Karlsruhe (TH), Karlsruhe, Germany, 2005.

[5]S. Deville, Freeze-Casting of Porous Ceramics: A Review of Current. Achievements and Issues, *Adv. Eng. Mater.*, **10**, 155-169 (2008).

[6]U.G.K. Wegst et al, Biomaterials by freeze casting, *Phil. Trans. R. Soc. A*, **368**, 2099-2121 (2010).

[7]S. Deville, E. Saiz, A.P. Tomsia, Ice-Templated Porous Alumina Structures, *Acta Mater.*, **55**, 1965-74 (2007).

[8]S. Roy, A. Wanner, Metal/Ceramic Composites from Freeze-Cast Ceramic Preforms: Domain Structure and Elastic Properties, Compos. *Sci. Technol.*, **68**, 1136-43 (2008).

[9]T. Ziegler et al, Elastic constants of metal/ceramic composites with lamellar microstructures: finite element modelling and ultrasonic experiments, *Compos. Sci. Technol.*, **69**, 620-626 (2009).

[10]T. Ziegler, A. Neubrand, R. Piat, Multiscale homogenization models for the elastic behaviour of metal/ceramic composites with lamellar domains,*Compos. Sci. Technol.*, **70(4)**, 664-670 (2010).

[11]S. Roy, J. Gibmeier, A. Wanner, In situ Study of Internal Load Transfer in a Novel Metal/Ceramic Composite Exhibiting Lamellar Microstructure Using Energy Dispersive Synchrotron X-ray Diffraction, *Adv. Eng. Mater.*, **11**, 471-477 (2009).

[12]M.E. Launey et al, A Novel Biomimetic Approach to the Design of High-Performance Ceramic/Metal Composites, *J. R. Soc. Interface*, **7**, 741-753 (2010).

[13]P. Kanoute et al, Multiscale Methods for Composites: A Review, *Arch. Comput. Method. Eng.*, **16**, 31–75, (2009).

[14]R. Piat et al, Minimal compliance design for metal–ceramic composites with lamellar microstructures, *Acta Materialia*, **59(12)**, 4835-46, (2011).

[15]H. Moulinec, P. Suquet, Intraphase strain heterogeneity in nonlinear composites: a computational approach, *J. Mech. A Solids*, **22**, 751-770 (2003).

[16]J. Qu, M. Cherkaoui, Fundamentals of Micromechanics of Solids, *Wiley*, New York, 2006.

[17]R. Piat, Y. Sinchuk, Nonlinear homogenization of metal-ceramic composites with lamellar microstructure, *PAMM Proc. Appl. Math. Mech.*, **11**, 545-546 (2011).

[18]M.P. Bendsoe, O. Sigmund, Topology optimization, *Springer*, Berlin, 2003.

[19]C.C. Swan, I. Kosaka, Voigt-Reuss topology optimization for structures with nonlinear material behaviors, *Int. J. Numer. Meth. Eng.*, **40**, 3785–3814 (1997).

[20]I. Lapczyk, J.A. Hurtado, Progressive damage modeling in fiber-reinforced materials, *Composites A*, **38**, 2333-41 (2007).

[21]S. Roy, B. Butz, A. Wanner, Damage evolution and domain-level anisotropy in metal/ceramic composites exhibiting lamellar microstructures, *Acta Mater.*, **58**, 2300-12 (2010).

[22]P. Pedersen, On optimal orientation of orthotopic materials, *Struct. Optim.*, **1**, 101–106 (1989).

[23]R.H. Byrd, M.E. Hribar, J. Nocedal, An Interior Point Method for Large Scale Nonlinear Programming, *SIAM J. Optim.* **9**, 877-900 (1999).

[24]Y. Sinchuk et al, Inelastic behavior of the single domain of metal-ceramic composites with lamellar microstructure, *PAMM Proc. Appl. Math. Mech.*, **11**, 285-286 (2011).

# MULTI-SCALE MODELING OF TEXTILE REINFORCED CERAMIC COMPOSITES

J. Vorel[a], S. Urbanová[a], E. Grippon[b], I. Jandejsek[a], M. Maršálková[c], M. Šejnoha[a]

[a] Department of Mechanics, Faculty of Civil Engineering, Czech Technical University in Prague, Thákurova 7, 166 29 Prague 6, Czech Republic
[b] LCTS, Univ. Bordeaux, 3 alle de la Botie, 33600 Pessac, France
[c] Department of Textile Materials, Technical University in Liberec, Studentská 2, 461 17 Liberec, Czech Republic

ABSTRACT

The present paper describes a two-step homogenization for the evaluation of effective elastic properties of textile reinforced ceramic composites. Attention is devoted to polysiloxane matrix based composites reinforced by plain weave textile fabrics. Basalt and carbon reinforcements are considered. X-ray microtomography as well as standard image analysis is utilized to estimate the volume fraction, shape and distribution of major porosity which considerably influences the resulting macroscopic response. The numerical procedure effectively combines the Mori-Tanaka averaging scheme and finite element simulation carried out on a suitable statistically equivalent periodic unit cell. The computational strategy employs the popular extended finite element method to avoid difficulties associated with meshing relatively complex geometries on the meso-scale. Comparison with the results obtained directly from the finite element simulations of available $\mu$CT projections together with experimental data derived from measurements of phase velocities of ultrasonic waves is also provided.

## INTRODUCTION

High strength and chemical stability of ceramics at elevated temperatures have been shown to be advantageous for many applications in high temperature and environmentally hostile surroundings. However, intrinsic brittleness of ceramics generally hindered their direct extension to large structural components which are expected to support larger tensile as well as compressive stresses. This drawback has led to the development of a variety of fibrous ceramic matrix composites with properties that are far superior to those of their monolithic counterparts. While the in-plane local stresses in individual plies are usually found to be below the allowable limits, the tensile out-of-plane stresses often lead to delamination due to a low strength of unidirectional fibrous laminates in this direction. Together with the need for manufacturing sufficiently thick structural units, this deficiency considerably contributed to the introduction of more complex three-dimensional fiber preforms with woven textile composites representing one particular class of such systems. See e.g.[1,2] for modeling initiatives in the field of ceramic composites.

Recently, a new class of plain weave textile reinforced ceramics derived from polysiloxane matrices has emerged showing applicability in many engineering areas. In view of biomechanical applications, a number of studies have already suggested acceptable mechanical properties of these "inexpensive" material systems comparable to human bones as well as their satisfactory bio- compatibility and bioactivity.[3] Their potential use in high temperature applications is discussed in.[4]

Regardless of their final use, the mechanical properties of these systems depend on the processing route that often leads to complex microstructures with a relatively large intrinsic porosity and a number of other fabrication related imperfections, including non-uniform layer widths, tow undulation, inter-layer shift and nesting. This invites development of sufficiently accurate geometrical models and feasible computational procedures. Since this constitutes our current research objectives we limit our attention to polymeric preforms of these systems expecting rather similar microstructural configuration with their ceramic counterparts.

Following our recent developments in the modeling of carbon-carbon textile composites[5,6] we organize the paper into two principal sections devoted to an experimental and a computational part of this project, both being considered equally important and thus supporting each other. In particular, we begin in the experimental part by briefly reviewing the essential steps needed in the formulation of a suitable computational model on the meso-scale (the level of yarns) including the prediction of yarn effective stiffnesses through micromechanical homogenization. The true structure of a porous phase is examined next with the help of computational microtomography ($\mu$CT). To support the predictive power of the proposed numerical approach we finally outline the derivation of effective properties of the laminate using the ultrasonic wave speed measurements. The numerical part begins with the standard finite element homogenization of a periodic unit cell and primarily mentions the difficulties associated with the mesh generation, and consequently promotes the extended finite element technique as a suitable alternative. We finally summarize all the presented results together with some concluding remarks.

LABORATORY MEASUREMENTS

The experimental program considers two types of composites with carbon and basalt plain weave textile reinforcements. These are manufactured by molding together eight layers of fabric first impregnated by the silicon resin Lukosil M130 (polymethylsilicone resin dissolved in xylene) and cured at 200°C for 6h to form the polymer precursors. Such composites would be further treated at temperatures exceeding 1100°C in vacuum to build ceramic matrix based composites. This step, however, goes beyond the present scope. Instead, we limit our attention to polymeric matrices and concentrate on the computational part of this investigation to propose and validate a suitable numerical approach to the derivation of effective elastic properties of such systems. Yet, the experimental part cannot be excluded and will comprise several steps including image analysis, X-ray microtomography to acquire basic information about the volume and distribution of major porosity and the derivation of effective properties by exploiting the ultrasonic wave speed measurements. Individual steps will be briefly described next.

Image analysis and yarn effective properties

This part of the experimental program is concerned with the determination of basic geometrical parameters of so-called two-ply Statistically Equivalent Periodic Unit Cell (SEPUC). The basic building block of this representative volume element (RVE) is represented by a single-ply periodic unit cell (PUC) displayed in Fig. 1(a). To allow for a mutual shift and ply nesting observed for real systems, see Fig. 2, the four basic parameters were supplemented by three possible translations $\Delta_1$, $\Delta_2$ and $\Delta_3$ of the two plies in the direction of the corresponding coordinate axes. A two-dimensional cut by the plane $X_2 = \pm a$ is plotted in Fig. 1(b).

The adopted model of the unit cell thus involves seven independent parameters

$$\mathbf{y} = [a, b, g, h, \Delta_1, \Delta_2, \Delta_3], \tag{1}$$

to be determined from available microstructural data. These are usually provided by two-

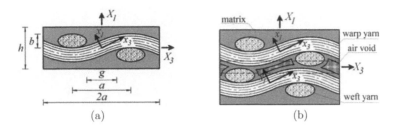

Figure 1. a) Single ply PUC,[5] b) two-ply SEPUC[6]

dimensional binary images taken either along the warp or weft direction, as seen in Fig. 2, for the two types of reinforcements. Matching, for example, the two-point probability functions of both the real system, e.g. Fig. 2(a), and SEPUC then allows for deriving the necessary parameters listed in Eq. (1). The elements of soft computing are typically employed to solve the associated minimization problem. Details can be found, e.g. in.[5,6] The resulting values adopted in the present study are shown in Table 1.

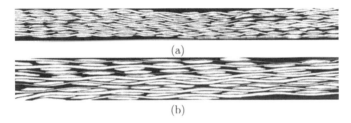

Figure 2. Binary images of eight-layer composites: a) Basalt, b) carbon fabric based textile composites

Table 1. Parameters of the periodic unit cell

| Parameter [$\mu$m] | | Basalt | Carbon |
|---|---|---|---|
| Yarn period | $(2a)$ | 1726 | 4072 |
| Yarn width | $(b)$ | 87 | 140 |
| Inter-yarn gap | $(g)$ | 312 | 490 |
| Layer height | $(h)$ | 183 | 314 |
| Horizontal shift | $(\Delta_{X_2}, \Delta_{X_3})$ | 417 | 14 |
| Vertical shift | $(\Delta_{X_1})$ | -53 | -63 |

It must be pointed out that so far the yarns in SEPUC as well as in binary images have been treated as a homogeneous material. Nevertheless, the yarn itself is a composite consisting of aligned transversely isotropic fibers bonded to an isotropic matrix. Images taken at the level of yarns, see Figs. 3(a)(c), suggest that such a composite can be considered as

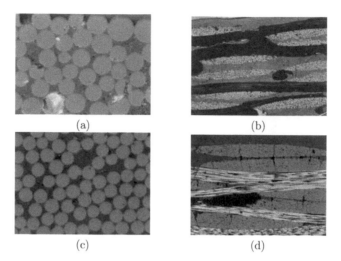

(a)                                (b)

(c)                                (d)

Figure 3. Images of yarns at various resolutions: a),b) Basalt, c),d) carbon fabric based composite

random with statistically uniform distribution of reinforcements having identical shape and orientation.

Without going into detail, we promote herein the Mori-Tanaka averaging scheme as a sufficiently accurate and reliable method to provide the estimates of effective properties of the yarn needed for the homogenization on the level of SEPUC. For a two-phase composite the $6 \times 6$ effective stiffness matrix of the yarn follows from

$$\mathbf{L}^{MT} = \mathbf{L}_m + c_f \left(\mathbf{L}_f - \mathbf{L}_m\right) \mathbf{T}_f \left(\mathbf{I} + c_m \mathbf{T}_f\right)^{-1}, \tag{2}$$

where $\mathbf{L}_r$ is the stiffness matrix of a given phase $r = m, f$ (subscripts $m, f$ stand for the matrix and fiber phase, respectively), $c_r$ refers to the volume fraction and $\mathbf{T}_f$ is the $6 \times 6$ concentration factor for the fiber phase written as

$$\mathbf{T}_f = \left[\mathbf{I} + \mathbf{SL}_m^{-1} \left(\mathbf{L}_f - \mathbf{L}_m\right)\right]^{-1}, \tag{3}$$

where $\mathbf{S}$ is the Eshelby tensor derived from a certain transformation problem of a single inclusion (an infinite cylinder of circular cross-section in this particular case) embedded in an infinite matrix.[7] Since this tensor depends on the matrix properties only, the method is fully explicit.

While image analysis may provide the volume fractions of individual phases $c_m$ and $c_f$ needed in Eqs. (2) and (3), the material parameters, if not supplied by the producer, have to be obtained either from literature or derived experimentally. The latter approach often employs nanoindentation.[8,9] Successful application of this measuring technique to carbon-carbon textile composites can be found e.g. in.[6] Table 2 contains the elastic moduli, Poisson's ratios and volume fractions of individual phases. Those found from nanoindentation are labeled by (*).

Table 2. Material parameters of individual phases (Basalt/Carbon fiber)

| | Material | $E_B/E_C$ [GPa] | $G_B/G_C$ [GPa] | $\nu_B/\nu_C$ [-] | $c_B/c_C$ [-] |
|---|---|---|---|---|---|
| fiber | longitudinal | 80/294 | 25/11.8 | 0.24 | 0.67/0.69 |
| | transverse | 18*/12.8* | 10/4.1 | 0.4 | |
| matrix | | 2.12* | 0.85 | 0.24 | 0.33/0.31 |

Next, inspecting Fig. 3(d) identifies a rather uniform distribution of inter-yarn cracks. Following[6] these are introduced independently in the second homogenization step using again the set of Eqs. (2) and (3) but in this case replacing the fiber phase by pores of zero stiffness and the matrix properties by those obtained from the first homogenization step. The volume fraction of pores equal to 0.01 was assumed for both fabric based composites.

X-ray microtomography

The X-ray microtomography served to render the three-dimensional phase information directly without performing a tedious and time consuming sectioning of two-dimensional images. The experimental setup used in the present study has successfully been applied to a number of material systems including bones[10,11] and carbon-carbon textile composites.[6] It consists of large-area flat panel sensor and microfocus X-ray source. Microfocus X-ray source L8601-01 (Hamamatsu Photonics K.K.) with tungsten anode, 5 $\mu$m spot size was used. For imaging, the C7942CA-22 X-ray detector (Hamamatsu Photonics K.K.) with the resolution of $2368 \times 2240$ pixels and physical dimensions of $120 \times 120$ mm was used; $2 \times 2$ pixel binning was used for better SNR (Signal-to-noise-ratio) and to reduce the model reconstruction time. The $\mu$CT scanning sequence assumed $360°$ sample rotation with projections taken each $0.5°$, i.e. 720 projections.

To obtain the spatial image data of the specimen, FBP cone-beam reconstruction was used. A spatial resolution of the resulting 3D images was 12 $\mu$m³. Two particular examples of the reconstructed 3D porosity are plotted in Fig. 4. The resulting porosity amounted to $\approx 10\%$ for basalt and to $\approx 12\%$ for carbon fabric based systems.

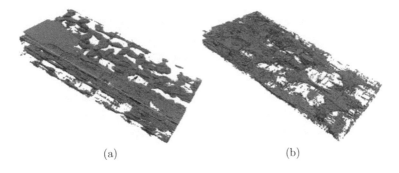

(a)             (b)

Figure 4. Reconstructed 3D porosity from micro-CT projections: a) basalt, b) carbon fabric based composite

Effective stiffnesses from ultrasonic wave speed measurements

The present section is concerned with the laboratory measurements of effective elastic properties to be compared with those derived later from numerical simulations. As is often the case, the desired effective properties are found indirectly from the measurements of, e.g. resonant frequencies[12] or ultrasonic wave speeds.[13,14] Both belong to the class of nondestructive techniques and can be utilized in the prediction of stiffness degradation with progressive damage.[14] Without giving any preference to a particular measuring technique we adopt here the latter approach as it allows for the determination of the entire stiffness matrix.

The evaluation of material stiffnesses by ultrasonic method consists of collecting speeds of waves propagating in different planes of the anisotropic solid. In this particular study, the acquisitions are made in three planes, two planes of symmetry $(X_1, X_2)$ and $(X_1, X_3)$ and one still perpendicular to the plane $(X_2, X_3)$ but rotated by the angle of 45° as seen in Fig. 5(a). For further reference this plane is denoted as $(X_1, X_{45°})$. The measurements within the $(X_2, X_3)$ plane are precluded by a relatively small thickness of the sample, see Table 3. Wave speed measurements are performed by comparing ultrasonic pulses propagating in water with pulses which are refracted by a plate-like sample immersed in water.

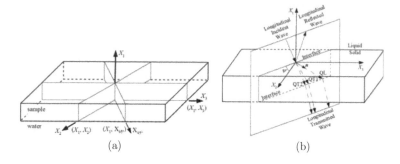

(a)                                    (b)

Figure 5. a) Selected incident planes, b) schematic diagram of the simple-transmission experiment, the incident plane is defined by the azimuthal angle $\psi$

It can be seen that these three incident planes are sufficient to derive all nine independent components of the stiffness matrix of an orthotropic material, to which the studied material systems belong, from the solution of the Christoffel equation. This equation, given by,

$$\left[ L_{ijkl} n_k n_l - \rho v^2 \delta_{ij} \right] P_j = 0, \tag{4}$$

relates the elastic stiffnesses $L_{ijkl}, i, j, k, l = 1, \ldots, 3$ of a homogeneous medium to the phase velocity $v$ of the elastic bulk wave propagating in the direction of the unit normal $\boldsymbol{n}$. The vector $\boldsymbol{P}$ is the corresponding polarization vector and $\rho$ is the effective density of the heterogeneous sample. In general, three modes can propagate along the direction $\boldsymbol{n}$ with different speeds and polarizations, see Fig. 5(b): one quasi-longitudinal mode (QL) and two quasi-transverse modes (QT1) and (QT2), e.g. the $(X_1, X_{45°})$ plane. If the incident plane is also the plane of symmetry $((X_1, X_2)$ and $(X_1, X_3)$ planes), only two waves are generated (QL1, QT1).

Knowing the elastic properties and the propagation direction, the corresponding velocities can be found from the solution of a cubic equation written as

$$\det\left[L_{ijkl}n_k n_l - \rho v^2 \delta_{ij}\right] = 0. \tag{5}$$

On the contrary, knowing the phase velocities for a given propagation direction $\boldsymbol{n}$ enables the determination of the stiffnesses $L_{ijkl}$ by inverting Eq. (5). This equation suggests that the $(X_1, X_2)$ incident plane allows for the determination of the $L_{11}, L_{22}, L_{12}, L_{66}$ stiffness components only, whereas the measurements in the $(X_1, X_3)$ plane may provide $L_{11}, L_{33}, L_{13}, L_{55}$. The third $(X_1, X_{45°})$ plane is therefore needed to assess the remaining two components $L_{23}, L_{44}$ of the stiffness matrix $\mathbf{L}$.

To avoid a direct inversion of Eq. (5), we identify the elasticity constants by minimizing the least square error of model responses $V(\mathbf{L}, \boldsymbol{n})$ and experimental values $V_e(\bar{v}(\boldsymbol{n}))$

$$L_{ij} \to \min[E(L_{ij})] = \sum_m \sum_{r=1}^N \left[V^{(X_1, X_m)}(\mathbf{L}, \boldsymbol{n}_r) - V_e^{(X_1, X_m)}(\bar{v}(\boldsymbol{n}_r))\right]^2, \tag{6}$$

$$V^{(X_1, X_m)}(\mathbf{L}, \boldsymbol{n}_r) = L_{ijkl}(\boldsymbol{n}_r)_k(\boldsymbol{n}_r)_l, \quad V_e^{(X_1, X_m)}(\bar{v}(\boldsymbol{n}_r)) = \rho\bar{v}^2(\boldsymbol{n}_r)\delta_{ij},$$

where $(X_1, X_m)$ stands for the incident plane ($m \equiv 2, 3, 45°$), $\bar{v}$ is the associated phase velocity in the propagation direction $\boldsymbol{n}_r$ and $N$ is the total number of measurements. The solution strategy based on genetic algorithms mentioned previously can be adopted again to run the minimization problem (6). A successful application to a similar problem of woven SiC/SiC composites can be found in.[14]

The actual measurements were carried out for four specimens. Two specimens cut along the warp and weft directions were considered for each type of composite. The basic geometrical parameters are listed in Table 3. In the present study, the warp direction is associated with the $X_3$ axis and the weft direction with the $X_2$ axis. The mass of dry samples and the corresponding densities needed in Eq. (6) are also provided in Table 3.

Table 3. Sample parameters

| Sample | In-plane dimensions [mm] | Thickness [mm] | Mass [g] | Density [g/mm³] |
|---|---|---|---|---|
| B-warp | 245×51.2 | 2.0 | 33.21 | 1.56 |
| B-weft | 213×50.0 | 2.03 | 29.70 | 1.54 |
| C-warp | 237×51.2 | 2.37 | 32.72 | 1.18 |
| C-weft | 209×51.3 | 2.19 | 26.14 | 1.24 |

The resulting elastic moduli derived from Eq. (6) are available in the first two rows of Tables 4 and 5. The results confirm the expected orthotropic symmetry of the specimens. A relatively large difference for in-plane elastic moduli $E_{22}$ and $E_{33}$ for basalt fabric based composites obtained for the two specimens can be attributed to a poor quality of these samples. Thus if relying solely on the experimental results, more measurements would be required to allow for a statistical evaluation. However, it should be kept in mind that the principal objective is to support the computational part of this study. In this regard, the available measurements are sufficient.

NUMERICAL CALCULATIONS

This section outlines the numerical evaluation of the effective properties of the two investigated systems. Attention is accorded to the meso-scale represented here by the Statistically

Equivalent Periodic Unit cell derived in the experimental part of this paper. The analysis is performed in the framework of the first-order homogenization theory which assumes the unit cell to be loaded, for example, by a macroscopically homogeneous strain vector $\boldsymbol{E}$. The local displacement field can be split into a linear homogeneous part $U_i = E_{ij}x_j$ and a fluctuation part $u_i^*$ being $Y$ periodic, where $Y$ represents the unit cell domain, to obtain

$$u_i(\boldsymbol{x}) = E_{ij}x_j + u_i^*(\boldsymbol{x}). \tag{7}$$

Next, writing the associated local strain as

$$\varepsilon(\boldsymbol{x}) = \boldsymbol{E} + \varepsilon^*(\boldsymbol{x}), \tag{8}$$

and introducing the local $\boldsymbol{\sigma}(\boldsymbol{x})$ and macroscopic $\boldsymbol{\Sigma}$ stress fields enables us to provide the Hill lemma in the form

$$\langle \varepsilon(\boldsymbol{x})^{\mathsf{T}} \boldsymbol{\sigma}(\boldsymbol{x}) \rangle = \boldsymbol{E}^{\mathsf{T}} \boldsymbol{\Sigma}, \tag{9}$$

where $\langle f \rangle$ represents the volume average of a given quantity $f$. Eq. (9) combined with the local and macroscopic constitutive laws then yields the effective stiffness matrix by solving six successive elasticity problems. To that end, the periodic unit cell is loaded, in turn, by each of the six components of $\boldsymbol{E}$, while the other five components vanish. The volume stress averages normalized with respect to $\boldsymbol{E}$ then furnish individual columns of the homogenized stiffness matrix $\boldsymbol{\mathsf{L}}^{\text{hom}}$. The required periodicity conditions (equal displacements $\boldsymbol{u}^*$ on opposite sides of the unit cell) can be accounted for through multi-point constraints, see e.g.[15] for details.

Three particular approaches are now examined to solve Eq. (9).

Direct Finite Element Analysis of SEPUC

We begin by considering the standard finite element method where the unit cell is meshed by, e.g. tetrahedral constant strain elements that confirm to material boundaries. Thus each element contains only one material phase. The resulting meshes are displayed in Fig. 6. Note that the meshes must be generated such as to allow for a simple introduction of periodic boundary conditions on opposite planes of SEPUC. Once having a reasonably fine mesh to accurately reflect the volume fractions of individual phases (matrix, yarns, voids) the analysis becomes straightforward proceeding along the steps discussed previously. The results are shown in the third row of Tables 4 and 5. Recall that the yarn properties were provided by the Mori-Tanaka homogenization steps described already in the experimental part of this paper.

Note that the porous phase, not specifically considered in the derivation of SEPUC, was created by coating each yarn with a layer of matrix such that the volume of coating complies with the respective volume fraction of the matrix phase. The remaining space of the total volume of SEPUC then devolves upon the porous phase, see[6] for further details.

Extended Finite Element Method

In Fig. 6 we may identify relatively sharp areas at yarn crossings that may lead to inappropriately distorted elements if attempting to match all geometrical boundaries. To avoid such complications we adopt the approach based on the extended finite element method (X-FEM), which allows us to treat complex geometries relatively easily. More specifically, the method enables an application of regular meshes, e.g. standard brick elements, which do not have to confirm to physical boundaries. These are captured by enriching the standard

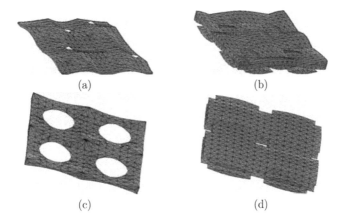

(a)  (b)

(c)  (d)

Figure 6. Standard finite element meshes for two SEPUCs: a),b) Basalt, c),d) carbon fabric based composite

approximation space of the displacements with a specific enrichment function $\psi(\boldsymbol{X})$ which renders the corresponding strains discontinuous along the material interface.

Following Moës et al.[16] the local displacement field can be written as

$$u_k(\boldsymbol{X}) = \sum_{i \in I} N^i(\boldsymbol{X}) r_k^i + \sum_{j \in I^*} N^j(\boldsymbol{X}) \psi(\boldsymbol{X}) a_k^j, \tag{10}$$

where $r_k^i$ are the standard nodal displacements, $N^i$ are the standard shape functions, $I$ represents the total number of finite element nodes in the analyzed domain, $I^* \subset I$ gives the number of nodes for which the support is split by the interface and $a_k^j$ are the additional degrees of freedom. The enrichment function $\psi(\boldsymbol{X})$ can attain the form

$$\psi(\boldsymbol{X}) = \sum_{i \in J} |\phi^i| N^i(\boldsymbol{X}) - \left| \sum_{i \in J} \phi^i N^i(\boldsymbol{X}) \right|, \tag{11}$$

where $\phi^i$ denotes the level set value in the node $i$, the signed distance of the element node to the interface with either a positive or a negative value depending on the material to which it belongs. $J$ stands for the number of nodes of the element for discretization containing the point $\boldsymbol{X}$. This function then locates interfaces implicitly as a union of points for which it attains a zero value (zero-level). Further details including numerical implementation are available in.[6]

Only sufficiently fine meshes which capture the location of individual interfaces and phase volume fractions reasonably well are mentioned here. The mesh consisting of $20 \times 20 \times 15$ 8-node brick elements was adopted for both material systems. The corresponding isosurfaces of the porous phase and yarns are plotted in Fig. 7. This refinement appears sufficient when compared to Fig. 6.

The actual evaluation of effective properties is identical to standard finite element analysis. Particular attention must be paid to the introduction of periodic boundary conditions for enhanced degrees of freedom, see.[6, 16] The predicted elastic moduli are stored in the fourth row of Tables 4 and 5.

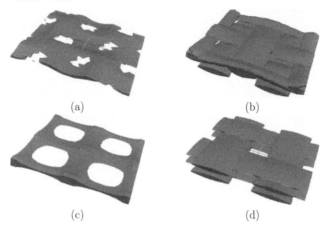

(a)

(b)

(c)

(d)

Figure 7. Isosurfaces of porous phase and yarns: a),b) basalt, c),d) carbon fabric based composite

So far all analyses have employed SEPUC as a representative volume element. On the other hand, one can directly exploit the reconstructed samples of real systems provided by computational micro-tomography, recall the voxel-type structure of reconstructed porosity in Fig. 4. At present, this is the only phase we may easily identify numerically. In view of this the yarns coated by the matrix phase are replaced by the new homogenized matrix found from the unit cell analysis free of pores. The new RVE (reconstructed samples consisting of the homogenized matrix weakened by real porosity) can then be subjected to homogenization. This analysis is easily carried out with the help of X-FEM.

To promote applicability of X-FEM in the analysis of real samples compare Figs. 8(a) and 8(b). Note that the black region in Fig. 8(b) represents the porous phase. It is clear that X-FEM is able to represent the real porosity relatively accurately. A three-dimensional X-FEM representation of porosity of both types of material systems used in the evaluation of effective properties is plotted in Fig. 9. The corresponding results are in the last row of Tables 4 and 5.

RESULTS AND DISCUSSIONS

Four particular approaches (three numerical and one experimental) were considered in the present study to estimate the effective elastic properties of multi-layered textile composites. To suggest sufficient generality of the proposed multi-scale first-order homogenization based computational scheme we examined two different types of textile reinforcements. Considerable differences between the basalt and carbon fabric based laminates can be observed for both the binary images of the two systems in Fig. 2 and the corresponding computational models (SEPUCs) plotted in Fig. 6 or Fig. 7. The associated effective elastic moduli are listed in Tables 4 and 5. In all cases, they were extracted from the components of the inverted stiffness matrix.

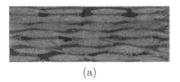
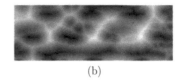

(a)                                                    (b)

Figure 8. a) 2D section of reconstructed $\mu$CT projections identifying position and shape of real porosity, b) X-FEM representation of porosity: level set function of real pores determined from 3D image of $\mu$CT projections (black = pores)

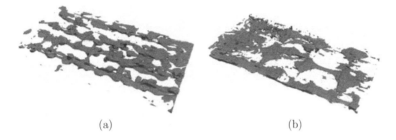

(a)                                                    (b)

Figure 9. X-FEM representation of porosity: a) Basalt, b) carbon fabric based composite

Though the adopted experimental method can be used alone as a tool for the derivation of all nine components of the stiffness matrix, it is considered here merely for the validation of numerical predictions. These on the other hand may serve to check the reliability of the experimental results. Here, this comparison may suggest the inadequacy of the experimentally derived in-plane shear moduli $G_{23}$, particularly for carbon fabric reinforced samples, which are unrealistically high when compared to the in-plane tensile moduli $E_{22}$ and $E_{33}$. See the first two rows of Table 5. The ultrasonic wave speeds propagating in the $(X_1, X_{45°})$ plane together with low thickness can make difficult to separate the quasi-longitudinal and the quasi-transverse modes, see Fig. 5(b). Therefore, the experimental identification of the $G_{23}$ component is delicate. The quality of examined samples may also play an important role. These questions are opened and will be the subject of our future research. Nevertheless, the actual in-plane shear modulus can for sure be bounded by the experimental and numerical values.

The predictions provided by the assumed numerical procedures are available in the last three rows of both tables. Except for the out-of-plane tensile modulus $E_{11}$ they show reasonable consistency. With reference to the experimental results, the last approach exploiting the reconstructed $\mu$CT projections appears to be the most accurate. The slight in-plane anisotropy, also observed experimentally but not predicted by unit cell models, can be associated with uneven distribution of pores, recall Fig. 4 or Fig. 9. On the contrary, the uniform distribution of pores in unit cell models may yield in-plane moduli slightly lower in comparison to the last numerical approach and experimental measurements. Nevertheless, given considerably less demanding computations, the concept of SEPUC in combination with X-FEM methodology appears promising and will be adopted in our next research effort

Table 4. Effective elastic moduli of laminate with basalt reinforcement

| Model | $E_{11}$ | $E_{22}$ | $E_{33}$ | $G_{23}$ | $G_{13}$ | $G_{12}$ |
|---|---|---|---|---|---|---|
| Experiment-warp | 3.8 | 16.3 | 16.3 | 11.5 | 1.7 | 1.7 |
| Experiment-weft | 3.2 | 6.9 | 7.9 | 11.9 | 1.8 | 1.6 |
| Simul-SEPUC-FEM | 0.3 | 12.8 | 12.8 | 1.7 | 0.2 | 0.2 |
| Simul-SEPUC-XFEM | 1.1 | 12.4 | 12.3 | 1.7 | 0.6 | 0.6 |
| Simul-CT-XFEM | 3.0 | 12.6 | 13.3 | 1.3 | 1.4 | 1.4 |

Table 5. Effective elastic moduli of laminate with carbon reinforcement

| Model | $E_{11}$ | $E_{22}$ | $E_{33}$ | $G_{23}$ | $G_{13}$ | $G_{12}$ |
|---|---|---|---|---|---|---|
| Experiment-warp | 3.5 | 41.6 | 42.2 | 35.9 | 0.7 | 0.7 |
| Experiment-weft | 3.5 | 47.7 | 41.6 | 32.5 | 1.2 | 1.2 |
| Simul-SEPUC-FEM | 1.9 | 39.4 | 39.4 | 1.6 | 0.7 | 0.7 |
| Simul-SEPUC-XFEM | 1.9 | 38.7 | 36.2 | 1.5 | 0.8 | 0.8 |
| Simul-CT-XFEM | 2.5 | 41.9 | 44.5 | 1.3 | 1.2 | 1.3 |

devoted to the damage modeling of ceramic composites.

ACKNOWLEDGMENTS
The financial support provided by the GAČR grant No. 105/11/0224 is gratefully acknowledged. We extend our personal thanks to Doc. Jiří Němeček from the Department of Mechanics of the Czech Technical University in Prague, for providing us with the phase properties obtained from nanoindentation tests and to Doc. Jan Zeman for providing us with the data of Statistically Equivalent Periodic Unit Cell.

REFERENCES
[1] G. Couégnat, E. Martin, J. Lamon, Multiscale modeling of the mechanical behavior of woven comosite materials, in: Proceedings of the 17th International Conference on Composite Materials (ICCM17), Edinburgh, Scotland, 2009.

[2] V. Herb, G. Couégnat, E. Martin, Damage assessment of thin SiC/SiC composite plates subjected to quasi-static indentation loading, Composites: Part A 41 (2010) 1766–1685.

[3] T. Suchý, K. Balík, M. Černý, M. Sochor, H. Hulejová, V. Pešáková, T. Fenclová, A composite based on glass fibers and siloxane matrix as a bone replacement, Ceramics-Silikáty 52 (1) (2008) 29–36.

[4] M. Černý, P. Glogar, P. Nekoksa, Z. Sucharda, Thermally resistant composites with R-glass fibres and polysiloxane-derived ceramic matrix, Bulletin of the Czech Society for Carbon Materials 1.

[5] J. Zeman, M. Šejnoha, Homogenization of balanced plain weave composites with imperfect microstructure: Part I – theoretical formulation, International Journal for Solids and Structures 41 (22–23) (2004) 6549–6571.

[6] J. Vorel, J. Zeman, M. Šejnoha, Homogenization of plain weave composites with imperfect microstructure: Part II–Analysis of real-world materials, International Journal for Multiscale Computational Engineering, Preprint avaialble at arXiv:1001.4063.

[7] G. J. Dvorak, Micromechanics of Composite Materials, 2013th Edition, Springer, 2012, ISBN-10: 9400741006, ISBN-13: 978-9400741003.

[8] J. Němeček, Creep effects in nanoindentation of hydrated phases of cement pastes, Material Characterization 60 (9) (2009) 1028–1034.

[9] J. Němeček, V. Šmilauer, L. Kopecký, Nanoindentation characteristics of alkali-activated aluminosilicate materials, Cement & Concrete Composites 33 (2011) 163–170.

[10] O. Jiroušek, P. Zlámal, D. Kytýř, M. Kroupa, Strain analysis of trabecular bone using time-resolved x-ray microtomography, Nuclear Instruments and Methods in Physics Research, Section A: Accelerators, Spectrometers, Detectors and Associated Equipment 633 (SUPPL. 1) (2011) S148–S151.

[11] O. Jiroušek, I. Jandejsek, D. Vavřík, Evaluation of strain field in microstructures using micro-ct and digital volume correlation, Journal of Instrumentation 6 (1).

[12] M. Černý, P. Glogar, L. Machota, Resonant frequency study of tensile and shear elasticity moduli of carbon fibre reinforced composites (CFRC), Carbon 38 (2000) 2139–2149.

[13] C. Aristegui, S. Baste, Optimal recovery of the elasticity tensor of general anisotropic materials from ultrasonic velocity data, J. Acoust. Soc. Am. 101 (2) (1997) 813–833.

[14] E. Grippon, S. Baste, E. Martin, C. Aristgui, G. Couegnat, Damage characterization of ceramic matrix composites, in: ECCM15 - 15th European Conference on Composite Materials, Venice, Italy, 24-28 June, 2012.

[15] J. Zeman, M. Šejnoha, Numerical evaluation of effective properties of graphite fiber tow impregnated by polymer matrix, Journal of the Mechanics and Physics of Solids 49 (1) (2001) 69–90.

[16] N. Moës, M. Cloirec, P. Cartraud, J.-F. Remacle, A computational approach to handle complex microstructure geometries, Computer Methods in Applied Mechanics and Engineering 192 (28-30) (2003) 3163–3177.

NUMERICAL ESTIMATION OF THE INFILTRABILITY OF WOVEN CMC PREFORMS

G. L. Vignoles[a], W. Ros[a,b], C. Germain[b]

[a]University of Bordeaux, LCTS, 3 allée de la Boétie, F-33600 Pessac, France
[b]University of Bordeaux, IMS, 350 Cours de la Libération, F-33410 Talence, France

ABSTRACT
    Preforms made of two different arrangements of woven SiC fibers were examined on their "infiltrability" (or ease to infiltrate) by chemical vapor infiltration (CVI) for the preparation of SiC/SiC composites after a first slurry impregnation/pyrolysis step. Based on X-ray CMT scans, numerical infiltration tools – previously validated on the CVI of pyrocarbon - were used to illustrate and to quantify the differences of infiltration behavior between these two samples.

INTRODUCTION
    The Chemical Vapor Infiltration (CVI) process is used, among others, for the fabrication of matrices in Ceramic-Matrix Composites (CMCs), considered for use in jet engine hot parts and in nuclear power engineering.[1] This process involves the transfer of gases (precursor, diluent, and by-products) inside and outside a fibrous architecture (the preform) and their chemical reactions — homogeneous and heterogeneous — eventually leading to the deposition of the solid matrix, thereby modifying the internal geometry of the porous medium. CVI allows preparing parts with outstanding quality since it preserves the fibers; however it is costly and difficult to control. Therefore process modeling appears as a very attractive tool for a better cost-effectiveness.[2] The present approach is intended to help the CMC materials designers account for "infiltrability", *i.e.* the ease of infiltration by gases in a preform, as a function of its geometrical parameters.

SIMULATION TOOLS AND STRATEGY
    The presented numerical approach allows dealing with simulations of CVI, treated as a diffusion/deposition problem in a slowly evolving porous medium, considering two scales.[3] The basic modeling tools are two pieces of software. The first code[4] features: (i) explicit treatment of gas/solid interfaces through a Simplified Marching Cubes[5] discretization, (ii) treatment of diffusion/reaction by random walks mimicking rarefied, intermediate, or continuum gas kinetics, (iii) surface movement upon deposition by a pseudo-VOF (Volume Of Fluid) method. The second code[6] is more adapted to the larger scales and features : (i) treatment of the porous medium as a continuum with fields of local properties such as pore volume fraction, internal surface area, tensorial diffusion coefficient, effective heterogeneous reaction constant (ii) random walks simulating Brownian motion, accounting for the local tensorial properties (i.e. the step sizes and orientations are biased by the local field values), (iii) simultaneous deposition inside the porous medium volume and as a seal-coat layer on its surface. These programs are capable of working in 3D images of actual media, as for instance X-ray Computerized Micro-Tomography data[7], but also on synthesized images. They have already been validated in the case of pyrocarbon CVI in needled carbon fiber fabrics.[8]
    The procedure used here, summarized in Figure 1, is the following: the materials are scanned at two distinct resolutions. The high-resolution tomographs are divided into sub-volumes in which the first code is used for the computation of properties of interest: internal surface area, pore volume fraction, tensorial diffusion coefficients, and effective reaction constants. The data are summarized as property-porosity correlations, featuring an "average law" and statistics on the deviations from the average.

**High-resolution tomograph**

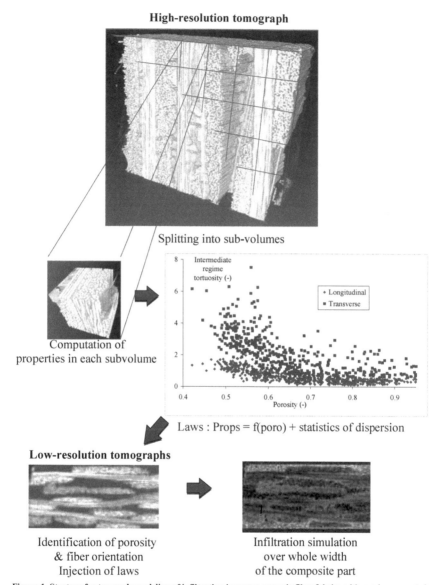

Splitting into sub-volumes

Computation of
properties in each subvolume

Laws : Props = f(poro) + statistics of dispersion

**Low-resolution tomographs**

Identification of porosity
& fiber orientation
Injection of laws

Infiltration simulation
over whole width
of the composite part

**Figure 1. Strategy for two-scale modeling of infiltration in woven ceramic fiber fabrics with pre-impregnated ceramic filler.**

Next, the low-resolution tomographs are processed in order to establish the relation between the local grayscale level and the local pore volume fraction and to detect the local fiber orientation. This step allows for the attribution of all simulation parameters in each voxel, respecting the local fiber direction and the statistical character of the property laws. The attribution algorithm is the following: (i) grayscale level/porosity correlation is made by adjusting the image histogram peak values to the void and solid phase respectively, interpolating the intermediate values, and leveling the values lying outside the inter-peak interval; (ii) fiber orientation is obtained by the eigenvalue and eigenvector analysis of the image Hessian (second derivative) matrix, (iii) surface area, transverse and longitudinal (with respect to fibers) diffusivity is obtained by first getting the "average law" value as a function of the local pore volume fraction, then by sampling a random value of the deviation so that the deviation histogram matches the experimental one. This last step is conveniently done by sampling a random number between 0 and 1 with uniform distribution, then by taking its image by the reciprocal function of the cumulative distribution function of the normalized deviations. The image is now ready for simulations by the large-scale diffusion/reaction code.

MATERIALS

The application case chosen here is CVI of SiC in a preform constituted by a woven fabric pre-densified with a ceramic obtained by Slurry Impregnation and Pyrolysis (SIP), defining a very complex network of pores with various sizes and shapes. Two series of preforms (hereafter M1 and M2) with different shapes are analyzed and compared to each other. The samples were embedded in epoxy resin. Their shape was rectangular with a small protuberance on one side for high-resolution scanning.

3D image data have been obtained by Synchrotron X-ray computerized microtomography at ID19 line of ESRF[9], using 20.5 keV quasi-monochromatic X-rays at two resolutions: 1.377 μm/voxel in high-resolution mode and 5.05 μm/voxel in low-resolution mode. The reconstructions from 1500 projections $2048^2$ square pixels in size were embedded in $2048^3$ cubic voxels blocks, suitably cropped outside the material for simulations. The resolution of the low-resolution images was further degraded by a factor 7 for M1 and 8 for M2, so that there exists a size ratio of ~25 between the large-scale and the small-scale voxels.

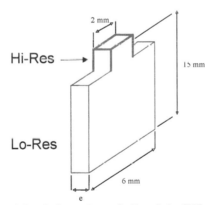

Figure 2. Sample shape, adequate for bi-resolution CMT scans

RESULTS AND DISCUSSION
    Figures 3 and 4 are plots of the pore volume fraction and effective transverse diffusion coefficient in materials M1 and M2 as the matrix volume amount increases, for 4 different choices of the deposition rate constant. Clearly, M1 is initially less porous and more penetrable than M2, but the situation gets reversed after 8-10% CVI matrix volume infiltration, since M1 undergoes a dramatic decrease in diffusion coefficient and consequently its porosity tends towards a given limit, indicating a premature pore closure. This effect is enhanced when the heterogeneous rate coefficient is high. On the other hand, M2 has a larger initial porosity and a lower initial transverse diffusivity, but we can see that at high infiltration rates this material does not suffer from premature plugging as fast as M1. Therefore, somewhat counter-intuitively, we can conclude that M2 has a better infiltration capability than M1.

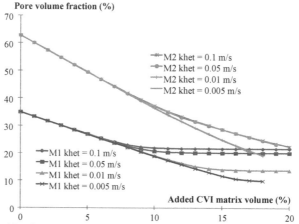
Figure 3. Interior pore volume fraction evolution as infiltration proceeds in materials M1 and M2.

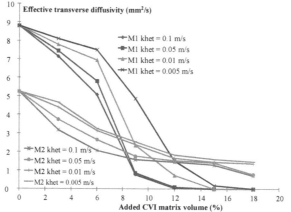
Figure 4. Effective transverse diffusivity as infiltration proceeds in materials M1 and M2.

CONCLUSION

We have presented a two scale CVI simulation method working on high/medium resolution 3D tomography data, and shown an application example on CMCs with woven fibrous reinforcements and pre-impregnated ceramic. We have shown that this numerical approach is of interest in the identification of preforms with better infiltration capabilities.

Moreover, it is also possible to use these tools on virtual images of fibrous architectures, and use them for the design of better materials. The infiltration computations also produce outputs that are suitable for further property computations, like mechanical or thermal FEM analysis; this completes a "virtual material toolbox" aimed at evaluating the potentialities of material and/or process variations without actually fabricating test samples.

ACKNOWLEDGEMENTS

This work was funded by Herakles (SAFRAN Group) and Aquitaine Region through a Ph. D. grant to W.R. as a part of the ARCOCE project.

REFERENCES

[1] J. C. Cavalier, I. Berdoyes, and E. Bouillon. Composites in aerospace industry. *Adv. Sci. Technol.* **50**: 153–162 (2006).

[2] G. L. Vignoles, "Modelling of CVI Processes", *Adv. Sci. Technol.* **50**: 97-106 (2006).

[3] G. L. Vignoles, C. Germain, C. Mulat, O. Coindreau, and W. Ros, Modeling of infiltration of fiber preforms based on X-ray tomographic imaging, *Adv. Sci. Technol.* **71**: 108-117 (2010).

[4] G. L. Vignoles, W. Ros, C. Mulat, O. Coindreau, and C. Germain. Pearson random walk algorithms for fiber-scale modeling of Chemical Vapor Infiltration, *Comput. Mater. Sci.* **50**(3):1157-1168 (2011).

[5] M. Donias, G. L. Vignoles, C. Mulat, C. Germain, and J.-F. Delesse, Simplified Marching Cubes: an efficient discretization scheme for simulations of deposition/ablation in complex media. *Comput. Mater. Sci.* **50**(3): 893-902 (2011).

[6] G. L. Vignoles, W. Ros, I. Szelengowicz, and C. Germain, 2011. A Brownian motion algorithm for tow scale modeling of chemical vapor infiltration, *Comput. Mater. Sci.* **50**(6): 1871-1878 (2011).

[7] G. L. Vignoles, C. Mulat, C. Germain, O. Coindreau, and J. Lachaud, Benefits of X-ray CMT for the modelling of C/C composites, Adv. Eng. Mater. **13**(3), 178–185 (2011).

[8] G. L. Vignoles, W. Ros, G. Chollon, F. Langlais, and C. Germain, "Quantitative validation of a multi-scale model of pyrocarbon chemical vapor infiltration from propane", , in "*Advanced Processing and Manufacturing Technologies for Structural and Multifunctional Materials and Systems VI (APMT6)*", T. Ohji, M. Singh, M. Halbig and S. Mathur, Editors, Ceramics Engineering and Science Proceedings **33**(8) 105-116 (2012).

[9] ESRF ID19 Industrial & Commercial Unit, *http://www.esrf.eu/Industry/files/brochure-3D.pdf* (consulted Dec. 2012)

MULTISCALE EXTRACTION OF MORPHOLOGICAL FEATURES IN WOVEN CMCs

C. Chapoullié[a,c], C. Germain[a], J-P. Da Costa[a], G. L. Vignoles[b], M. Cataldi[b,c]

[a]University of Bordeaux, IMS, 350 Cours de la Libération, F-33410 Talence, France
[b]University of Bordeaux, LCTS, 3 allée de la Boétie, F-33600 Pessac, France
[c]SAFRAN Herakles, Les 5 Chemins, F-33187 Le Haillan, France

ABSTRACT
    Woven CMCs are key materials in aeronautic industry. Their characterization is often carried out on two-dimensional microscopy data and very few methods aim at the 3D description of their real inner structure. We propose to explore the structure of woven CMCs in 3D using high resolution tomography. In particular, we present two segmentation algorithms combining differential geometry and mathematical morphology. The first one performs at low resolution. It allows extracting yarn envelopes using local orientations. The second one allows fiber identification at a higher resolution. It is based on the extraction of fiber axes using fiber local orientation. After this two-scale segmentation, various morphological features are estimated such as: diameter distribution, diameter variations along fibers, fiber orientation distribution, intra-yarn fiber density. Our algorithms are tested on X-ray tomographs of a fibrous preform at the resolution of 1.377 µm per voxel.

INTRODUCTION
    Woven ceramic-matrix composites (CMC) are choice materials in aeronautic industry[1] because of their thermal and mechanical properties. The characterization of the properties of woven CMC properties is crucial in order to understand their behavior and defects. These properties can then be used as inputs for mechanical simulation algorithms.
    More precisely, a critical point consists in analyzing weave, yarn and fiber geometry. The difficulty of such analysis resides in the multi-scale nature of these materials. Regarding the geometric characterization of woven CMC, two main categories of studies exist. The first one concerns mechanical simulation. Durville's approach[2] consists in applying mechanical constraints to virtual yarns composed of fibers in order to simulate the weaving process. The analysis of geometric and mechanical properties is then performed on the simulated material.
    The second type of studies relies on image analysis. Approaches found in literature concern local analysis of materials in 2D[3]. These methods allow from a 2D microscopic image to extract characteristic parameters like fiber diameter distribution[4] or fiber orientation mapping and classification[5,6]. Improvements in acquisition techniques have also allowed the development of some 3D fiber analysis methods. For instance, 3D microtomography has been used to perform fiber segmentation[7], fiber orientation distribution calculation[8] or fiber axis detection[9]. Besides, literature also contains methods to study yarns or weave geometry. 2D-micrography images have been processed to analyze woven materials at meso and macro-scales[10] and synthesize virtual 3D materials. X-ray tomography has been exploited to characterize woven yarns and extract their mechanical[11] or geometrical[12] parameters with the aim of describing the material at meso and macro-scale. Application to SiC-fiber reinforced ceramics at yarn scale has been reported recently[13]. Geometrical information has been given under the form of yarn centroids and area, aspect ratio, and cross sections orientations. Averages and distributions of stochastic deviations have been produced.
    Up to now, while woven material analyses concerning fibers are only local, yarn studies remain global without considering fiber arrangements within yarns. The purpose of this paper is to present a double-scale approach for woven CMC analysis based on the processing of tomography images combining yarn (weft/warp classification and geometry characterization) and fiber analysis (3D description of the geometry, evolution and topology of fibers within yarns). The approach hinges on

two main steps: segmentation and characterization. For the first step, two segmentation algorithms sharing the same basic principles but working at different scales are described. The first one allows extraction of yarn envelopes while the second one identifies individual fibers. Once the yarn and fibers are identified, the characterization stage provides parameters such as fiber diameter distribution, diameter variations along fibers, fiber orientation distribution, and intra-yarn density.

This paper is organized as follows: the next paragraph provides a description of the data used for the study. Then, we will present the two algorithms used to segment yarns and fibers. A review of some characterization results will then be addressed. Finally, we will present a few conclusions and prospects.

INPUT DATA

The approach has been conducted on a 2D woven CMC before matrix reinforcement, embedded in epoxy resin. 3D image data have been obtained by Synchrotron X-ray computerized tomography at a resolution of 1.377 micron per voxel. The data of interest represent a CMC sample of size $1.4mm \times 0.99mm \times 1.4mm$, i.e. $1041 \times 721 \times 1024$ pixels. Two data blocks have been used. On both blocks, the direction normal to the weaving plane is vertical. In the first block (visible in Figure 1) weft and warp directions are more or less orthogonal to the front and to the right view respectively. In the second one, weft and warp directions are rotated by 45° around the vertical axis.

The resolution and size of these tomographic images allow identifying simultaneously yarns at coarse scale and fibers at fine scale. Images show a good contrast between fibers and resin. The cylindrical shape of fibers is clearly visible. Their radius is about 5 pixels. Figure 1 provides a 3D view of the first data block and three 2D views: front, right and top views.

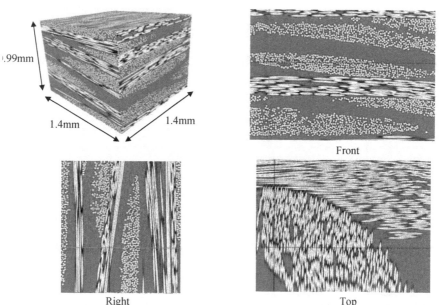

Figure 1: Tomographic data of a woven CMC at the resolution of 1,377micron/voxel. 3D view and 2D sections (front, right and top views)

YARN SEGMENTATION ALGORITHM

Algorithm overview

As shown in the block diagram of Figure 2, the yarn segmentation algorithm proposed herein consists of two main stages from the initial 3D data block to the final segmentation result. The first stage, represented as a green rectangle, is inspired by previous work[14]. It provides weft and warp masks. The second stage, represented as a red rectangle, aims at individual yarn identification.

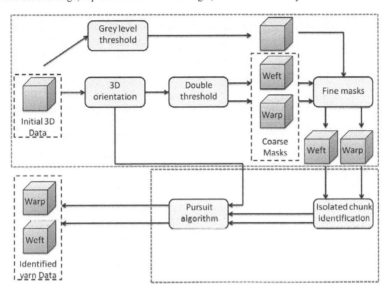

Figure 2 : Block diagram of the yarn segmentation algorithm. The algorithm is divided in two stages: warp and weft yarn separation (top green rectangle) and individual yarn identification (bottom red rectangle)

Weft and warp yarn separation

The first stage of this algorithm aims at generating two masks separating respectively warp yarns and weft yarns. To get these masks, three other primary masks have to be calculated. The first one is obtained by basic intensity grey level thresholding to separate CMC and resin. The other two are obtained by calculating local 3D orientation vectors using the structure tensor[15] and classifying them into weft and warp. Two coarse masks are thus obtained. After that, weft and warp coarse masks are multiplied with the first all-yarn mask and, with a few morphological operations (dilation and erosion to smooth the envelope and fill porosity) fine weft and warp masks are ready to use.

Individual yarn identification

The second stage addresses individual yarn identification. It proceeds in two steps: isolated chunk identification and yarn pursuit. By "isolated chunk" we mean the yarn sections without any contact with neighboring yarns. An area-based criterion is used to decide whether the yarn is in contact with another one or not. This criterion also allows elimination of yarns intersected by data block borders.

The isolated chunks are then used by the pursuit algorithm : they are grown inside the mask following 3D orientation vectors. Figure 3 shows isolated chunks (with color labels) before (on the left) and after (on the middle) the pursuit algorithm.

After spreading labels inside entire yarns, a post-treatment is applied to paste together chunks belonging to the same yarn. This final step results in the identification of all individual yarns (see Figure 3, right).

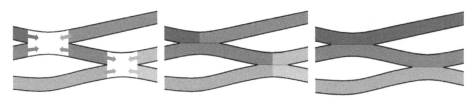

Figure 3 : Yarn longitudinal sections showing chunks with color labels. Left: isolated chunks with arrows representing spreading directions. Center: chunks after pursuit. Right: final results showing individual yarns.

Retrieval of the entire yarn from at least one isolated chunk is thus possible, even if the chunk is only one pixel thick, without mixing up yarns in contact. Figure 4 shows the 3D resulting block with three entire yarn envelopes superimposed to the initial image data. The interest of this method is that no hypothesis on the yarn section shape is made whatsoever.

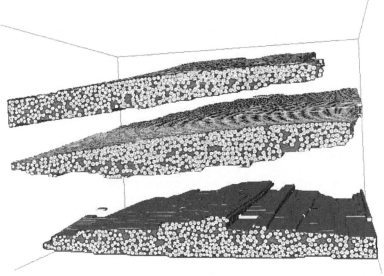

Figure 4 : 3D representation of yarn envelopes with superimposed initial data

FIBER SEGMENTATION ALGORITHM

The algorithm that labels fibers is inspired by the previous one, but operates at a finer scale.

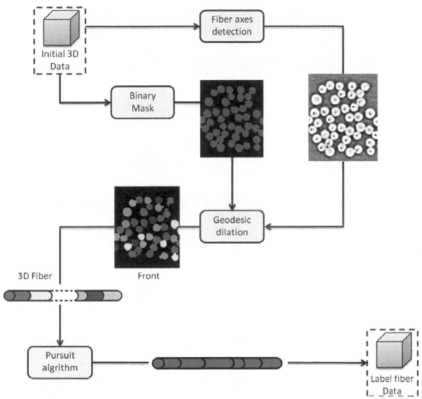

Figure 5 : Fiber segmentation algorithm block diagram

The first step of this algorithm is the detection of individual fiber centers, *i.e.* voxels belonging to fiber axes. The approach of Mulat *et al.* [9,13], based on differential geometry, has been chosen.

The second major step of the algorithm is the identification and labeling of fiber segments. This is done by associating neighboring individual axes voxels previously detected and by operating a geodesic dilation[16] inside the fiber binary mask (see Figure 5). This step results in a 3D block were fibers are split into small cylindrical segments.

The last step is the fiber pursuit algorithm. Indeed, after center detection, labels are not the same along a fiber. Interruptions can also occur due to interruptions of the center extraction procedure itself (often in high density parts of the yarn). To solve this problem, a version of the pursuit algorithm adapted to high resolution is used. It consists in using local 3D orientation vectors to follow fibers and associating labels along them. This method allows unambiguous labeling of the fibers: every fiber has a single label and there is no spreading of this label into neighboring fibers.

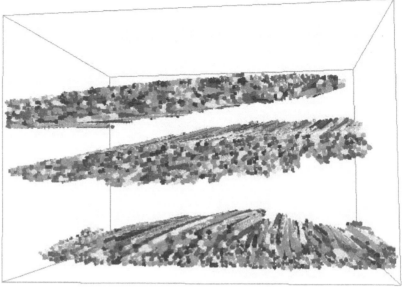

**Figure 6 : 3D representation of final label fiber in the three previously extracted entire yarns**

RESULTS

Segmentation result

The application of the algorithms presented above to the CMC data block yielded satisfactory results: weft and warp yarns were well retrieved and discriminated. Inside the yarns, except in high density regions where fiber envelopes are indiscernible, fibers were well segmented and labeled, from one end of the block to the other. Some errors can be found in high density regions of the yarns and also in blurred areas of the 3D data block.

Figure 6 shows a 3D view of the final segmentation result. Only the yarns orthogonal to the front view are represented. Color labels are given to fibers.

Once the data block was segmented into fibers, a number of properties have been extracted.

Diameter distribution

For each orthogonal slice, fiber diameters $d$ are deduced from fiber section areas $s$, considering that a fiber section can be seen as a disk with diameter $= 2\sqrt{\dfrac{s}{\pi}}$. Figure 7(left) shows the distribution of fiber section diameters throughout the data block. It can be seen here that the diameter distribution was unimodal and symmetrically distributed around its mean value.

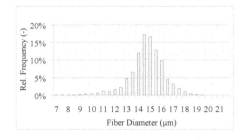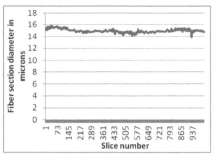

Figure 7 : Diameter distribution in the data block (on the left) and variation along a particular fiber (on the right)

Diameter variation along fiber:

Individual fibers being identified, it is also possible to look at diameter variations along fibers. Figure 7 shows such variations along one specific fiber. A smoothing operation was performed by averaging along five consecutive slices. It is seen that the diameter is approximately constant along a fiber (at least within a block of 1.4mm width).

Fiber orientation distribution:

Regarding fiber orientation information, two angles have been extracted for each fiber section. In the case of a fiber belonging to a yarn oriented along the Oz axis, the yaw angle $\beta$ between the fiber (in yellow) and the plane yOz (in red) and the pitch angle $\alpha$ between the fiber and the plane xOz (in blue) are computed.

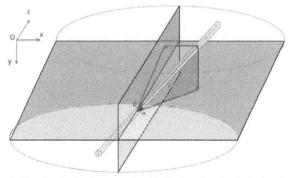

Figure 8 : Fiber (in yellow) angle computation with yOz plane (in red) and xOz plane (in blue)

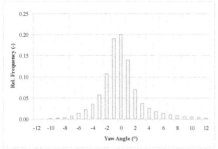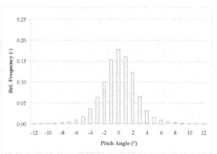

Figure 9 : Angle distribution on a yarn. Left: yaw, Right: pitch.

Figure 9 shows the distributions of these angles throughout one specific yarn within the CMC block, with a $\pm 4°$ standard deviation in yaw and $\pm 3°$ in pitch. Such a quantification of the inner orientation variability of yarns helps identifying whether a yarn is under tensile stress or not. Low tensile stresses result in scattered orientations whereas strong stresses – in the case of high density yarn parts – result in tight distribution.

Intra-yarn fiber density:
Knowing yarn and fiber mask allows extracting the intra-yarn fiber density. To do that, for each pixel in a yarn, we count, within a fixed-diameter circle centered on the pixel (circle diameter is about three times the mean fiber diameter i.e. about 43 microns), the ratio between the number of labeled pixels (fiber pixels) and the total number of pixels in the circle. An example of density map on a 2D slice of a yarn is given in Figure 10. We can distinguish zones with higher density close to the yarn "skin", resulting from compaction by compression against a transverse yarn.

Figure 10 : Intra-yarn fiber density, high density in red, low density in blue, outside yarn in black

CONCLUSION
We have presented a two scale segmentation method to process high resolution 3D tomography data of 2D woven materials before matrix reinforcement. In short, this method:
- provides individual yarn envelopes avoiding confusions between neighboring yarns, and not making any assumption on their shape;
- identifies individual fibers within yarns
- extracts characteristic parameters: fiber diameter distribution, fiber diameter variation along fibers, fiber orientation distribution and intra-yarn fiber density.
It has been tested on two data blocks with a resolution of 1.377 microns by voxel. Segmentation results are satisfactory. Characterization results seem promising in the sense that they provide relevant information for material studies in industrial applications: indeed, this approach provides a thorough description of the inner structure of the weave both at yarn scale and at fiber scale.

One of the limits of the algorithm is input data quality. Indeed, for orientation estimation and fiber axis detection, high contrast between fibers and resin is required. In blurred parts of the tomographic data, such conditions may not be verified. As a result, some fiber axes may not be detected and orientations may be poorly estimated, hindering the pursuit process.

The present study was carried out in the precise context of a 2D woven CMC before matrix reinforcement, embedded in epoxy resin. One can thus legitimately wonder whether the algorithms can be generalized to the case of other types of weaves, for instance weaves with 3 directions of yarns or needled 2D weaves. As the pursuit algorithm only needs 3D orientation vectors and binary masks as inputs, we can reasonably think that it remains applicable in various cases. Of course, this should be verified. It seems interesting to test the algorithm in other contexts, as soon as the following double constraint is satisfied: (i) image data can be segmented and (ii) it has a high enough resolution to allow orientation estimation.

Moreover, we have demonstrated the capabilities of the algorithm by extracting a few different parameters (diameters, orientation, density). Yarns and fibers being identified, many other morphological features could be calculated. For instance, it may be interesting to produce 3D maps of fiber diameters, orientations and curvature in order to help visualizing the material inner structure and understanding material behavior.

Finally, the output of this work can also be used as a primary step of a synthesis approach. Extracted morphological parameters are indeed possible inputs for any simulation algorithm aiming at the synthesis of virtual materials. They could also be considered as references for the final validity control of the produced virtual materials.

ACKNOWLEDGEMENTS

This work was funded by Herakles (SAFRAN Group) and DGA (French Ministry of Defense) through a Ph. D. grant to C.C..

REFERENCES
[1] J.C. Cavalier, I. Berdoyes and E. Bouillon, Composites in aerospace industry, *Adv. Sci. Technol.* **50**, 153–162 (2006).
[2] D. Durville, A Finite Element Approach of the Behaviour of Woven Materials at Microscopic Scale, in *Mechanics of Microstructured Solids*, J.-F. Ganghoffer and F. Pastrone, editors, *Lecture Notes in Applied and Computational Mechanics* **46**, 39–46 (2009).
[3] M. Coster, J.-L. Chermant, Image analysis and mathematical morphology for civil engineering materials, *Cement & Concrete Composites* **23**(2), 133–151 (2001).
[4] S. Yang, A. Tewari, A.M. Gokhale, Modeling of non-uniform spatial arrangement of fibers in a ceramic matrix composite, *Acta Mater.* **45**(7), 3059–3069 (1997).
[5] R. Blanc, Ch. Germain, Fiber orientation measurements in composite materials, *Composites: Part A*, **37**(2), 197–206 (2006).
[6] N.C. Davidson, A.R. Clarke, G. Archenhold, Large-area, high-resolution image analysis of composite materials, *J. Microsc.* **185**(2), 233–242 (1997).
[7] G.L. Vignoles, Image segmentation for phase-contrast hard X-ray CMT of C/C composites, *Carbon* **39**(2), 167–173 (2001).
[8] J.C. Tan, I.A. Elliott, W. Clyne, Analysis of tomography images of bounded fiber networks to measure distributions of fiber segment length and fiber orientation, *Adv. Eng. Mater.* **8**(6), 495–500 (2006).
[9] C. Mulat, Axis detection of cylindrical objects in 3-D images, *J. of Electr. Imaging* **17**(3), 0311081–0311089 (2008).

[10] B. Piezel, L. Laiarinandrasana, A. Thionnet, A multilevel finite element analysis of a textile composite, *Procs. 17th Intl. Conf. on Composite Materials*, 27-31 July 2009, Edinburgh, Scotland, W. M. Banks & M. R. Wisnom editors, ID11.1, 8 pp. (2009).

[11] P. Badel, E. Vidal-Sallé, E.Maire, L. Bigorgne, P. Boisse, Simulation and tomography analyses of textile composite reinforcement, *Procs. 17th Intl. Conf. on Composite Materials*, 27-31 July 2009, Edinburgh, Scotland, W. M. Banks & M. R. Wisnom editors, C1.2, 10 pp. (2009).

[12] G. Hivet, A. Wendling, E. Vidal-Sallé, B. Laine, P. Boisse, Modeling strategies for fabrics unit cell geometry - Application to permeability simulations, *Int. J. Mater. Forming* **3**(1), 727–730 (2010).

[13] H. Bale, M. Blacklock, M. R. Begley, D. B. Marshall, B. N. Cox, R. O. Ritchie, Characterizing Three-Dimensional Textile Ceramic Composites Using Synchrotron X-Ray Micro-Computed-Tomography. *J. Amer. Ceram. Soc.* **95**(1), 392–402 (2012).

[14] C. Mulat, M. Donias, P. Baylou, G. Vignoles, Ch. Germain, Optimal orientation estimators for detection of cylindrical objects, *Signal, Image & Video Processing* **2**(1), 51–58 (2008).

[15] H. Knutsson, Representing local structure using tensors, in *Proc.s 6th Scand. Conf. on Im. Anal.*, Oulu, Finland, June 1989, J. Roening and M. Pietikainen, editors, pp. 244–251 (1989).

[16] J. Serra, *Image analysis and mathematical morphology I*, Academic Press, London (1982).

# Materials for Extreme Environments: Ultrahigh Temperature Ceramics and Nanolaminated Ternary Carbides and Nitrides

# INFLUENCE OF PRECURSORS STOICHIOMETRY ON SHS SYNTHESIS OF $Ti_2AlC$ POWDERS

L. Chlubny, J. Lis, M.M. Bućko
AGH - University of Science and Technology, Faculty of Material Science and Ceramics
Department of Technology of Ceramics and Refractories, Al. Mickiewicza 30, 30-059
Cracow, Poland

## ABSTRACT

In the Ti-Al-C system can be found very interesting group of ternary compounds called MAX-phases. These compounds, such as $Ti_2AlC$ and $Ti_3AlC_2$, are characterised by heterodesmic layer structure consisting of covalent and metallic chemical bonds. These specific structure leads to their semi-ductile features locating them on the boundary between metals and ceramics. These features may result in wide range of potential applications, for example as a part of ceramic armour. One of potential effective and efficient methods of obtaining these materials is Self-propagating High-temperature Synthesis (SHS), basing on exothermal effect of reaction.

The objective of this work was to apply SHS method to obtain sinterable powders of $Ti_2AlC$ and to examine influence of different stoichiometry of various precursors such as elementary metallic powders, intermetallic powders in Ti-Al system and titanium carbide on final product of the reaction.

## INTRODUCTION

Among many covalent materials, such as carbides or nitrides, a group of ternary and quaternary compounds, referred in literature as H-phases, Hägg-phases, Novotny-phases or thermodynamically stable nanolaminates, can be found. These compounds have a $M_{n+1}AX_n$ stoichiometry, where M is an early transition metal, A is an element of A groups (mostly IIIA or IVA) and X is carbon and/or nitrogen. Heterodesmic structures of these phases are hexagonal, P63/mmc, and specifically layered. They consist of alternate near close-packed layers of $M_6X$ octahedrons with strong covalent bonds and layers of A atoms located at the centre of trigonal prisms. The $M_6X$ octahedral, similar to those forming respective binary carbides, are connected one to another by shared edges. Variability of chemical composition of the nanolaminate is usually labeled by the symbol describing their stoichiometry, e.g. $Ti_2AlC$ represents 211 type phase and $Ti_3AlC_2$ – 312 type. Differences between the respective phases consist in the number of M layers separating the A-layers: in the 211's there are two whereas in the 321's three M-layers [1-3]. The layered, heterodesmic structure of MAX phases leads to an extraordinary set of properties. These materials combine properties of ceramics such as high stiffness, moderately low coefficient of thermal expansion and excellent thermal and chemical resistance with low hardness, good compressive strength, high fracture toughness, ductile behavior, good electrical and thermal conductivity which are characteristic for metals. They can be used to produce ceramic armor based on functionally graded materials (FGM) or as a matrix in ceramic-based composites reinforced by covalent phases [4]. Very wide spectrum of latest information about properties, synthesis and applications of MAX phases materials in Ti-Al-C system can be found in complex and comprehensive review by X.H. Wang and Y.C. Zhou [5].

The Self-propagating High-temperature Synthesis (SHS) is a method applied for obtaining numerous materials such as carbides, borides, nitrides, oxides, intermetallic compounds and composites. The principle of this method is utilization of exothermal effect of chemical synthesis. This synthesis can proceed in a powder bed of solid substrates or as filtration combustion where at least one of the substrates is in gaseous state. To initiate the process an external source of heat has to be used and then the self-sustaining reaction front is propagating through the bed of substrates until all of them are consumed. This process could be initiated by the local ignition or by the thermal explosion. The final form of the synthesized material may depends on kind of precursors used for synthesis and the technique applied. Low energy consumption, high temperatures obtained during the process, high efficiency and simple apparatus are the features which are characteristic for this type of synthesis. The lack of control of the process is the disadvantage of this method [6].

The main purpose of this work was optimization of SHS synthesis applied to obtain sinterable powders of Ti$_2$AlC nanolaminate materials and investigation of influence of stoichiometry of precursors used during reaction on final products' phase composition. The final objective was to obtain powder with highest content of ternary phase (Ti$_2$AlC) and lowest content of TiC impurities, which strongly affects properties of the dense, sintered material, decreasing its pseudo-plastic behaviour.

The possibility of obtaining powders with high content of Ti$_2$AlC by SHS synthesis was proved by authors in their previous work, where use of intermetallic material, namely TiAl together with elemental titanium powder and carbon resulted with powder containing 95.2 wt% of Ti$_2$AlC and 4.8 wt. % of TiC [7].

Also other authors are confirming possibilities of synthesis of powders or bulk material characterised with purity of the MAX phase around 95%. In the paper by X.H. Wang and Y.C. Zhou a vast part concerning the state of art of latest Ti$_2$AlC and Ti$_3$AlC$_2$ synthesis method can be found [5]. In their work P. Wang et.al shows possibility of synthesizing good quality Ti$_2$AlC contaminated by other MAX phase, namely Ti$_3$AlC$_2$ by hot-pressing method [8]. Also Y. Bai et.al in their papers presents possibility of SHS synthesis with Pseudo Hot Isostatic Pressing were dominating phase of a product is Ti$_2$AlC accompanied by minor amounts of TiAl [9, 10]. Possibility of synthesis of Ti$_2$AlC was also reported by A.G. Zhou et.al where synthesis was conducted from elemental powders and mixture of both MAX phases was obtained as a result [11]. Possibility of using intermetallic materials in the MAX phases in Ti-Al-C system SHS synthesis were proved in paper of Lopacinski et.al [15]. Different system of ignition, namely thermal explosion was applied by Y. Khoptiar and I. Gotman [12]. Also works by Liu et.al and Guo et.al provides new information about SHS synthesis of Ti$_2$AlC [13, 14].

The deeper insight into the tendencies of reaction to form Ti$_2$AlC as a result of SHS synthesis with various stoichiometries of reactants may help in further optimization of synthesis and elimination of TiC impurities in the powder.

PREPARATION

Following the experience gained during previous synthesis of ternary materials such as Ti$_3$AlC$_2$, Ti$_2$AlN and also Ti$_2$AlC, various materials such as elementary powders, intermetallic materials in the Ti-Al system and titanium carbide powder were selected to be used as precursors for synthesis of Ti$_2$AlC powders [7, 15, 16, 17, 18, 19]. Mostly the intermetallic powders were used in the experiments, basing on the results achieved in the mentioned researches. Due to the relatively low availability of commercial intermetallic powders in the Ti-Al system, it was decided to synthesize them in the first stage of experiment. TiAl and Ti$_3$Al powders were synthesized by SHS method [17]. Titanium hydride powder, TiH$_2$, and metallic aluminium powder with grain sizes below 10 μm were used as sources of titanium and aluminium. The mixture for SHS synthesis had a molar ratio of 1:1 and 3:1 respectively (equations 1-2).

$$TiH_2 + Al \rightarrow TiAl + H_2 \qquad (1)$$

$$3TiH_2 + Al \rightarrow Ti_3Al + 3H_2 \qquad (2)$$

Powders were initially homogenized in dry isopropanol using a ball-mill. The homogenized and dried mixtures were placed in a graphite crucible which was heated in a graphite furnace up to 1200°C, at this temperature SHS reaction was initiated by the thermal explosion. Then obtained products were crushed in a roll crusher to the grain size ca. 1 mm and afterwards powders were ground to the grain size ca. 10 μm for 8 hours in the rotary-vibratory mill in dry isopropanol, using WC balls as a grinding medium [23]. Other powders used during synthesis of ternary compound were commercially available aluminium powder (grain size below 6.7 μm, +99% pure), graphite powder used as a source of carbon (Merck no. 1.04206.9050, 99.8% pure, grain size 99.5% < 50μm), titanium powder (AEE TI-109, 99.7% pure, ~100 mesh) and titanium carbide powder (AEE TI-301, 99.9% pure, ~325 mesh).

All of the Ti$_2$AlC syntheses were conducted by SHS method with a local ignition system and with use of various precursors' stoichiometry. The same experimental procedure was applied for researches on stoichiometry influence of other MAX phase, namely Ti$_3$AlC$_2$ [20]. The mixtures of substrates for the SHS synthesis were set in assumed ratios and homogenized for 12 hours. The SHS reactions stoichiometries are presented in equations 3-10 respectively. The 1.05, 1.1, 1.2 etc. corresponds to 5, 10 and 20 wt.% of excess precursor respectively.

$$2Ti + Al + C \rightarrow Ti_2AlC \qquad (3)$$
$$TiAl + Ti + C \rightarrow Ti_2AlC \qquad (4)$$
$$Ti + Al + TiC \rightarrow Ti_2AlC \qquad (5)$$
$$2Ti + 1.1Al + C \rightarrow Ti_2AlC \qquad (6)$$
$$2Ti + 1.2Al + C \rightarrow Ti_2AlC \qquad (7)$$
$$1.05TiAl + Ti + C \rightarrow Ti_2AlC \qquad (8)$$
$$1.1TiAl + Ti + C \rightarrow Ti_2AlC \qquad (9)$$
$$1.2TiAl + Ti + C \rightarrow Ti_2AlC \qquad (10)$$

After the synthesis obtained products were ground and the XRD analysis method was applied to determine phase composition of the synthesised materials. The data for quantitative and qualitative phase analysis were acquired from ICCD [21]. Quantities of the respective phases were calculated according to the Rietveld analysis [22]. The measurements were made within an accuracy of 0.5%.

RESULTS AND DISCUSSION
The X-ray diffraction analysis of SHS synthesis in the Ti-Al system proved that TiAl synthesised by SHS method was almost pure and contained only 5% of Ti$_3$Al impurities (Figure 1), while Ti$_3$Al powder did not contain any of the other phases (Fig. 2) [23].

In the case of the ternary phase synthesis, the highest amount of Ti$_2$AlC phase (84 wt.%) was achieved in the reaction 8. This dominating phase was accompanied by the other MAX phase that can be found in this system, namely Ti$_3$AlC$_2$ (5.9 wt.%), relatively low content of TiC (8.7 wt.%) and minor amount of Ti$_3$AlC (1.4%). The XRD pattern of obtained powder is presented on Figure 3. The peak corresponding to the graphite that can be seen on the XRD pattern is a result of impurities introduced by the graphite combustion boat in which the synthesis was conducted and is not included in final results. This procedure was applied to all of the experimental results. In the case of reaction 9 where 10% of excess TiAl was used as a precursor, the lowest level of TiC impurities (4 wt.%) was achieved. Results of phase quantities analysis are presented in Table 1.

Table I. Products of SHS synthesis of Ti$_3$AlC$_2$ phase composition.

| No. | Chemical reaction | Ti$_2$AlC [wt.%] | Ti$_3$AlC$_2$ [wt.%] | TiC [wt.%] | Al$_4$C$_3$ [wt.%] | Ti$_3$AlC [wt.%] |
|-----|-------------------|---------|-----------|------|---------|---------|
| 3 | 2Ti + Al + C → Ti$_2$AlC | 64.8 | 26.1 | 9.1 | - | - |
| 4 | TiAl + Ti + C → Ti$_2$AlC | 49.5 | 18.4 | 15.5 | 16.5 | - |
| 5 | Ti + Al + TiC → Ti$_2$AlC | 57.1 | 18.3 | 23.1 | - | 1.5 |
| 6 | 2Ti + 1.1Al + C → Ti$_2$AlC | 55.5 | 10.4 | 8.1 | 26.0 | - |
| 7 | 2Ti + 1.2Al + C → Ti$_2$AlC | 17.6 | 37.3 | 6.3 | 38.8 | - |
| 8 | 1.05TiAl + Ti + C → Ti$_2$AlC | **84.0** | 5.9 | 8.7 | - | 1.4 |
| 9 | 1.1TiAl + Ti + C → Ti$_2$AlC | 79.1 | 7.8 | 4.0 | 9.1 | - |
| 10 | 1.2TiAl + Ti + C → Ti$_2$AlC | 65.8 | 12.1 | 10.2 | 12.0 | - |

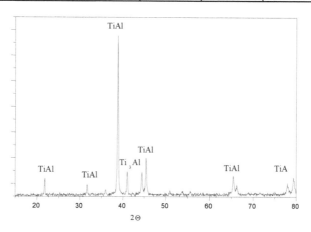

Figure 1. XRD pattern of the TiAl powders obtained by SHS [23]

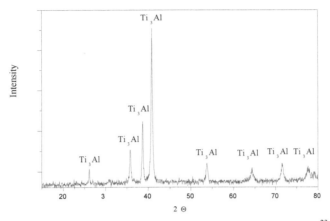

Figure 2. XRD pattern of the Ti$_3$Al powders obtained by SHS [23]

It is worth to notice that excess of TiAl precursor leads to increase of the Ti$_2$AlC content in the final product at cost of Ti$_3$AlC$_2$ and TiC. Different situation was observed in case of reactions were pure metallic Al powder was used as a precursor. Excess amounts of Al lead to decrease of both MAX phases content and promote formation of Al$_4$C$_3$ which reaches almost 39 wt.% in the case of reaction with 20% of excess aluminium precursor. Relatively high content of MAX phases was obtained in reaction were TiC was used as a precursor (reaction 5), but content of unreacted titanium carbide impurities in final product is too high and might affect properties of the dense materials produced from such powder.

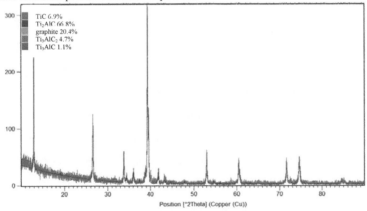

Figure 3. XRD pattern of SHS derived Ti$_2$AlC powder
with highest content of MAX phase.

CONCLUSION

Powders with high content of Ti$_2$AlC and relatively low level of TiC impurities can be obtained by Self-propagating High-temperature Synthesis (SHS). This method is not only possible but also effective and efficient.

The best results were achieved when excess amount (5 wt.%) of TiAl was used as a precursor for SHS synthesis of Ti$_2$AlC. In this case final product contained 84 wt.% of Ti$_2$AlC. Very good results were also achieved when 10% of excess TiAl was used were lowest contamination with titanium carbide was achieved. The high amounts of MAX phases in obtained powders combined with low level of TiC impurities may result in interesting properties of dense, sintered materials made of these powders.

Increasing of excess Al during the SHS reaction promotes formation of Al$_4$C$_3$. Use of titanium carbide as one of precursors results also in quite high amount of MAX phase in the final product, but the high content of unreacted TiC may affect properties of the sintered material. Similar observations were done during SHS synthesis of Ti$_3$AlC$_2$ powders [20].

The difference between the powder obtained in previous experiments after SHS synthesis of TiAl + Ti + Al (in which 95.2 wt.% were obtained ) and powders obtained in latest experiments may be result of a different source of titanium powder and some alteration to the SHS device since the previous experiments [7]. Also the volumes of mixture undergoing SHS reaction were significantly different.

Careful control of the chemical reaction stoichiometry seems to play important role during SHS synthesis of Ti$_2$AlC powders and further more detailed researches will be conducted.

Also investigation on influence of source of substrates (grain size, purity etc.) on SHS synthesis of MAX phases in Ti-Al-C system as well as effect of scaling up process will be conducted to optimize the synthesis conditions.

Sintering conditions of obtained powders as well as their mechanical properties will be examined.

ACKNOWLEDGMENTS

This work was supported by the National Science Centre under the grant no. 2472/B/T02/2011/40

REFERENCES

[1] W. Jeitschko, H. Nowotny, F.Benesovsky, Kohlenstoffhaltige ternare Verbindungen (II-Phase). *Monatsh. Chem.* 94, 1963, p 672-678

[2] H. Nowotny, Structurchemie Einiger Verbindungen der Ubergangsmetalle mit den Elementen C, Si, Ge, Sn. *Prog. Solid State Chem.* 2 1970, p 27

[3] M.W. Barsoum: The MN+1AXN Phases a New Class of Solids; Thermodynamically Stable Nanolaminates- *Prog Solid St. Chem.* 28, 2000, p 201-281

[4] L. Chlubny, J. Lis "Properties of Hot-pressed Ti$_2$AlN Obtained by SHS Process", *Proceedings of the 36th International Conference on Advanced Ceramics and Composites"*, CD-ROM, volume editors Michael Halbig, Sanjay Mathur, 2012, Wiley

[5] X.H. Wang and Y.C. Zhou "Layered Machinable and Electrically Conductive Ti$_2$AlC and Ti$_3$AlC$_2$ Ceramics: a Review", *J. Mater. Sci. Technol.,* 26(5), 2010, p 385-416

[6] J. Lis: Spiekalne proszki związków kowalencyjnych otrzymywane metodą Samorozwijającej się Syntezy Wysokotemperaturowej (SHS) - *Ceramics 44* : (1994) (*in Polish*)

[7] L. Chlubny, J. Lis, M.M. Bucko "Sintering and Hot-Pressing of Ti$_2$AlC Obtained By SHS Process", *Advances in Science and Technology,* 63, 2010, p 282-286

[8] P. Wang, B. Mei, X. Hong, W. Zhou "Synthesis of Ti₂AlC by hot pressing and its mechanical and electrical properties", *Trans. Nonferrous Met. Soc. China*, 17, 2007, p 1001-1004

[9] Y. Bai, X. He, C. Zhu, G. Chen " Microstructures, Electrical, Thermal, and Mechanical Properties of Bulk Ti₂AlC Synthesized by Self-Propagating High-Temperature Combustion Synthesis with Pseudo Hot Isostatic Pressing ", *J. Am. Ceram. Soc.*, 95 [1], 2012, p 358–364

[10] Y. L. Bai, X. D. He, Y. B. Li, C. C. Zhu, and S. Zhang, "Rapid Synthesis of Bulk Ti₂AlC by Self-Propagating High Temperature Combustion Synthesis with a Pseudo-Hot Isostatic Pressing Process," *J. Mater. Res.*, 24 [8], 2009, p 2528–35

[11] A.G. Zhou, C.A. Wang, Z.B. Ge, L.F. Wu "Preparation of Ti₃AlC₂ and Ti₂AlC by Self propagating High temeprature Synthesis" *J. Mater. Sci. Lett.*, 21, 2001, p 1971

[12] Y. Khoptiar, I. Gotman, "Ti₂AlC Ternary Carbide Synthesized by Thermal Explosion," *Mater. Lett.*, 57 [1], 2002, p. 72–76

[13] G. Liu, K.X. Chen, J.M. Guo, H.P. Zhou, J.M.F.Ferreira "Layered Growth of T₁₂AlC and Ti₃AlC₂ in Combustion Synthesis", *Mater. Lett.*, 61, 2007, p 779

[14] J. M. Guo, K. X. Chen, Z. B. Ge, H. P. Zhou, and X. S. Ning, "Combustion Synthesis Ternary Carbide Ti₂AlC₁₋ₓ" *Rare Metal Mat. Eng.*, 32 [12], 2003, p 1029-32

[15] M. Lopacinski, J. Puszynski, and J. Lis, "Synthesis of Ternary Titanium Aluminum Carbides Using Self-Propagating High-Temperature Synthesis Technique," J. Am. Ceram. Soc., 84 [12], 2001, p 3051–3053

[16] J. Lis, L. Chlubny, M. Lopacinski, L. Stobierski, M. M. Bucko " Ceramic Nanolaminates - Processing and Application", *Journal of the European Ceramic Society*, 28, 2008, p 1009–1014

[17] L. Chlubny, M.M. Bucko, J. Lis "Intermetalics as a precursors in SHS synthesis of the materials in Ti-Al-C-N system" *Advances in Science and Technology*, 45, 2006, p 1047-1051

[18] L. Chlubny, M.M. Bucko, J. Lis "Phase Evolution and Properties of Ti₂AlN Based Materials, Obtained by SHS Method" Mechanical Properties and Processing of Ceramic Binary, Ternary and Composite Systems, *Ceramic Engineering and Science Proceedings*, Volume 29, Issue 2, 2008, Jonathan Salem, Greg Hilmas, and William Fahrenholtz, editors; Tatsuki Ohji and Andrew Wereszczak, volume editors, 2008, p 13-20

[19] L. Chlubny, J. Lis, M.M. Bucko: Preparation of Ti₃AlC₂ and Ti₂AlC powders by SHS method MS&T Pittsburgh 09: Material Science and Technology 2009, 2009, p 2205-2213

[20] L. Chlubny, J. Lis: Influence of Precursors Stoichiometry on SHS Synthesis of Ti₃AlC₂ Powders, Ceramic Transactions *(in print)*

[21] "Joint Commitee for Powder Diffraction Standards: International Center for Diffraction Data"

[22] H. M. Rietveld: "A profile refinement method for nuclear and magnetic structures." J. Appl. Cryst. **2** (1969) p. 65-71

[23] L. Chlubny: New materials in Ti-Al-C-N system. - PhD Thesis. AGH-University of Science and Technology, Kraków 2006. *(in Polish)*

# XRD AND TG-DSC ANALYSIS OF THE SILICON CARBIDE-PALLADIUM REACTION

M. Gentile[a], P.Xiao[b], T. Abram[a]

[a]Centre for Nuclear Energy Technology (C-NET), School of Mechanical, Aerospace and Civil Engineering, The University of Manchester, Manchester M13 9PL, UK

[b]Materials Science Centre, School of Materials, The University of Manchester, Manchester M13 9PL, UK

ABSTRACT

The attack of palladium on silicon carbide was investigated using high purity powders of $\alpha$-SiC and palladium blended to produce a mixture with composition of 95 at.% SiC and 5 at.% Pd and cold pressed to pellets.

The palladium silicon carbide reaction was studied by thermogravimetry (TG) and differential scanning calorimentry (DSC), whereas the phase composition of the specimens was studied using X-ray diffractometry (XRD) before and after thermoscans. X-ray powder analysis (XRD) was employed in combination with Rietveld method for a quantitative phase analysis. Thermoscans of $\alpha$-SiC-5at. %Pd pellet present three exothermic peaks at 773 K, 1144 K and 1615 K that indicate the presence of three reaction stages. XRD patterns of $\alpha$-SiC-5at. %Pd pellets show the presence of $Pd_xSi$ phase that developed after the thermal treatment.

This work provides a better understanding the palladium-silicon carbide reaction mechanism, in particular analyses the interaction between palladium and alpha silicon carbide.

## INTRODUCTION

Silicon carbide has many applications in a wide range of industries. Due to its high thermal conductivity[1], wide band gap and large electron mobility[2] it is used in high-power and high temperature electronic device. In addition due to its chemical inertness and low neutron absorption cross-section it is used as coating in tristructural isotropic (TRISO) fuel[3,4] and it is seen as alternative to zirconium alloys for a new generation of nuclear fuel cladding.

In nuclear application silicon carbide provides an effective barrier to the diffusion of gaseous and metallic fission products[5,6,7]. Therefore, the integrity of silicon carbide material is crucial to the safety and the performance of nuclear power plants.

Nuclear plants produce energy using the heat that occurs during nuclear fission when an atom splits into two smaller parts forming two new elements and releases energy. A significant amount of palladium is formed by the fission process of nuclear fuel. Palladium unlike other fission products such as rare-earth elements, causes corrosion of the silicon carbide layer at various kernel compositions in tristructural-isotropic (TRISO) nuclear fuels particles[8].

TRISO nuclear fuel particles consist in a fuel kernel, typically $UO_x$, coated with four layers of three isotropic materials that are carbon, dense pyrolitic carbon (PyC) and silicon carbide (SiC).

In particular, silicon carbide has the function to retain fission products at elevated temperatures and to give to TRISO particle more structural integrity.

Thus, for the optimization of the diffusion barrier in next-generation TRISO fuels it is important to understand the silicon carbide interaction with different fission products, in particular with palladium.

Most of the published results found that the fuel temperature mainly governs the palladium silicon carbide interaction. Therefore in this paper the palladium silicon carbide interaction is analysed through thermal analysis in a temperature range comprised between 293 K and 1773 K in order to better explain the influence of the temperature on the palladium attack of silicon carbide. The studied temperature spectrum covers the operating temperature of diverse nuclear power reactors such as boiling water reactors (BWR) and very high temperature reactors (VHTR).

Thermogravimetry and differential scanning calorimetry studies are coupled with X-ray diffraction analysis in order to characterize the phases present in the specimens. The morphology of specimen surface after heat treatments was investigated through secondary electron microscopy (SEM).

EXPERIMENTAL WORK

Specimens were produced from high purity powders of alpha silicon carbide (Goodfellow, mean particle size < 1 $\mu m$, 99.9% trace metal basis) and palladium (Sigma-Aldrich, mean particle size < 1 $\mu m$, ≤ 99.9% trace metal basis).

Powders containing 95 at. % SiC and 5 at. % Pd were blended for 12h in an automatic machine. After blending the powder mixture was uniaxially cold-pressed under a pressure 0.98 MPa in hard steel dies using a hydraulic press to form compacts of cylindrical pellets having 5 mm diameter.

Thermogravimetry (TG) and differential scanning calorimentry (DSC) tests were carried out through NETZSCH STA 449 F1 thermal analiser. Thermoscans were conducted in 20-ml/min argon flow from 293 K to 1773 K at a temperature increment of 5 K/min.

The phase composition of the specimens before and after thermoscans was identified through x-ray diffraction (XRD) and quantified through Rietveld method. X-ray diffraction measurements were performed using a PANalytical X'Pert MPD X-ray diffractometer. The tube voltage and current were 40 kV and 40 mA, respectively. The tube anode was CuK$\alpha_1$ ($\lambda$ = 0.154 06 nm). The XRD data were analysed using PANanalytical X'Pert HighScore Plus software. Rietveld refinement was performed using the structural models from ICSD database with TOPAS 4.2 software.

The specimens were embedded in copper-loaded resin and ground/polished through successive grades of SiC paper, followed by finishing to 1 $\mu m$ diamond. The surface of mechanical polished specimens was characterized by scanning electron microscopy (FEG-SEM JEOL 6300, Tokyo, Japan).

RESULTS

**Thermogravimetry and differential scanning calorimetry (TG – DSC)**

Simultaneous thermogravimetry and differential standard calorimetry analysis were carried out on pure silicon carbide powders and mixed silicon carbide-palladium powders in order to study the silicon-carbide palladium reaction. All the measurements were carried out in argon gas.

Figure 1 reveals the dynamical TG curves of pure $\alpha$ −SiC and $\alpha$ −SiC-5at. %Pd pellets in a temperature range comprised from 293 K to 1773 K.

The diagram shows that pure silicon carbide specimen did not change his weight up to the temperature of 1638 K. After reaching this temperature the specimen's weight slowly decreased of the 0.42%.

On the contrary for the SiC-5at. %Pd pellet, the specimen started to gain weight at about 723 K reaching the maximum value of weight change at 1408 K, after this temperature the percentage of weight change decreased of 2.18%.

Figure 2 shows DSC curves of pure $\alpha$ −SiC and $\alpha$ −SiC-5at. %Pd pellets in a temperature range comprised from 293 K to 1773 K.

The curve of pure silicon carbide specimen reveals an exothermic peak in the temperature range comprised between approximately 923 K and 1263 K. Conversely the plot representing the differential scanning calorimetry analysis of the palladium silicon carbide specimen shows three exothermic peaks, which indicate the presence of three reaction stages. The first peak was in the temperature range comprised between 673 K and 873 K, while the second and third peak occurred from 1063 K to 1253 K and from 1357 K to 1773 K respectively.

The sudden rise of the TG and DSC values registered in the initial part of the curves in Figure 1 and Figure 2 was due to buoyancy.

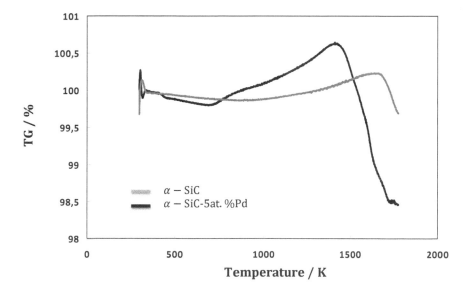

**Figure 1** TG curves of pure $\alpha$ −SiC and $\alpha$ −SiC-5at. %Pd pellets in a temperature range comprised from 293 K to 1773 K.

**X-ray diffraction (EDX) and secondary electron microscopy (SEM) analysis**

Phases developed in pure silicon carbide and palladium silicon carbide compact pellets at diverse temperatures during thermoscans were investigated through x-ray diffraction (XRD). The diffraction patterns were analysed using the PANanalytical X'Pert HighScore Plus software.

XRD analysis was carried out on specimens before and after the heat treatment. In Figure 2 letters A, B, C, D, E in the DSC plot indicate the temperatures at which the analyses were carried out.

XRD spectra of specimens before heat-treatment revealed that the alpha silicon carbide powders were moissanite 6H and 4H polytypes.

Figure 3 shows XRD spectra of pure $\alpha$ −SiC and $\alpha$ −SiC-5at. %Pd specimens after annealing up to 1773 K at a temperature increment of 5 K/min. The XRD spectra corresponding to the temperature of 1773 K show that at the end of the heat treatment $SiO_2$, $Pd_2Si$ peaks were produced in the palladium silicon carbide pellet.

Secondary electron microscopy (SEM) was employed to distinguish the metal phase from the oxide and ceramic powder. The examination was carried out in regions near the surface of the annealed specimens. Figure 4 presents secondary electron micrographs of mechanical polished pure $\alpha$-SiC and $\alpha$-SiC-5at. %Pd after thermoscans.

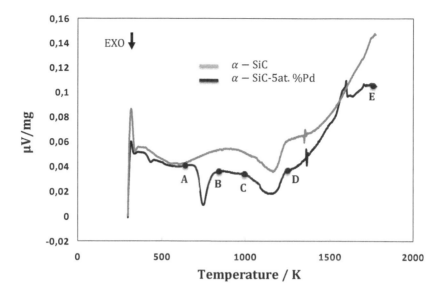

**Figure 2** DSC curves of pure $\alpha$ −SiC and $\alpha$ −SiC-5at. %Pd pellets in a temperature range comprised from 293 K to 1773 K.

Secondary electron microscopic observation on a polished surface of SiC-5at. %Pd shows that shiny palladium metal spots are still present in the compact at the end of the thermoscan (Figure 4(b)). This indicates that only part of the palladium in the compact reacted to form di-palladium silicides (Pd$_2$Si). Conversely dark spots on the surface of SiC-5at. %Pd pellet are associated with the formation of silicon oxide or carbon area, although due to the size of the powder it was not possible to carry out Energy Dispersive X-ray spectroscopy (EDX) analysis.

Silicon carbide palladium reaction in the temperature range comprised between 293 K and 1773 K was checked by XRD studies. Figure 5 displays a series of XRD spectra at diverse temperatures representing the phase evolution SiC/Pd compacts containing 5at. % Pd during thermoscans.

XRD spectra were carried out at the temperatures indicated by the points A, B, C, D, and E in the DSC plot of Figure 2. The point A, B, C, D and E indicate respectively the following temperatures 623 K, 873 K, 1063 K, 1253 K, and 1773 K.

According to estimates made analysing the XRD spectra through PANanalytical X'Pert HighScore Plus software, in the first reaction stage occurring from 623 K to 873 K the phase Pd$_3$Si phase was formed.

All the Pd$_3$Si phase was converted in Pd$_2$Si phase during the second reaction stage that occurred in the temperature range comprised between 1063K and 1253 K. XRD spectra do not detect the presence of Pd$_3$Si phase in specimens annealed to temperatures higher than 1253 K.

The XRD spectrum carried out at the final temperature show the presence of Pd$_2$Si and SiO$_2$ indicating that the third exothermic peak is associated to the active oxidation of silicon carbide.

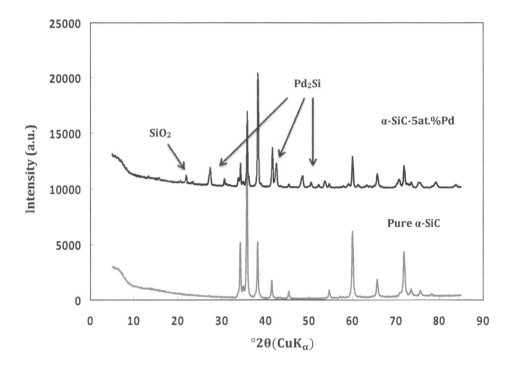

**Figure 3** XRD spectra of pure $\alpha$ −SiC and $\alpha$ −SiC-5at. %Pd specimens after annealing up to 1773 K at a temperature increment of 5 K/min.

**Figure 4** Secondary electron micrograph of mechanical polished silicon carbide and palladium powders after the thermaoscans carried out up to 1773 K: a) pure α-SiC; b) α-SiC-5at. %Pd.

In order to quantify the phases developed during the heat treatment, Rietveld refinement was performed on XRD spectra from specimens that had been annealed up to the temperatures of 600 K and 980 K. These are the temperatures at the end of the first reaction stage and the second reaction stage respectively. Table 1 shows the results of the phase quantification.

**Table 1**
Phase quantification using Rietveld method for specimens annealed up to 600 K and 980 K.

| Phase | Phase composition at 600 K [at.%] | Phase composition at 980 K [at.%] |
|---|---|---|
| $\alpha - SiC$ | 87.32 | 81.83 |
| $Pd$ | 2.79 | 8.95 |
| $PdO$ | 0.79 | 1.42 |
| $Pd_2Si$ | - | 7.8 |
| $Pd_3Si$ | 9.09 | - |

DISCUSSION

Palladium silicon carbide interaction limits the use of silicon carbide as coating of nuclear fuel particles and in high power and high temperature electronic device. Previous studies have found that palladium causes degradation of silicon carbide properties because it reacts with silicon carbide forming palladium silicides that create weak points in the structure of silicon carbide decreasing the strength of the material.

In addition, it has been observed that silicon carbide and palladium form periodic-layered structures with $Pd_xSi$ layer adjacent to carbon layer[9,10] due to the diffusion of palladium species.

In particular a number of studies have shown that the temperature play a key role in the palladium silicon carbide reaction.

Therefore in this study thermal analysis was carried out on mixed powders of silicon carbide and palladium in the temperature range comprised between 293 K and 1773 K.

Thermogravimetry analysis registered a gain in weight due to oxidation of silicon carbide in the temperature range comprised between 723 K and 1408 K. The weight gain was followed by weight loss associated with the active oxidation of silicon carbide.

On the other side, differential scanning calorimetry analysis coupled to X-ray diffraction identified three exothermic reaction stages.

In the first reaction stage occurring from 623 K to 873 K, palladium reacted with silicon carbide forming $Pd_3Si$, while in the second reaction stage occurring between 1023 K and 1253 K $Pd_3Si$ was converted in $Pd_2Si$. XRD analysis confirmed that there was not $Pd_3Si$ phase in specimen annealed at temperatures higher of 1253 K. The third exothermic peak is associated to the active oxidation of silicon carbide.

Figure 6 presents Pd-Si-C phase diagrams at the temperatures of 973 K and 1273 K, which are the final temperatures of the first reaction stage and the second reaction stage respectively. An analysis of the Pd-Si-C phase diagram at the temperatures of 973 K and 1273 K reveals that the formation of Pd₂Si is thermodynamically favored at the end of each reaction stage. Thus, the XRD results at temperatures lower than 1253 K are in contrast with thermodynamic data.

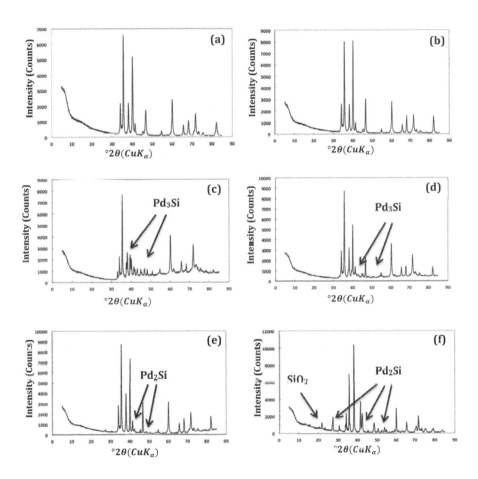

**Figure 5** XRD spectra at diverse temperatures showing the phase development in α −SiC-5at. %Pd compact during thermoscans:(a) specimen before annealing; (b) specimen annealed up to 623 K (Letter A); (c) specimen annealed up to 873 K (Letter B); (d) specimen annealed up to 1063 K (Letter C); (e) specimen annealed up to 1253 K (Letter D); (f) specimen annealed up to 1773 K (Letter E).The letters associated to each plot indicate in Figure 2 the final temperature of the specimens' heat treatment.

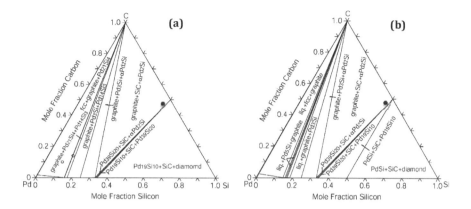

**Figure 6** Isothermal section of Pd-Si-C system at different temperature: (a) 973 K; (b) 1273 K.[11]

However, previous research work revealed that Pd₃Si phase is generated in specimens annealed at temperatures around 698 K[12,13], while contrasting results were observed in specimens annealed at temperature higher than 1073 K.

After annealing at temperature higher than 1073 K some specimens presented Pd₃Si and Pd₂Si phases while other specimens presented only Pd₂Si phase[14,11].

These contrasting results can be associated to the chemical composition of the gas in which the heat treatment was carried out. Previous results[14] shows that in specimens annealed in air, Pd₃Si phase was formed at lower temperature and with lower palladium content than in specimens annealed in vacuum

The results presented in this paper confirm these findings. The weight gain registered by the thermogravimetry plot indicates the formation of silicon oxide due to residual oxygen present in the argon gas. The formation of silicon oxide starts together with the first reaction stage in which Pd₃Si are formed. This indicates that probably the oxidation of silicon carbide plays a role in the palladium silicide interaction. Further studies are necessary to identify the influence of oxygen on the mechanism of interaction between palladium and silicon carbide.

CONCLUSIONS

A thermal analysis of the palladium silicon carbide interaction revealed three reaction stages in the temperature range comprised between 293 K and 1773 K. The three reaction stages were exothermic. During the first reaction stage, occurring from 673 K to 873 K, palladium interacted with silicon carbide forming Pd₃Si phase, while in the second reaction stage from 1063 K to 1253 K Pd₃Si phase disappeared forming Pd₂Si phase. The third exothermic reaction from 1357 K to 1773 K was due to the active oxidation of silicon carbide. Thermogravimetry analysis revealed that silicon oxide was created together with Pd₃Si phase. Comparison of the present results with previous studies suggests that the oxidation of silicon carbide plays a role in the palladium silicon carbide interaction.

REFERENCES

1. Slack, G. A. *Journal of Applied Physics* **1964**, *35*, 3460;
2. Barret, D. L.; Campbell, R. B. *Journal of Applied Physics* **1967**, *38*;
3. Ford, L. H.; Hibbert, N. S.; Martin, D. G. *Journal of Nuclear Materials* **1973**, *45*, 139 149.
4. Snead, L. L.; Nozawa, T.; Katoh, Y.; Byun, T.; Kondo, S.; Petti, D. A. *Journal of Nuclear Materials* **2007**, *371*, 329 377.
5. Minato, K.; Fukuda, K.; Ishikawa, A.; Mita, N. *Journal of Nuclear Materials* **1997**, *246*, 215 222.
6. Lauf, R. J.; Lindemer, T. B.; Pearson, R. L. *Journal of Nuclear Materials* **1964**, *120*, 30.
7. Minato, K.; Ogawa, T.; Fukuda, K.; Sekino, H.; Miyanishi, H.; Kado, S.; Takahashi, I. *Journal of Nuclear Materials* **1993**, *202*, 47 53.
8. Minato, K.; Ogawa, T.; Kashimura, S.; Fukuda, K.; Shimizu, M.; Tayama, Y.; Takahashi, I. *Journal of Nuclear Materials* **1990**, *172*, 184 196.
9. Bhanumurthy, K.; Schmid-Fetzer, R. *Composites: Part A* **2001**, *32*, 569 574.
10. Lopez-Honorato, E.; Fu, K.; Meadows, P. J.; Tan, J.; Xiao, P. *Journal of American Ceramic Society* **2010**, *93*, 4135 4141.
11. Du, Z.; Guo, C.; Yang, X.; Liu, T. *Intermetallics* **2006**, 14, 560 569.
12. Chen, L.; Hunter, G. W.; Neudeck, P. G.; Bansal, G.; Petit, J. B.; Knight, D. *Journal of Vacuum Science Technology A* **1997**, *15*, 1228.
13. Roy, S.; Basu, S.; Jacob, C.; Tyagi, A. K. *Applied Surface Science* **2002**, *202*, 73 79.
14. Suzuki, H.; Iseki, T.; Imanaka, T. *Journal of Nuclear Science and Technology* **1977**, *14*, 438 442.

# MODELLING DAMAGE AND FAILURE IN STRUCTURAL CERAMICS AT ULTRA-HIGH TEMPERATURES

M. Pettinà[1], F. Biglari[1], D. D. Jayaseelan[2], L. J. Vandeperre[2], P. Brown[3], A. Heaton[3] and K. Nikbin[1]

[1]Department of Mechanical Engineering, Imperial College London, London, UK
[2]Centre for Advanced Structural Ceramics, Imperial College London, London, UK
[3]Dstl Porton Down, Salisbury, Wiltshire SP4 0JQ

## ABSTRACT

This paper considers the state of the art in both testing and modelling for ultra-high temperature ceramics and identifies the material variables that need to be derived for modelling structural failures in ceramics at extreme temperatures. By focusing on monolithic $ZrB_2$ ultra-high temperature ceramic, for which a reasonable set of material creep data is available, the paper identifies a damage model appropriate for the failure mechanism of the material at temperatures above 1800K. Failure mapping for this material has shown that power law creep deformation, similar to metals at lower temperatures, is the dominant mechanism at temperatures greater than 1800K and at stresses above 100 MPa. As an example a representative three point bend geometry, which is planned to be tested, is modeled using the relevant creep constitutive properties that have been found for $ZrB_2$. The creep properties, over the temperature range, 1800-2500K are used in a multiaxial ductility exhaustion damage model to predict cracking in the three point bend sample model. In this way the essential properties required to develop predictive damage simulations are investigated, underlining the importance of having accurate and appropriate materials test data.

## INTRODUCTION

Assessment of damage and failure of structural ceramics at extreme temperatures is an area that needs substantial development. In order to produce a robust tool for predictive modelling of failure in ultra-high temperature ceramic (UHTC) materials it is important to assess what material and mechanical properties can be reasonably derived to input in the models. Engineering applications involving exposure of components to extreme thermal and chemical environments require materials capable of withstanding such harsh conditions. UHTC materials are considered among the best candidates for extreme temperature applications, due to their high melting point and excellent combination of chemical stability, high electrical and thermal conductivities and resistance to corrosion/erosion.[1] $ZrB_2$ and $HfB_2$ are part of this class of materials and are mainly used in combination with SiC to create particle-reinforced ceramics with improved oxidation resistance and mechanical properties.

Table I. Properties for $ZrB_2$ available in the literature

| Property | Temperature | Value | Unit | Reference |
|---|---|---|---|---|
| Melting temperature | | 3518 | K | 2 |
| Density | RT | 5.49 | $g/cm^3$ | 3 |
| Bend strength | RT | 300–580 | MPa | 4–6 |
| | 2273K | 100 | MPa | 4–6 |
| Elastic modulus | RT | 489–493 | GPa | 7 |
| | 1673K | 220–360 | GPa | 8 |
| Hardness | RT | 19–29 | GPa | 9–11 |
| | 2273K | 0.61 | GPa | 12 |
| Fracture toughness | RT | 3.5–4.2 | $MPa \cdot m^{1/2}$ | 1 |
| Thermal conductivity | 523K | 55 | W/mK | 3 |
| | 2273K | 44 | W/mK | 3 |

  In order to assess damage and failure of UHTC components in service conditions consistent and verifiable material properties ranging from room temperature to ultra-high temperatures (above 2200K) are needed. However, performing experiments at very high temperatures is costly and challenging, and often test data available are insufficient to fully characterize the material at those temperature regimes. This paper focuses on monolithic $ZrB_2$ which seems to be the most investigated material of the class of UHTCs. A substantial amount of data on mechanical properties for this material is reported in the literature.[1-12] A summary of the properties of interest for this paper is provided in Table I. It is clear that for temperatures above 2200K the material properties data are incomplete and difficult to find. It is therefore important to be able to model the material in the ultra-high temperature region using the best available data and take into account the level of accuracy and scatter in any data used. The work described in this paper attempts to predict damage and cracking in pre-cracked geometries in order to recommend improved testing methods for an experimental program.

BACKGROUND FOR FRACTURE MECHANICS METHODS UNDER CREEP CONDITIONS

  In this section a brief overview of the relevance of fracture mechanics to creep damage and cracking behaviour in components is discussed. The basic concept is the assumption that the tensile and creep properties of the material determine the stress and strain rate distributions ahead of a cracked component and by determining these properties it is possible to evaluate the stress state at the crack tip and hence the rate of damage and crack growth in the component. This should be true to a whole range of materials from the very brittle where elastic properties dominate to the extreme ductile conditions where elastic/plastic/creep conditions prevail.

In general, metals at temperatures greater than half their melting points undergo time dependent creep deformation. The behaviour can be described by a creep curve (a plot of creep strain vs. time) which consists of three regions, namely the primary, secondary and tertiary creep regimes, as shown in Fig. 1. Parameter of interest is the secondary steady state region in Fig. 1 which is described by the well-known Norton's power law creep law:

$$\dot{\varepsilon}_c = A\sigma^n \tag{1}$$

where $\dot{\varepsilon}_c$ is the steady state creep strain rate, $\sigma$ is the applied stress, $A$ is a temperature dependent material constant and $n$ is the power law creep stress exponent. Furthermore to account for the three regions in Fig. 1 the average creep strain rate, $\dot{\varepsilon}_A$, may be employed, defined by the ratio of the uniaxial failure strain, $\varepsilon_f$, to the rupture time in a uniaxial creep test, $t_r$. This may also be written in the power law form:

$$\dot{\varepsilon}_A = A\sigma_e{}^n = \frac{\varepsilon_f}{t_r} \tag{2}$$

where $\sigma_e$ is the equivalent stress.

Therefore, based on the power law creep behaviour response of the material, the stress/strain rate distributions can be estimated using fracture mechanics methods. In such a case the crack tip stress and strain rate fields in a cracked geometry can be described in terms of the $C^*$ parameter[13-15] and by the HRR stress/strain rate field distribution,[16-18] given by:

$$\sigma_{ij} = \sigma_0 \left(\frac{C^*}{\dot{\varepsilon}_0 \sigma_0 I_n r}\right)^{\frac{1}{n+1}} \tilde{\sigma}_{ij}(\theta, n) \tag{3}$$

$$\dot{\varepsilon}_{ij} = \dot{\varepsilon}_0 \left(\frac{C^*}{\dot{\varepsilon}_0 \sigma_0 I_n r}\right)^{\frac{n}{n+1}} \tilde{\varepsilon}_{ij}(\theta, n) \tag{4}$$

where $r$ is the radial distance from the crack tip and $\theta$ is the crack tip angle. In Eqs. 3 and 4, $C^*$ is the amplitude of the crack tip fields, which in an experiment may be estimated from the remote load line displacement rate, $I_n$ is a non-dimensional function of $n$, and $\tilde{\sigma}_{ij}$ and $\tilde{\varepsilon}_{ij}$ are dimensionless functions of $n$ and $\theta$. The values of the functions, $I_n$, $\tilde{\sigma}_{ij}$ and $\tilde{\varepsilon}_{ij}$ depend on whether plane stress or plane strain conditions prevail near the crack tip and solutions are available for these functions in Ref.[18]

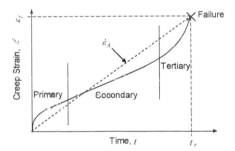

Figure 1. Schematic for a creep curve showing definition of the average creep strain rate $\dot{\varepsilon}_A$

The $C^*$ parameter may be determined experimentally for crack growth tests from the creep load line displacement rate measurements, $\dot{\Delta}^c$, using the relation:[14,15]

$$C^* = \frac{P\dot{\Delta}^c}{B_n b} F \tag{5}$$

where $P$ is the applied load, $b$ is the remaining ligament ahead of the crack, $B_n$ is the net thickness between side-grooves, and $F$ is a factor which depends on crack length, specimen geometry and creep stress index, $n$. In the case of a three point bend test, the creep displacement rate at the load line will be measured and converted into $C*$ using Eq. 5 which should then correlate the rate of crack initiation and growth in the geometry.

When a steady state creep stress and damage distribution is achieved ahead of the crack tip a power law relationship between the creep crack growth (CCG) rate, $\dot{a}$, and the $C*$ parameter is expected to be in the form:[13-15]

$$\dot{a} = DC^{*\phi} \tag{6}$$

where $D$ and $\phi$ are temperature dependent material constants. The value of $\phi$ is typically a fraction close to but less than unity. Similarly for a 3-point bend $ZrB_2$ specimen the cracking behaviour in the ultra-high temperature range might be expected to follow Eq. 6.

## FAILURE MECHANISMS IN $ZrB_2$ IN THE ULTRA HIGH TEMPERATURE RANGE

Recently a deformation mechanism map for $ZrB_2$ has been developed by Wang et al.[12] to identify its failure mechanisms and the evolution of its toughness with temperature. The map combines new data on the hardness from the same authors with data available in the literature[9-11] and shows the dominant mechanisms as a function of the conditions, i.e. strain rate, stress and temperature. In the low temperature regime (<1800K) and low stresses (<100 MPa) glide is controlled by lattice resistance, resulting in brittle fracture rather than plastic deformation. As temperatures reaches 1800K and higher the predominant mechanism becomes power law creep, and this is consistent with a time-dependent permanent creep deformation rather than elastic brittle fracture that dominates ceramics at low temperatures. Diffusion creep is the mechanism controlling deformation at higher temperatures and lower stresses (<100 MPa) whilst dislocation driven creep deformation controls damage at higher stresses.

Same parameters used to generate the deformation mechanism map for $ZrB_2$ were then used by Wang et al.[12] to plot constant-temperature lines on a shear strain rate vs. shear stress map. By doing this a comparison with creep data generated by Rhodes et al.[6] was made, showing good agreement and validating the approach.

For stresses greater than 100 MPa the value of $n$ is about 7 for the different temperatures suggesting a typical creep deformation behaviour for $ZrB_2$, which might be assumed to be similar to that of metallic alloys. Relevant creep constitutive properties calculated from Wang et al.[12] for different temperature and stresses are summarized in Fig. 2 and tabulated in Table II according to Eq. 2.

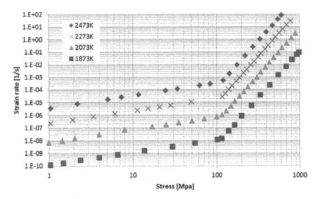

Figure 2. Power law creep curves calculated based on Wang et al.[12]

This paper makes use of the fact that power law creep properties given in Table II for $ZrB_2$ are similar to those of metallic alloys at high temperatures. Therefore it is postulated that the stress/strain rate response at ultra-high temperatures in a $ZrB_2$ specimen can be modeled by a power law relation, and can subsequently be used as input in a remaining multiaxial ductility exhaustion damage model to predict damage and cracking. The model described briefly in the next section is used in a subroutine written in FORTRAN code to predict failure of components under multiaxial stress states. By coupling this with the finite element (FE) software ABAQUS[19] the damage evolution due to creep can be monitored over the temperature range of interest. A cracked 3-point bending specimen (described in detail in the next section) has been chosen as an initial example to test the methodology.

Table II. Power law creep properties extrapolated from Figure 2

| Temp [K] | <100 MPa $n$ | <100 MPa $A\ [\frac{1}{MPa^n \cdot s}]$ | >100 MPa $n$ | >100 MPa $A\ [\frac{1}{MPa^n \cdot s}]$ |
|---|---|---|---|---|
| 2473 | 1 | 3.00E-06 | 7.3 | 3.00E-19 |
| 2273 | 1 | 2.00E-07 | 7.4 | 2.00E-20 |
| 2073 | 1 | 7.00E-09 | 7.3 | 5.00E-22 |
| 1873 | 1 | 1.00E-10 | 7.4 | 9.00E-24 |

FINITE ELEMENT MODELLING

A predictive creep fracture modelling procedure using the aforementioned methodology in a finite element method[16,18] has been developed to model creep crack growth rates in geometries containing cracks. It has been found that uncoupled continuum damage methods can be used to predict creep crack growth within a numerical framework. In this section, a brief review of the approach is presented.

Material Model

In most metallic materials the properties needed to develop the stress and strain rates at the crack tip are elastic-plastic-creep as in Eq. 7. The plastic strains are understood to be independent of strain rate giving the total strain as:

$$\varepsilon = \varepsilon_{el} + \varepsilon_{pl} + \varepsilon_c \tag{7}$$

where $\varepsilon_{el}$, $\varepsilon_{pl}$ and $\varepsilon_c$ are elastic, plastic and creep strains respectively. However, for more brittle materials the material behaviour could be described by elastic-creep deformation rates as:

$$\dot{\varepsilon} = \dot{\varepsilon}_{el} + \dot{\varepsilon}_c \tag{8}$$

The creep response is described by a secondary creep law using the average creep properties as in Eq. 2.

Continuum Damage Modelling

Continuum damage methods have been used to predict failure of metallic components at high temperature, focusing on the prediction of crack initiation and rupture life.[20-22] In this work, an uncoupled damage-based approach is used to simulate crack growth from the transient to the steady state regime. It follows the approach in Refs.[23,24] whereby nodes ahead of the crack tip are released when the damage reaches a critical value, simulating the formation of a sharp crack. A similar node-release approach to model crack growth under creep conditions has recently been employed,[25] in which the crack growth rate has been assumed a priori, based on experimental data, and has not been determined within the analysis.

Following Ref.[23], a damage parameter, $\omega$, is defined such that $0 \leq \omega \leq 1$. Failure at a material point occurs when $\omega = 1$. The rate of damage accumulation is related to the equivalent creep strain rate, $\dot{\varepsilon}_c$, by:

$$\dot{\omega} = \frac{\dot{\varepsilon}_c}{\varepsilon_f^*} \tag{9}$$

where $\varepsilon_f^*$ is the multiaxial creep ductility. The total damage at any instant is the integral of the damage rate in Eq. 9 up to that time:

$$\omega = \int_0^t \dot{\omega} \, dt \tag{10}$$

Thus, failure occurs in the vicinity of the crack tip when the local accumulated strain reaches the local multiaxial creep ductility. The multiaxial creep ductility, $\varepsilon_f^*$, can be obtained from a number of available void growth models. Rice and Tracey[17] have developed the following expression for rigid-plastic deformation:

$$\frac{\varepsilon_f^*}{\varepsilon_f} = 0.521/\sinh\left(\frac{3}{2}\frac{\sigma_m}{\sigma_e}\right) \tag{11}$$

where $\sigma_m$ is the hydrostatic stress. Alternatively, Cocks and Ashby[26] have proposed to represent the multiaxial creep ductility of the material as:

$$\frac{\varepsilon_f^*}{\varepsilon_f} = \sinh\left[\frac{2}{3}\left(\frac{n-1/2}{n+1/2}\right)\right] / \sinh\left[2\left(\frac{n-1/2}{n+1/2}\right)\frac{\sigma_m}{\sigma_e}\right] \tag{12}$$

For most practical purposes the value of $\varepsilon_f^*$ is taken as the uniaxial failure strain $\varepsilon_f$ (plane stress conditions) or $\varepsilon_f/30$ (plane strain conditions).[27] The value of $n$ in Norton's law lies within 5 to 10 for most engineering materials. In these cases, therefore, $(n-0.5)/(n+0.5)$ would be between 0.818 and 0.905, which can be assumed as a constant in Eq. 12, allowing an approximation to be written as follows:[28]

$$\frac{\varepsilon_f^*}{\varepsilon_f} = 0.61/\sinh\left(\sqrt{3}\frac{\sigma_m}{\sigma_e}\right) \tag{13}$$

where $(n-0.5)/(n+0.5)$ is approximated as $\sqrt{3}/2 \approx 0.866$.

The comparison between Eqs. 11-13 is shown in Fig. 3. It can be observed that the reduction of creep ductility based on Eq. 12 gives little dependence on the value of $n$ except for $n = 2$ which is an unrealistic value for engineering materials. Therefore it could be said that the creep index independent Eq. 13 gives a suitable estimation of Eq. 12, whilst Eq. 11 gives upper bound of Eq. 12 except for $n = 2$. The simplified relationship in Eq. 13 effectively describes the importance of constraint on the inverse effect of appropriate multiaxial failure strain that is needed for damage to develop at the crack tip.

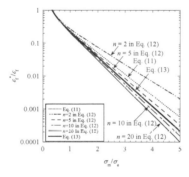

Figure 3. Comparison of the influence of stress state on creep ductility as a function of creep index $n$ for different void growth models

## CREEP CRACK GROWTH PREDICTION MODELS

As a brief summary, this section highlights the analytical modelling of failure in metals under creep loading conditions. The NSW models[23,24] have been widely used to predict the steady state cracking rate using uniaxial creep data. In these models the creep crack growth is predicted to occur when a critical quantity of damage, measured by creep ductility exhaustion approach, is attained at a characteristic distance, $r_c$, ahead of the crack tip. A modified version of the NSW model, known as NSW-MOD, has been derived in Ref.[28] to predict the CCG rate under steady state creep conditions by considering the dependency of creep strain on the crack tip angle, $\theta$, and the power law creep stress exponent, $n$, in addition to the stress state. The CCG rate prediction from the NSW-MOD model, $\dot{a}_{NSW-MOD}$, is given by:

$$\dot{a}_{NSW-MOD} = (n+1)\left(\frac{C^*}{I_n}\right)^{\frac{n}{n+1}}(Ar_c)^{\frac{1}{n+1}}\left.\frac{\tilde{\varepsilon}_e(\theta,n)}{\varepsilon_f^*(\theta,n)}\right|_{max} \tag{14}$$

where $\tilde{\varepsilon}_e$ is a non-dimensional function of $\theta$ and $n$, and $I_n$ is a non-dimensional function of $n$. The values of $\tilde{\varepsilon}_e$ and $I_n$ are tabulated in Refs.[16,18] The creep process zone, $r_c$, is usually taken as the material's average grain size. The NSW-MOD model needs the plane stress uniaxial creep failure ductility of the material and the calculation of the parameter $C^*$ to predict cracking in cracked geometries. Similar to what is done for metals, by using this model for ceramics in the ultra-high temperature regime where the deformation mechanism is described by power law creep, it may be possible to predict the upper/lower bounds for creep crack growth once the parameter $C^*$ and the failure strain $\varepsilon_f$ are known. These can be derived from experimental tests which are not completed at present.

FINITE ELEMENT FRAMEWORK FOR UHTC MATERIALS

All finite element analyses were performed using ABAQUS.[19] A 3D geometry of a single edge notch 3-point bend (SENB) specimen is analyzed. With reference to Figure 4, specimen geometry is $W = 2$ mm, $L = 22$ mm, $B = 2.5$ mm and $a_i/W = 0.1$, and the applied load, $P$, is 20 N. Both plane stress and plane strain analyses have been carried out. Crack growth was modeled using a nodal-release technique, as presented in Ref.[23] When damage parameter $\omega$ reaches unity at two integration points ahead of the crack tip, the node at the crack tip is released. Regular square elements were used in the vicinity of the crack tip so that the crack grows through a region of uniform elements. It is assumed in the analysis that the crack grows in the plane of the initial crack front. The mesh size at the crack tip is approximately 20 μm.

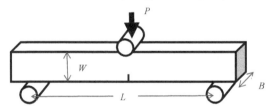

Figure 4. Schematic illustration of 3-point bending test of a rectangular cross-section beam

Because of load and shape symmetry only a quarter of the beam was modeled in the FE simulations. A preliminary analysis was carried out using both coarse and fine meshes. The finite element mesh for the quarter of the beam is shown in Figure 5.

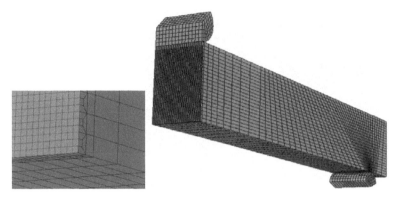

Figure 5. Finite element mesh for three point bending analysis

The finite element mesh consists of 22723 elements and 26954 nodes. Due to the high stress intensity around the crack tip and loading points an innovative mesh refinement scheme has been employed. This meshing scheme allows for an element refinement ratio of 1 to 9 on the symmetry line were an initial crack is introduced. Checks were made to ensure mesh is adequately refined at the loading points. A symmetry boundary condition has been specified on the front face along x direction. This face contains a crack at the bottom with a separated area from the symmetry line. To accommodate for a smooth FE solution for crack growth at this face, a high density mesh has been employed.

Figure 6 shows the stress distribution perpendicular to the crack surface (S11). During creep crack propagation, an increase in stress levels of more than two times the initial values can be seen in Figures 6a to 6d.

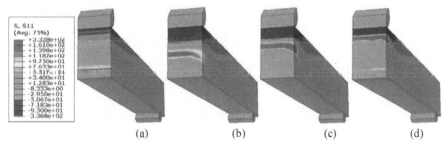

Figure 6. Stress distribution at (a) start and (d) end of creep crack propagation simulation

Figure 7 illustrates the damage development during creep crack growth. Material constant $D$ in Eq. 6 is used as a scalar damage index to degrade material stiffness. This material degradation causes the separation of the crack front from the face with a x-symmetry boundary condition, which has been specified for this face of the mesh. Consequently, the crack growth can propagate freely in three dimensional mesh domains without any restriction. It can be seen that, the damage is more dominant in the inner area.

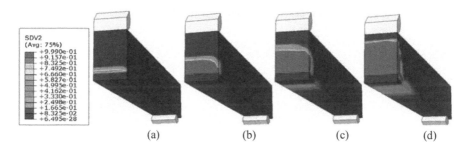

(a)                    (b)                    (c)                    (d)

Figure 7. Stages in Creep damage and crack propagation simulation.

Figure 8 shows damage predictions for temperatures 1873K and 2473K. Due to lower ductility of the ceramic at 1873K, the development of damage takes more time to spread than at the higher temperature 2473K. The material parameter $A$ (shown in Table II) of Eq. 2 is highly dependent on temperature and values of $9 \times 10^{-24}$ at T = 1873K and $3 \times 10^{-19}$ at T = 2473K show a staggering increase of creep strain rate dependency to the load. In Figure 8 the creep time (hours) is shown for a roughly similar amount of creep damage. Since the material is more ductile at higher temperature, creep damage develops faster.

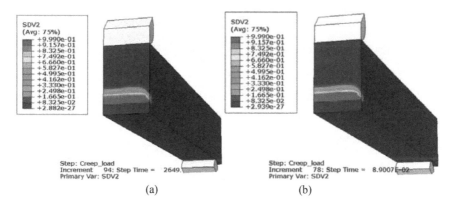

(a)                                        (b)

Figure 8. Creep damage distribution at temperatures (a) 1873K and (b) 2473K.

CONCLUSIONS

The state of the art in both testing and modelling for ultra-high temperature ceramics suggests that considerable development is needed in developing a robust methodology to predict damage and failure in UHTC components. A continuum damage model using multiaxial ductility exhaustion has been proposed for use in the failure prediction of UHTC materials which exhibit a power law creep deformation mechanism at extreme temperatures. Once power law creep material properties and $C^*$ creep parameter are known, the stress and strain rate field distributions in the component can be calculated using a fracture mechanics-based approach. $C^*$ parameter can be determined via experiments or FE methods as a function of loading conditions and creep load line displacement rate. Finally, using $C^*$ and the multiaxial creep ductility, which can be estimated using one of the models available, cracking rate in the component can be predicted. By focusing on monolithic $ZrB_2$ ultra-high temperature ceramic, for which a reasonable set of material creep data is available, the paper identifies a damage model appropriate for the failure mechanisms of the material at temperatures above 1800K. Failure mapping for this material has shown that power law creep deformation, similar to metals at lower temperatures, is the dominant mechanism at temperatures higher than 1800K and at stresses above 100 MPa. As an example, a representative three point bend geometry is numerically modeled using $ZrB_2$ relevant creep constitutive properties in a multiaxial ductility exhaustion damage model to predict cracking in the component. Future work will focus on evaluating numerical results from FE analyses and comparing them with theoretical models predictions and experiments that are planned to be performed.

ACKNOWLEDGEMENTS

The authors thank the UK's Defence Science and Technology Laboratory (DSTL) for providing the financial support for this work under contract number DSTLX-1000064072.

NOMENCLATURE

| | |
|---|---|
| $\dot{a}$ | crack growth rate |
| $A$ | coefficient in the power law creep strain rate expression |
| $B$ | specimen thickness |
| $B_n$ | net section thickness between side-grooves |
| $C^*$ | steady state creep fracture mechanics parameter |
| $D$ | material constant for creep crack growth correlation with $C^*$ |
| $F$ | geometric function for experimental estimates of $C^*$ |
| $I_N$ | amplitude function in the HRR stress field distribution |
| $n$ | power law creep stress exponent |
| $P$ | applied load |
| $t_r$ | time to rupture |
| $W$ | specimen width |
| $\Delta, \dot{\Delta}$ | load line displacement, load line displacement rate |
| $\dot{\Delta}_c$ | component of the load line displacement rate directly associated with the accumulation of creep strains |
| $\varepsilon_f, \varepsilon_f^*$ | uniaxial and multiaxial creep ductility |
| $\dot{\varepsilon}_c, \dot{\varepsilon}_A$ | creep strain rate, average creep strain rate |
| $\tilde{\varepsilon}_{ij}$ | non-dimensional strain tensor |
| $\phi$ | material constant creep crack growth correlation with $C^*$ |
| $\sigma_{ref}$ | reference stress |
| $\sigma_e, \sigma_m$ | equivalent stress and mean (or hydrostatic) stress |
| $\tilde{\sigma}_{ij}, \tilde{\sigma}_e$ | non-dimensional stress tensor, non-dimensional equivalent stress |
| $\omega, \dot{\omega}$ | creep damage parameter, rate of creep damage accumulation |

REFERENCES

[1]W. G. Fahrenholtz, G. E. Hilmas, I. G. Talmy, and J. A. Zaykoski, "Refractory Diborides of Zirconium and Hafnium," *Journal of the American Ceramic Society*, **90** [5] 1347–1364 (2007).

[2] J. Justin and A. Jankowiak, "Ultra High Temperature Ceramics: Densification, Properties and Thermal Stability," *The Onera Journal - Aerospace Lab*, [3] 1–11 (2011).

[3] R. Loehman, E. L. Corral, H. P. Dumm, P. Kotula, and R. Tandon, "Ultra High Temperature Ceramics for Hypersonic Vehicle Applications," *Sandia Report SAND 2006-2925*, (2006).

[4] S. Guo, "Densification of $ZrB_2$-based composites and their mechanical and physical properties: A review," *Journal of the European Ceramic Society*, **29** [6] 995–1011 (2009).

[5] J. W. Zimmermann, G. E. Hilmas, and W. G. Fahrenholtz, "Thermal shock resistance of $ZrB_2$ and $ZrB_2$–30% SiC," *Materials Chemistry and Physics*, **112** [1] 140–145 (2008).

[6] E. V Clougherty, R. J. Hill, W. H. Rhodes, and E. T. Peters, "Research and Development of Refractory Oxidation-Resistant Diborides. Part II.," in *Processing and Characterization*, Cambridge, MA: *ManLabs Inc.*, 1970.

[7] A. L. Chamberlain, W. G. Fahrenholtz, G. E. Hilmas, and D. T. Ellerby, "High-Strength Zirconium Diboride-Based Ceramics," *Journal of the American Ceramic Society*, **87** [6] 1170–1172 (2004).

[8] W. Li, F. Yang, and D. Fang, "The temperature-dependent fracture strength model for ultra-high temperature ceramics," *Acta Mechanica Sinica*, **26** [2] 235–239 (2009).

[9] L. Bsenko and T. Lundstrom, "The high-temperature harndess of $ZrB_2$ and $HfB_2$," *Journal of Less-Common Metals*, **34** 273–278 (1974).

[10] V. Bhakhri, J. Wang, N. Ur-rehman, C. Ciurea, F. Giuliani, and L. J. Vandeperre, "Instrumented nanoindentation investigation into the mechanical behavior of ceramics at moderately elevated temperatures," *Journal of Materials Research*, **27** [01] 65–75 (2011).

[11] Y. Xuan, C. Chen, and S. Otani, "High temperature microhardness of $ZrB_2$ single crystals," *Journal of Physics D: Applied Physics*, **98** 98–101 (2002).

[12] J. Wang, F. Giuliani, and L. J. Vandeperre, "A deformation mechanism map for $ZrB_2$," unpublished.

[13] K. M. Nikbin, D. Smith, and G. A. Webster, "Influence of Creep Ductility and State of Stress on Creep Crack Growth," in *Advances in Life Prediction Methods at Elevated Temperatures*, D. Woodford and J. Whitehead, Eds. *ASME*, pp. 249–258 (1983).

[14] G. A. Webster and R. A. Ainsworth, High Temperature Component Life Assessment, *Springer*, 1994.

[15] ASTM, "E1457-07-Standard Test Method for Measurement of Creep Crack Growth Times in Metals," (2013).

[16] H. Riedel and J. Rice, "Tensile cracks in creeping solids," *Fracture Mechanics: 12th Conference*, 112–130 (1980).

[17] J. Rice and D. Tracey, "On the ductile enlargement of voids in triaxial stress fields," *Journal of the Mechanics and Physics of Solids*, **17** 201–17 (1969).

[18] C. Shih, "Tables of HRR Singular Field Quantities," Brown University, Providence, (1983).

[19] D. Hibbitt, B. Karlsson, and P. Sorensen, ABAQUS/Standard User's Manual, Version 6. Providence, RI: *ABAQUS ltd.*, 2012.

[20] D. Hayhurst, P. Dimmer, and C. Morrison, "Development of continuum damage in the creep rupture of notched bars," *Philosophical Transactions of the Royal Society of London. Series A, Mathematical and Physical Sciences*, **311** [1516] 103–129 (1984).

[21] T. Hyde, W. Sun, and A. Becker, "Creep crack growth in welds: a damage mechanics approach to predicting initiation and growth of circumferential cracks," *International Journal of Pressure Vessels and Piping*, **78** 765–771 (2001).

[22] I. Perrin and D. Hayhurst, "Continuum damage mechanics analyses of type IV creep failure in ferritic steel crossweld specimens," *International Journal of Pressure Vessels and Piping*, **76** [9] 599–617 (1999).

[23]M. Yatomi, K. M. Nikbin, and N. P. O'Dowd, "Creep crack growth prediction using a damage based approach," *International Journal of Pressure Vessels and Piping*, **80** [7–8] 573–583 (2003).

[24]M. Yatomi, N. P. O'Dowd, and K. M. Nikbin, "Computational Modelling of High Temperature Steady State Crack Growth Using a Damage-Based Approach," in *Application of Fracture Mechanics in Failure Assessment*, pp. 5–12 (2003).

[25]L. G. Zhao, J. Tong, and J. Byrne, "Finite element simulation of creep-crack growth in a nickel base superalloy," *Engineering Fracture Mechanics*, **68** [10] 1157–1170 (2001).

[26]A. Cocks and M. Ashby, "Intergranular fracture during power-law creep under multiaxial stresses," *Metal science*, pp. 395–402 (1980).

[27]M. Tan, N. Celard, K. M. Nikbin, and G. Webster, "Comparison of creep crack initiation and growth in four steels tested in HIDA," *International Journal of Pressure Vessels and Piping*, **78** [11–12] 737–747 (2001).

[28]M. Yatomi and K. M. Nikbin, "Numerical prediction of constraint effect on creep crack growth using a simplified void growth models," to be published (2013).

# INFLUENCE OF PRECURSOR ZIRCONIUM CARBIDE POWDERS ON THE PROPERTIES OF THE SPARK PLASMA SINTERED CERAMIC COMPOSITE MATERIALS

Nikolai Voltsihhin[1], Irina Hussainova[1], Simo-Pekka Hannula[2], Mart Viljus[1]
[1]Department of Materials Engineering, Tallinn University of Technology, Ehitajate tee 5, 19086 Tallinn, Estonia
[2]Department of Materials Science and Engineering, Aalto University, P.O.Box 16200, FI-00076 Aalto, Finland

ABSTRACT
    This study is aimed at developing, processing and characterize ZrC - based ceramics for high temperature applications. Two grades of ZrC were used as matrixes, and partially stabilized zirconia, titanium carbide and molybdenum were used as a second phase and were spark plasma sintered to >98% of theoretical density at temperatures between 1700 and 2000 °C and pressure of 50 MPa. The volume fraction of additives varied from 20 to 40% in the precursor powder blend and different composites were produced. The presence of the retained tetragonal zirconia in ZrC-ZrO$_2$ composite prospects this material for increased fracture toughness characteristics. Increasing of TiC content from 25 to 40vol% results in hardness increment of 4GPa. ZrC-TiC composite is dependant on the sintering regimes being used for fabrication.

## INTRODUCTION

    Advanced engineering ceramics have many industrially attractive characteristics/properties such as high hardness and stiffness, excellent thermo- and chemical stability, etc, while their brittleness often restricts them from many structural applications. In order to overcome the problem, incorporation of particulates, flakes and short/long fibers into ceramics matrix, as a second phase, to produce toughened ceramic materials has been an eminent practice for decades.[1] Thus materials design approaches imparting a substantial degree of toughness to ceramics can obviously have a significant effect on in-service damage of the materials.

    High temperature ceramics (HTC) can be applied in various conditions and environments that are subjected to high temperatures, chemical attack, abrasive wear, etc. Zirconium carbide (ZrC) belongs to a class of high temperature ceramics and meets mostly the HTC requirements. Zirconium carbide exhibits high hardness ( -25.5GPa) and modulus of elasticity (~ 440GPa) as well as high electrical conductivity (78 x 10$^{-6}$ $\Omega$/cm) combined with a high melting point at about 3440°C. The high melting point and strong covalent bonding between atoms of ZrC makes it difficult to sinter with conventional techniques and without additional pressure during process. One of the advanced and novel sintering techniques for consolidation of ceramics is spark plasma sintering (SPS). SPS approach enhances the densification kinetics and decreases the sintering time producing a fine-grained microstructure that is often non-achievable by conventional methods, such as vacuum sintering and hot-pressing.[3]

    Although designed materials characteristics can be attractive theoretically the properties of the final product are highly dependent on the raw or input materials being used.[4,5] The quality of the raw materials, in turn, depends on the method of their production. There are several ways to produce ZrC powders, such as self-propagating high-temperature synthesis, sol-gel technique and others.[2,6] Each of the processes has its advantages and disadvantages, thus affecting the sintering kinetics and properties of the material obtained.

    To improve mechanical performance and to increase the sinterability of ZrC some additives should be used. Among mostly studied there are ZrC-Mo refractory cermets and ZrC-ZrO$_2$ composites produced by different routines.[7-10] Tetragonal to monoclinic phase transformation and the associated volume change of ZrO$_2$ is one of the challenges when it is used to modify other ceramics. Transformability of stabilized zirconia can be affected by the variety of factors, such as stabiliser type, stabiliser amount, grain size and grain size distribution and residual stresses.[11-13] Thermal residual stresses during sintering process and during high temperature loads are the most critical parameters influencing transformability of tetragonal $ZrO_2$ particles located in the carbide material matrix. However zirconia can undergo polymorphic transition at room temperature if the suficcient stress is applied.[14] Wide variety of experiments have been made on zirconia as a toughening component with other ceramic materials, including $TiB_2$, TiC, WC, ZrC and others.[15,16] Only few of those however were consolidated with SPS.

The aim of this study is to exploit the potential of the SPS method attempting the consolidation of the ZrC based composites produced from the different grades of zirconium carbide powders, by adding various additions in different amounts and to compare the densification behavior and mechanical properties of the final products, and to see how ZrC will behave in role of matrix material for different additives.

EXPERIMENTAL

Precursors and composites

Two different commercially available ZrC powders were used for the production of different composites. One ZrC powder was supplied by ABO Swiss Co. Ltd. (S1 – supplier 1. The powder purchased was reported to have average particle size (APS) of 50 nm and approximately 97-98% purity. A secondary ZrC powder was purchased from Strategic Metal Investments Ltd. (S2 – supplier 2). APS of this powder was about 3.5μm and approximately 99% pure. Other powders sourced included TiC supplied by Pacific Particulate Materials with an APS of 2-3 μm and 99% pure; nano $ZrO_2$ stabilized by 3mol% of $Y_2O_3$ with APS of 30nm, produced and supplied by Tosoh; and Mo with an APS of 2-3 μm. All of these compounds and elements possess high melting points above 2500°C and, therefore composites of these materials would be inherently HTC. Supplied powders were designed into composites named in Table I. Composites were produced in double amount. First contain mixed-size and second micron-size ZrC as a basis.

Table I. Composites designed for further study.

| Component/Grade→ | ZA(-S2) | ZB-S2 | ZC(-S2) | ZD(-S2) | ZE(-S2) | ZF(-S2) |
|---|---|---|---|---|---|---|
| Percentage % by volume (vol%) | | | | | | |
| ZrC (S1,S2) | 65 | 70 | 60 | 75 | 75 | 60 |
| Mo | 25 | 30 | - | - | - | - |
| $ZrO_2$ | 10 | - | 40 | 25 | - | - |
| TiC | - | - | - | - | 25 | 40 |
| Percentage % by weight (wt%) | | | | | | |
| ZrC(S1,S2) | 60 | 60 | 62 | 76 | 80 | 67 |
| Mo | 32 | 40 | - | - | - | - |
| $ZrO_2$ | 8 | - | 38 | 24 | - | - |
| TiC | - | - | - | - | 20 | 33 |

Mixing and sintering parameters and chracterization

All the composites were mixed with the help of planetary ball mill device (Fritsch, Pulverisette 6 classic) at 300rpm in ethanol environment for 6 hours. Grinding media was chosen depending on which materials are being processed to minimize the contamination effect. In case of ZrC-TiC composite WC balls were used for mixing and in case of ZrC-$ZrO_2$ composites $ZrO_2$ balls were used. In case of ZC grade the transformability issue of $ZrO_2$ was examined by blending the powders in dry and wet environments.

Sintering was done by SPS (FCT GmbH) with a vertically located pyrometer temperature control and operating at computer controlled regimes that allows data and sintering processes (such as densification behavior) analyzing. Firstly, parameters for the sintering process were chosen theoretically following the high melting points of the compounds. After first experiments it was decided to change the parameters of sintering for some composites and to see the effect of change on densification behavior. The production of Ø20mm samples in graphite mould was performed at 50MPa. Parameters of sintering experiments can be found in Table 2.

Table II. Sintering parameters of the composed materials.

| Grade | Heating rate; °C/min | Dwell T;°C | Dwell time; min | Cooling rate; °C/min | Pressure; MPa | Environment |
|---|---|---|---|---|---|---|
| ZA(-S2) | 100 | 1750 | 5 | 300 | 50 | Vacuum |
| ZB-S2-1 | 100 | 1750 | 5 | 300 | 50 | Vacuum |
| ZB-S2-2 | 100 | 1750 | 10 | 300 | 50 | Vacuum |
| ZC-(S2)-1 | 100 | 1900 | 10 | 300 | 50 | Vacuum |
| ZC-S2-2 | 100 | 1600 | 10 | 300 | 50 | Vacuum |
| ZD-S2-1 | 100 | 1600 | 10 | 300 | 50 | Vacuum |
| ZD-S2-2 | 100 | 1600 | 6 | 300 | 50 | Vacuum |
| ZE(-S2)-1 | 75 | 1900 | 10 | 300 | 50 | Vacuum |
| ZE(-S2)-2 | 75 | 2000 | 10 | 300 | 50 | Vacuum |
| ZF-S2-1 | 75 | 1900 | 20 | 300 | 50 | Vacuum |
| ZF(-S2)-2 | 75 | 1900 | 10 | 300 | 50 | Vacuum |
| ZF(-S2)-3 | 75 | 2000 | 10 | 300 | 50 | Vacuum |

Densities of the specimens were measured using Archimedes method with distilled water as an immersing media. The Vickers hardness was measured using Indentec 5030 SKV at the load of 10 kg according to ISO 6507. Modulus of elasticity was measured according to EN ISO 14577 using indentation modulus technique with the help of hardness testing machine ZHU zwicki-Line Z2.5.[18,19]

Phase compositions of the powders' mixtures and bulk samples were analyzed with the help of XRD (Philips PW3830 X-ray Generator, 4 kW, Cu-Anode). Samples were irradiated with CuKa radiation at 40 kV and 30 mA, in a $\theta - 2\theta$ scan with a step size of 0.02° and a count time of 0.4 s. Microstructures of the polished bulk surfaces were observed by the scanning electron microscopy (Hitachi S-4700 and EVO Zeis) equipped with EDS.

RESULTS AND DISCUSSION

Precursor ZrC powders
Powders were first analyzed for grain size and chemical composites and presence of additives. Particles size and chemical composite were examined by SEM and EDS with results shown in Figure 1. Micron-size powder consists of ZrC particles of 1.5-5 μm in size and some insignificant amount of fine particles of 0.5 – 1.5 μm in size. The image b on Figure 1 stands for ZrC considered as nanosize powder by specification and looks much finer than the micron-size powder. However, it cannot be concluded as true nanosize powder as there is a wide range of particles, of a micron size range and hence this powder can be determined as mixed-size particle

powder. In some places agglomerates of ZrC particles can be observed, but half of the micron size particles are whole monolith grains. Therefore, it is possible that powder consist of two different-size powders: nano and micro.

EDS analyze has revealed presence of some unexpected elements inside mixed-size ZrC powder while the same analyze of micron size powder indicates its purity. Presence of such elements as oxygen, chlorine and fluorine made this powder questionable and additional XRD observation of this powder was done. Some unidentified peak together with ZrC and free carbon peaks were measured by XRD (Figure 2). After inserting of the chlorine and fluorine to the search engine of XRD database this peak could be determined as $ZrCl_2$. This compound can

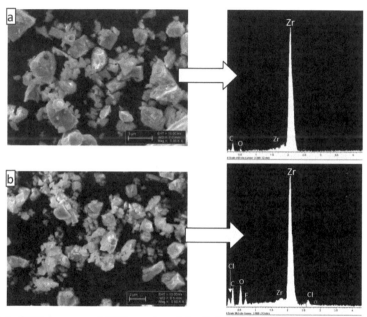

Figure 1. SEM images and EDS spectra of the ZrC powders a) micron-size ZrC powder; b) mixed-size ZrC.

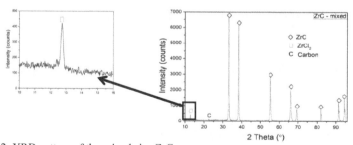

Figure 2. XRD pattern of the mixed size ZrC.

come from zirconocene dichloride ($C_{10}H_{10}Cl_2Zr$) and this type of organometallic compounds is possible for laser pyrolysis powders production technique.

### XRD of the blended ZrC-ZrO₂ composite powders

Milling of the ZC and ZC-S2 mixtures in dry environment has revealed spontaneous transformation of initial tetragonal $ZrO_2$ crystals into monoclinic ones. It is important to retain the tetragonal state of zirconia particles and to induce this phase transformation in a controlled manner and not as an unexpected result discovered after some of the fabrication stages. The powder milled in ethanol showed the undesired phase transformation partially whereas dry milled powder was fully transformed into monoclinic (Figure 3). Presence of new peaks can be observed compare with that mixed in ethanol. Dry environment at high rotating speeds creates conditions for severe plastic deformation on microscale and thus the stress level is sufficient for transformation of tetragonal phase into monoclinic.

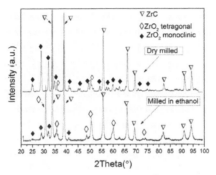

Figure 3. Planetary dry and ethanol ball milled ZC composite.

### Sintering behavior and properties of the ZA, ZA-S2 and ZB-S2 composites

ZA and ZA-S2 composites containing 25vol%Mo and 10vol%ZrO₂ but with different ZrC grades were sintered at the same conditions, while densification curves of their sintering cycles differ. The densification curves of ZA and ZA-S2 materials are represented in Figure 4.

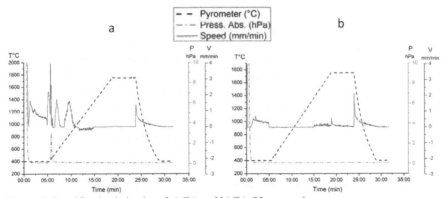

Figure 4. Densification behavior of a) ZA and b) ZA-S2 composites.

There is a sudden increase of pistons movement speed curve and absolute pressure curve at about 450°C on Figure 4a for ZA composite. Pressure increase during sintering can be explained by evaporation of some volatile species from the powder. If the powder did contain Cl or F species, as suspected from XRD and EDS results (Figure 1b and 2), it can be predicted that evaporation of these chemicals is taking place at this moment. To eliminate possible evaporation, other composite series were additionally given a pre-sintering in argon environment at 700°C for 1 hour before SPS.

ZB-S2 composite was sintered at the same conditions and sintering curve of this composite is very similar to ZA-S2 (Figure 4b). However measured density of this material was very low and it was decided to sinter it once again with a longer sintering time of 10 minutes. Densification and hardness results of these materials are showed in Table 3. Low density and poor sinterability of ZB-S2 grade even at longer dwell period can be attributed to the high melting point of both Mo and ZrC and insufficient temperature of sintering. As the result of low densification hardness of ZB composites is weak. However addition of 10vol% of zirconia improves sinterability and densification behavior of this material and allows sintering of dense specimen. On the other hand the sinterability and densification is always dependant on the precursor powders being utilized. ZA density and hardness values are noticeably lower than ZA-S2 which is obviously the result of presence of contaminations in the starting ZrC powder. That is confirmed by XRD and EDS observations given in Figures 1 and 2.

Table III. Density and hardness values.

| Grade | Dwell temperature °C | Dwell time; min | Theoretical density g/cm$^3$ | Measured density g/cm$^3$ | Desnification % | Hardness HV10 |
|---|---|---|---|---|---|---|
| ZA | 1750 | 5 | 7,42 | 7,23 | 97,4 | 1206±90 |
| ZA-S2 | 1750 | 5 | 7,42 | 7,41 | 100 | 1700±46 |
| ZB-S2-1 | 1750 | 5 | 7,65 | 7 | 91,5 | 560±22 |
| ZB-S2-2 | 1750 | 10 | 7,65 | 7,23 | 94,4 | 692±34 |

Sintering behavior and properties of ZC, ZD and ZC-S2, ZD-S2 composites

Sintering behavior demonstrated in Figure 4 was inherent for all of the composites. Densification curves of all doubled composites differed similarly depending on the ZrC grade used as a basis. Densification of materials composites containing mixed-size ZrC starts at much lower temperatures compared to micron-size ZrC containing composites. Nevertheless the densification behavior of the composites with micron-size ZrC is more stable and predictable, even though the consolidation temperature is higher. Densification steps are examined more closely on the ZC-S2-1 composite densification curve (Figure5).

Previously ZrC-ZrO$_2$ composites were fully densified by the mean of pressureless sintering in different environmental conditions at temperatures higher than 2000°C.[8] In SPS technique the current is applied directly to the material, therefore local current densities are very high and as the result the measured sintering temperatures are lower than the conventional methods. For this reason the composites were firstly sintered at 1900°C. However, observation of the sintering process and analysis of the densification curve after the experiment has revealed much lower dwell temperature needed for densification to occur. Figure 5 demonstrates displacement rate of the upper electrode against sintering time. When the pressure is applied at room temperature the densification displacement is positive and the powder is compressed to its green density (stage 1). In the heating stage due to the expansion of the graphite moulds the

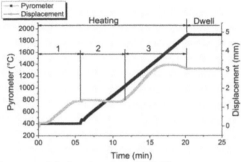

Figure 5. Change of piston speed during sintering of ZC-S2.

displacement is in the opposite direction (stage 2). In the next step densification occurs and piston movement increases rapidly between 1000 and 1640 °C (stage 3).

During the first step densification is going on due to rearrangement of particles and plastic deformation under the applied pressure of the pistons. The densification step has a large rise because of the nanosize of the zirconia particles which tend to decrease their surface area and rearrange as soon as distance between them is decreased. In second step the maximum pressure is reached and pistons movement ends. Also some negative movement can be observed because of the thermal expansion of the powder and graphite material. Third step is the most important process characterization moment, indicating densification induced by current and concomitant joule heating. Densification starts between 1000 and 1100°C. Particles of $ZrO_2$ and ZrC located very close to each other after the first step now start to interact threw high heating and pressure energies. Densification now occurs by mass transfer mechanism and necks between $ZrO_2$ particles start to form. Neck formation leads to initiation of grain boundaries. There is some negative slope after major densification is finished indicating thermal expansion of the material.

According to the graph the major densification of the powder is between 1600°C and 1700°C. Hence next sintering was made at 1600°C temperature which is 300°C lower than previous sintering.

ZD-S2 composite was decided to sinter at the same dwell temperature. No big difference could be observed between sintering behavior of ZD-S2 and ZC-S2, even though ZD-S2 has higher ZrC concentration. To optimize the process it was decided to sinter the ZD-S2 composite at 1600°C for 6 minutes and to study the effect of dwell time on the final properties of the composite.

In case of ZC material, which was based on the mixed size ZrC powder densification temperature was noticeably lower. This material starts to densify at about 650°C, which is about 350°C lower than that of ZC-S2.

It should be noted that among all composites' sintering graphs ZC and ZC-S2 have lowest densification temperatures and thus it can be concluded that $ZrO_2$ influences the densification process by lowering the consolidation temperature. The same behavior of adding the $ZrO_2$ to ZrC was observed in [8].

Difference in sintering behavior resulted in difference between the properties of ZC-1, ZC-2 and ZC-S2-1, ZC-S2-1materials. Figure 6 demonstrates these differences. First series of property bars are for ZC-1 sample based on mixed-size ZrC. It has poor densification, hardness and

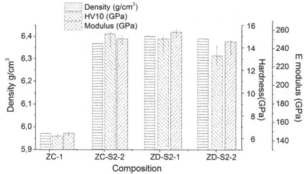

Figure 6. Properties of the ZrC-ZrO$_2$ composites.

modulus of elasticity in comparison with micron-size based samples. As in case of ZA composite this kind of result is explained by impurities presence in the material.

Powders based on micron-size ZrC could be consolidated to 99% of theoretical density by SPS, which is significantly high considering refractory nature of the carbide. The hardness of the composites is between 12 and 15GPa. Theoretically hardness of the ZD-S2 specimen should be at least 17GPa due to higher content of hard ZrC phase. The graph shows reverse dependence and it can be seen that hardness of the lower ZrC containing specimen (ZC-S2-1) is higher than ZD-S2-2 specimen. The reason for such behavior can be poor sinterability of the high ZrC containing material and short dwell time.

Elastic modulus is similar for all micron-size ZrC containing samples and is between 245 and 260GPa while maximum is for lower ZrC containing specimen. Still the modulus is far from theoretical, which should be 320GPa for ZC-S2 and 350GPa for another composite. The lowest modulus is again for 6 min sintered ZD-S2 specimen indicating not sufficient sintering time.

From analysis of the properties it can be concluded that increasing of zirconia content improves sinterability and homogeneity in properties, while specimens with higher ZrC content do not demonstrate any advantages in hardness for example. ZrC is difficult to sinter alone and therefore an additional phase such as nano sized zirconia enhances sinterability noticeably and lowers the sintering temperature of ZrC.

The inequalities in mechanical properties between mixed-size ZrC and microns-size ZrC containing composites can be explained more clearly by the microstructure characterization and indication of chemical composite. This can be observed from the Figure 7. XRD patterns of the ZC(S2) composites depicted on Figure 7 c and d show presence of two phases: ZrC and ZrO$_2$ (whether tetragonal or monoclinic). On the SEM micrographs dark regions are ZrO$_2$ and lighter regions are ZrC. In case of ZC-S2 microstructure zirconia has completely lost particulate appearance and takes shape of a long thick agglomerated sleeve between ZrC islands. Such a big agglomerates of zirconia could be the reason for lowered hardness of this composite.[20] From Figure 7a it is difficult to define the amount of phases. Gray background is definitely ZrC and dark spots should be ZrO$_2$ as it can be concluded from neighbor (Figure 7b) image. Thereby, it becomes unclear what are the white regions on the image. XRD does not show any additional phases. Based on some literature it can be predicted that white regions are zironium oxicarbide formations, however it is not detected by XRD, or it just unnoticeable on the background of ZrC peaks as it was demonstrated in other srudies.[8,21] Clear distinction of phases,

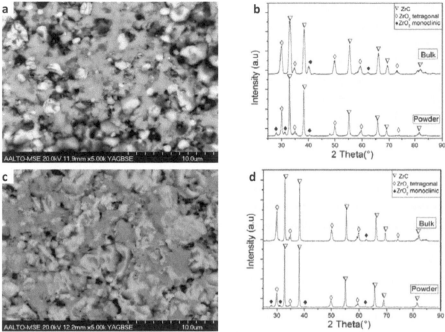

Figure 7. SEM images and XRD patterns of the: a, b) ZC; c, d) ZC-S2 composites.

low amount of pores and absence of additional phases show superiority of micron-size ZrC containing materials under mixed-size ZrC containing composites and explain big difference in mechanical properties of the ZrC-ZrO$_2$ bulk bodies based on different ZrC grades.

Sintering behavior and properties of the ZrC-TiC composites

Densification behavior of the composites containing 40 and 25vol% of TiC was very similar to the composites described before. Two short densification steps at different temperatures (750°C and 1200°C) were observed for the ZE material while ZE-S2 demonstrated one long pistons movement section from 1200°C up to 1900°C. Again, the influence of precursor ZrC powder on the densification and sintering behavior is observed and as the result properties of the bulk bodies vary noticeably. It can be seen from Figure 8a that composites with mixed-size ZrC demonstrate poor densification and mechanical properties in comparison with micronsize ZrC containing specimens. There is also dependence of mechanical properties on the TiC content in ZrC. For example hardness bar of ZF-S2-1 sample containing 40vol%TiC is noticeably higher than hardness of ZE-S2-1 containing 25vol%TiC. This is expected result because TiC has hardness of about 32GPa and thus enhances ZrC hardness. Nevertheless modulus of elasticity is between 340 and 360GPa for both ZF-S2-1 and ZE-S2-1 samples. However it can be higher for 40vol%TiC containing composite as modulus of elasticity of TiC is also higher than modulus of ZrC. After additional sintering experiments much higher modulus could be obtained for ZF-S2 material. Figure 8b demonstrates dependence of ZF-S2 composite

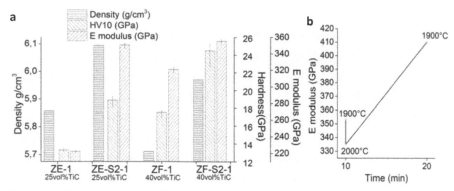

Figure 8. a) Comparative bars of TiC containing composites based on different grades of ZrC; b) dependence of ZF-S2 composite properties on the sintering parameters.

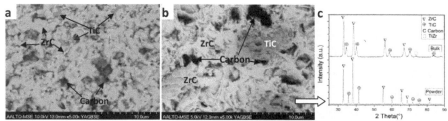

Figure 9. SEM images of the a) ZF-1 and b) ZF-S2-1 composites c) XRD pattern of the ZE-S2-1

modulus on sintering parameters. It can be noticed that sintering at higher temperature of 2000°C does not give better results, but on the contrary lowers modulus of elasticity for 20GPa. Increasing sintering time from 10 to 20 minutes results in higher modulus of elasticity for 50GPa.

Figure 9 represents images of the ZE and ZE-S2 materials. Here the dark regions are TiC and light regions are ZrC. XRD of the micron-size ZrC containing bulk specimen (Figure 9c) has shown presence of one more additional phase, which is TiZr. It is expected because of the TiC-ZrC mutual solubility especially at temperatures near 2000°C.[21] According to phase diagram of ZrC-TiC system composed by Gusev TiC and ZrC will form a (Zr,Ti)C and this meets the results of XRD.[22] As the result of TiC and ZrC decomposition free carbon is produced and it can be seen as black spots. Those spots are the main reason for wide range of standard deviation in hardness and lowered density of this material. Peaks of TiC and ZrC are somehow insignificantly shifted compare to as mixed powder. Most probably because of the solubility of Ti and Zr the crystal lattices are rearranged and transformed tensions inside the crystalline affect the XRD result.

Dark TiC areas are about 4-7μm long and 1-3μm wide for ZF-S2 specimens' microstructure. TiC areas as well as ZrC areas of ZF specimen are smaller and more homogeneously distributed. This indicates that mixed-size ZrC has positive influence on the formation of composite' microstructure. However, this microstructure did not result in high mechanical properties because of low densification degree (Figure 8a). ZF-S2-1 specimen on the other hand has quite high hardness values of 24GPa units even with coarse grain microstructure and precipitated free carbon and undersintered voids in a view of dark spots.

CONCLUSION

Various composites containing two types of ZrC were produced with the help of SPS technique. Some of them were successfully densified to theoretical density. Zirconium carbide was investigated as a matrix material in combination with different ceramics and interactions between those compounds have been observed. Transformability of tetragonal zirconia into monoclinic during mixing process has been found to take place while mixing in dry environment, due to severe plastic deformation. However the tetragonal zirconia phase was retained in bulk specimen and thus makes it possible to enhance fracture toughness of ZrC.

XRD and EDS measurements of the precursor ZrC powders revealed presence of some chemical impurities in raw ZrC powder. Image analysis of the as received powder by SEM showed that ZrC specified as nanosize particle powder can be determined as mixed-size particles powder. Sintering diagrams has shown that densification temperature of the mixed-size ZrC containing composites can be 300-400°C lower than for micron-size ZrC containing composites. Another point derived from sintering graphs observation is good sinterability of $ZrO_2$ containing composites, which sintering temperature was lowered from 1900°C to 1600°C without any loss in densification and mechanical properties. Addition of 10vol% of $ZrO_2$ enhances densification of ZrC-Mo composite up to theoretical density.

In case of all composites specimens based on micron size ZrC has shown better properties. As a result hardness of the same material pairs based on various ZrC grades showed differences in hardness of up to 8,5GPa. Micron-size ZF-S2-1 composite has demonstrated the highest among all composites hardness values of up to 24GPa.

ACKNOWLEDGEMENTS

The authors would like to thank MSc. Ekin Cura for his help and support in doing experiments. This work has been partially supported by graduate school "Functional materials and technologies" funded from the European Social Foundation under the project 1.2.0401.09-0079 in Estonia and Finnish Academy of Science under grant N. 259596.

REFERENCES

[1] A. G. Evans, Perspective on the Development of High-Toughness Ceramics, *J. Am. Ceram. Soc.*, **73**,187–206 (1990).

[2] H. O. Pierson, Handbook of Refractory Carbides and Nitrides, *Noyes*, Westwood, 1996.

[3] J. E. Garay, Current-activated, Pressure-assisted Densification of Materials, *Annu. Rev. Mater. Sci*, **40**, 445-68 (2010).

[4] D. Yung, L. Kollo, I. Hussainova, A. Zikin, Reactive Sintering of Zirconium Carbide Based systems, In: Otto T, editor. *Proceedings of 8th International DAAAM Baltic Conference*, Tallinn, 2012, 783-788.

[5] S. G. Huanga, K. Vanmeensela, L. Lib, O. Van der Biesta, J. Vleugelsa, Influence of Starting Powder on the Microstructure of WC–Co Hardmetals Obtained by Spark Plasma Sintering, *Mater. Sci. Eng. A*, **475**,87-91(2008).

[6] M. Umalas, V. Reedo, A. Lõhmus, I. Hussainova, K. Juhani, Synthesis of ZrC-TiC Blend by Novel Combination of Sol-gel Method and Carbothermal Reduction, In: Otto T, editor. *8th International DAAAM Baltic Conference Industrial engineering*, Tallinn, 2012, 753-758.

7 S. E. Landwehr, G. E. Hilmas, W. G. Fahrenholtz, I. G. Talmy, Processing of ZrC–Mo Cermets for High-temperature Applications, part I: Chemical Interactions in the ZrC–Mo system, *J. Am. Ceram. Soc.*, **90**, 1998-02 (2007).

[8] E. Min-Haga, W. D. Scott, Sintering and Mechanical Properties of ZrC-ZrO$_2$ Composites, *J. Mater. Sci.*, **23**, 2865-70 (1988).

[9] D. Sciti, S. Guicciardi, M. Nygren, Spark Plasma Sintering and Mechanical Behavior of ZrC-Based Composites, *Scr. Mater.*, **59**, 638-41 (2008).

[10] S. E. Landwehra, G. E. Hilmasa, W.G. Fahrenholtz, I. G. Talmy, S. G. DiPietroc, Microstructure and Mechanical Characterization of ZrC–Mo Cermets produced by hot isostatic pressing, *Mater. Sci. Eng. A.*, **497**, 79-86 (2008).

[11] R. H. J. Hannink, M. V. Swain, Progress in transformation toughening of ceramics, *Annu. Rev. Maler. Sci.*, **24**, 359-408(1994).

[12] A. B. Yuichiro Morikawa, M. Kawahara, M. J. Mayo, Fracture Toughness Of Nanocrystalline Tetragonal Zirconia With Low Yttria Content, *Acta. Mater.*, **50**, 4555-62 (2002) .

[13] A. Krella, A. Teresiakb, D. Schläferb, Grain size dependent residual microstresses in submicron Al$_2$O$_3$ and ZrO$_2$, *J. Eur. Ceram. Soc.*, **16**, 803-811, (1996).

[14] R. H. J. Hannink, P. M. Kelly, B. C. Muddle, Transformation Toughening in Zirconia-Containing Ceramics, *J. Am. Ceram.Soc.*, **83**, 461-87,(2000).

[15] J. Vleugels, O. Van der Biest, Development and Characterization Of Y$_2$O$_3$-stabilized ZrO$_2$ (Y-TZP) Composites with TiB$_2$, TiN, TiC, and TiC$_{0.5}$N, *J. Am. Ceram. Soc.*, **82**, 2717-20 (1999).

[16] I. Hussainova, M. Antonov , N. Voltsihhin, Assessment of Zirconia Doped Hardmetals as Tribomaterials, *Wear*, **271**, 1909-1915 (2011).

[17] E. Kimmari, I. Hussainova, A. Smirnov, R. Traksmaa, Irina Preis, Processing and microstructural characterization of WC-based cermets doped by ZrO$_2$, *Est. J. Eng.*, **15**, 275-82 (2009).

[18] W. C. Oliver, G. M. Pharr, Improved Technique for Determining Hardness and Elastic Modulus Using Load and Displacement Sensing Indentation Experiments, *J. Mater. Res.*, (1992).

[19] W. C. Oliver and G. M. Pharr, Measurement of Hardness and Elastic Modulus by Instrumented Indentation: Advances in Understanding and Refinements to Methodology, *J. Mater. Res.*, **19**, 3-20 (2004).

[20] A. Teber, F. Schoenstein, F. Tetard, M. Abdellaoui, N. Jouini, Effect of SPS Process Sintering on the Microstructure and Mechanical Properties of Nanocrystalline TiC for Tools Application, *Int. J. Refract. Met. Hard. Mater.*,**30**. 64-70 (2012).

[21] A. Markström, K. Frisk, Experimental and Thermodynamic Evaluation of the Miscibility Gaps in MC Carbides for the C–Co–Ti–V–W–Zr System, *Calphad*, **33**, 530-8, 2009.

[22] A. I. Gusev, Calculation of Phase Diagrams of Pseudobinary Systems Based on High-Melting Carbides of Titanium, Zirconium, Hafnium, and Vanadium, *Inorg Mater*, **20**, 976-81 (1984).

# Second Annual Global Young Investigator Forum

# DIELECTRIC AND PIEZOELECTRIC PROPERTIES OF Sr AND La CO-DOPED PZT CERAMICS

Volkan Kalem[1] and Muharrem Timucin[2]

[1]Metallurgical and Materials Engineering, Selcuk University, Konya, Turkey
[2]Metallurgical and Materials Engineering, Middle East Technical University, Ankara, Turkey

## ABSTRACT

PZT based piezoelectric ceramics doped with $Sr^{2+}$ and $La^{3+}$, designated as PSLZT, were prepared by conventional processing techniques. The effect of the Zr/Ti ratio on the structural, dielectric and piezoelectric properties was investigated. XRD results showed that all PSLZT compositions had perovskite structure in which decreasing Zr/Ti ratio increased the tetragonality. The morphotropic phase boundary (MPB) appeared as a region extending from Zr/Ti ratio of 52/48 to 58/42. The PSLZT ceramic with a Zr/Ti ratio of 54/46 exhibited a remarkably high piezoelectric strain coefficient of 640 pC/N with attendant parameters of dielectric constant (1800), electromechanical coupling coefficient (0.56), mechanical quality factor (70), and Curie temperature (272 °C). This soft PSLZT ceramic composition is a good candidate for electromechanical applications where quick switching and high operation temperature are needed.

## INTRODUCTION

Lead-zirconate-titanate (PZT) ceramics are extensively studied and used in electromechanical applications due to their superior ferroelectric and dielectric properties. The dielectric and ferroelectric characteristics were modified for specific applications by doping with small amounts of other oxides and by varying the Zr/Ti ratio. The Zr/Ti ratio has been given an important consideration in the production of PZT ceramics since the compositions in the vicinity of the morphotropic phase boundary (MPB) exhibit enhanced piezoelectric activity. The effects of Zr/Ti ratio and those of different additives on the structural and electrical properties of PZT ceramics were investigated in various earlier studies[1 5].

The present study is concerned with the influence of the Zr/Ti ratio on PSLZT ceramics which have dual incorporation of Sr and La into the PZT. Sr is an isovalent additive substituting for Pb on the A site in the perovskite structure. It has entered into the formulation of a large number of PZT ceramics, including both the soft and the hard types. Sr additions exhibited improved dielectric permittivity, $\varepsilon$, and larger piezoelectric strain coefficient, $d_{31}$, with slight increase in the electromechanical coupling factor, $k_p$. These effects were believed to be due to the straining of the PZT lattice since the size of Sr is smaller than that of the parent Pb[6,7].

The size of the $La^{3+}$ ion (0.136nm) is compatible with that of $Pb^{2+}$ (0.149nm). $La^{3+}$ has been used as an additive to PZT in manufacturing two different types of PLZT ceramics. The compositions with high $ZrO_2$ content have been used in electrooptic applications due to their transparency and high photostriction. A notable composition developed for this purpose was a PLZT ceramic represented as 9/65/35, the ratio standing for the atomic percentages of La/Zr/Ti[8]. The compositions of the second group of PLZT ceramics lie closer to or within the MPB region. Large improvements were obtained in dielectric permittivity and piezoelectric strain coefficient when small additions of $La_2O_3$ were made as a donor dopant to various PZT compositions[9-11]. The highest piezoelectric strain coefficient reported in the literature for a MPB type pure PLZT ceramic ($d_{33}$=455 pC/N) was achieved in the composition $Pb_{0.97}La_{0.03}(Zr_{0.53}Ti_{0.47})O_3$[12].

Although the literature on the dielectric and piezoelectric properties of PSZT and PLZT type ceramics is wealthy, similar information on the PSLZT ceramics is rather scarce. The influence of La substitution on the dielectric permittivity and the radial piezoelectric strain coefficient was reported by Kulcsar[13] for a fixed Zr/Ti ratio. Mehta et al.[14] examined the change in tetragonality index as a function of co-doping in $Pb_{0.94}Sr_xLa_y(Zr_{0.52}Ti_{0.48})O_3$. But no data was provided on electromechanical properties. The study of Ramam and Chandramouli[15] on the effects of Sr substitution into a La and Mn doped PZT was limited to a ceramic of fixed composition. The effect of variations in the Zr/Ti ratio was not examined. In view of the scarcity of information on electromechanical properties of PSLZT ceramics, the focus of the present study was to examine in detail the effect of Zr/Ti ratio on the microstructure, dielectric and piezoelectric properties of PZT ceramics.

EXPERIMENTAL

Ceramic Preparation

High purity powders of PbO, $SrCO_3$, $La_2O_3$, $TiO_2$ (Merck) and $ZrO_2$ (SEPR-CS10) were used to prepare the desired compositions of PSLZT ceramics by the mixed-oxide method. The oxide constituents were weighed in accordance with the molecular formula $Pb_{0.94}Sr_{0.05}La_{0.01}(Zr_xTi_{1-x})_{0.9975}O_3$ (x=0.5 to 0.6). Each powder batch was blended thoroughly in an agate mortar and pestle under ethanol. The dried mixture was compacted as a slug in a hardened steel die and then calcined at 820 °C for 2 h. The product was ball milled for 12 h; it was dried and then given a second calcination at 850 °C for 2 h. The final slug was crushed and ball milled with stabilized zirconia balls in a medium of ethanol in a plastic vial for 16 h.

Green ceramic discs, each measuring 13 mm in diameter and 1 mm in thickness, were prepared from the PSLZT powders plasticized by 3 wt% PEG addition. The discs were compacted by uniaxial pressing in a hardened tool steel die under a load of 150 MPa. After binder burnout in air at 600 °C, the discs were sintered by soaking at 1240 °C for 2 h in a closed saggar assembly which contained PSLZT bedding to inhibit PbO evaporation from the ceramic samples.

Characterization

The densities of sintered ceramics were determined by the liquid displacement technique based on the Archimedes' principle. The microstructural studies were conducted mainly on polished sections under a FEI Nova Nano 430 FEG scanning electron microscope (SEM). The polished surfaces were given a thermal etch at 1000 °C for 1 h in enclosures containing PZT bedding to compensate for PbO loss.

XRD analyses were performed for monitoring the powder synthesis process, for determination of the lattice parameters, and quantification of the tetragonal, T, and rhombohedral, R, phase fractions present in sintered ceramics. For electromechanical measurements, the flat surfaces of the sintered discs were lapped and then metalized with a silver paste. Painted silver electrodes were fired at 750 °C for 20 min. For the poling process, the electroded discs were exposed to a DC electric field of 3 kV/mm for 30 min in a silicon oil bath at 120 °C.

The piezoelectric strain coefficient ($d_{33}$) of each disc was measured 24 h after poling, by a Berlincourt D33-meter. $k_p$ and $Q_m$ were determined by the resonance/anti-resonance method according to the IRE standards using a HP4194A impedance analyzer. The dielectric properties, free dielectric constant ($K_{33}^T$) and loss tangent (tanδ), of the poled ceramics were calculated from the capacitance and dissipation factor values measured at 1 kHz and sample dimensions. Hysteresis behavior was characterized with a modified Sawyer-Tower circuit at 25 °C using a 50 Hz driving field. The Curie temperature, $T_C$, of poled discs was determined by establishing the variation of dielectric permittivity with temperature in the range 100 to 350 °C. These measurements were done at 1 kHz frequency by placing the samples in a small pot furnace in which the data on permittivity were taken upon cooling the ceramics from high temperature at a rate of 3 °C/min.

RESULTS AND DISCUSSION

The PSLZT ceramics were based on PZT having 1 at% La, 5 at% Sr. The ceramics were produced with Zr/Ti ratio ranging between 50/50 to 60/40. The data related to various physical properties are summarized in Table 1. The porosity in the PSLZT ceramics was rather small; the values of %TD which is an indicator of the degree of densification achieved in the sintering process were in excess of 97% of the theoretical. SEM observations on fracture surfaces also revealed limited porosity. The average grain size in the PSLZT ceramics, evaluated from micrographs shown in Figure 1 revealed that the change in the Zr/Ti ratio had little effect on the size of the polycrystals; the average grain size in all the samples was about 3.5 μm. However, the size distribution was rather wide; small grains filling the interstices among the larger ones may be the reason for better densification.

Table I. Physical and electrical properties of $Pb_{0.94}Sr_{0.05}La_{0.01}(Zr_xTi_{1-x})O_3$ ceramics.

| x | 0.50 | 0.52 | 0.54 | 0.56 | 0.58 | 0.60 |
|---|------|------|------|------|------|------|
| $d_{33}$ (pC/N) | 393 | 480 | 640 | 560 | 427 | 380 |
| $k_p$ | 0.39 | 0.42 | 0.56 | 0.53 | 0.42 | 0.40 |
| $Q_m$ | 96 | 83 | 70 | 84 | 99 | 115 |
| $K_{33}^T$ | 972 | 1475 | 1800 | 1138 | 608 | 487 |
| tanδ (%) | 1.15 | 1.46 | 1.55 | 1.89 | 2.22 | 2.34 |
| Density as %TD | 98.12 | 97.98 | 98.29 | 98.36 | 97.23 | 97.48 |
| Average Grain Size , μm | 3.31 | 3.44 | 3.62 | 3.64 | 3.64 | 3.70 |

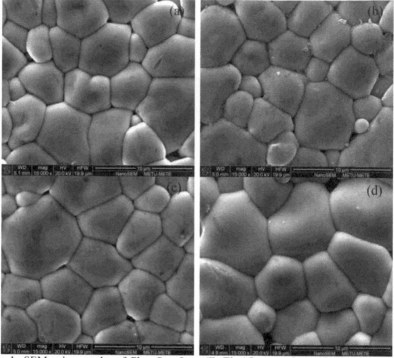

Figure 1. SEM micrographs of $Pb_{0.94}Sr_{0.05}La_{0.01}(Zr_xTi_{1-x})O_3$ ceramics with Zr content of (a) x=0.52 , (b) x=0.54, (c) x=0.58, (d) x=0.60.

Figure 2 shows XRD patterns of the PSLZT ceramics with varying Zr/Ti ratio. All of the samples had pure perovskite structure. The variations in the morphology of the peaks in the XRD patterns indicated that, for the present study, the MPB in the PSLZT ceramics covered a wide range of Zr/Ti ratio. The patterns revealed the co-existence of T and R modifications. In order to resolve the overlapping triplet peaks from the (200)T, (200)R, and (002)T planes in the $2\theta$ range $43°$ to $46°$, Peakfit v4.0 software was used to deconvolute the peaks assuming Lorentzian line shape. The relative volume fractions of rhombohedral and tetragonal phases, %R and %T, were estimated by the integrated intensities of XRD peaks through the following relationships:

$$\%R = \frac{I_{(200)R}}{I_{(200)T} + I_{(200)R} + I_{(002)T}} , \quad \%T = 100 - \%R \tag{1}$$

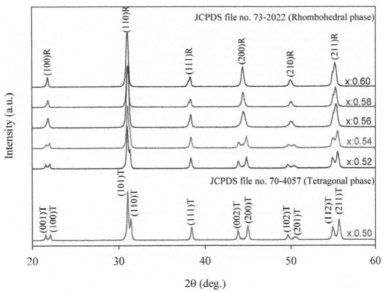

Figure 2. The XRD patterns of the $Pb_{0.94}Sr_{0.05}La_{0.01}(Zr_xTi_{1-x})O_3$ ceramics.

The calculations on the relative proportions of the T and R phases in PSLZT ceramics in the composition range Zr/Ti = 50/50 to Zr/Ti = 60/40 resulted in the quantitative distribution shown in Figure 3. The MPB region was located within the range of Zr/Ti extending from 52/48 to 58/42. Inside this region, the change in the lattice constants and tetragonality with the Zr/Ti ratio is shown in Figure 4. The tetragonality index of 1.021 and the $a_R$ value of 4.067 Å at the Zr/Ti ratio of 52/48 were consistent with the values determined by Mehta et al.[14] for the $Pb_{0.94}Sr_xLa_y(Zr_{0.52}Ti_{0.48})O_3$ ceramics with La/Sr ratio similar to the present work.

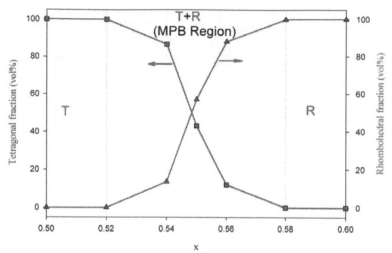

Figure 3. Variations in relative contents of the T and R modifications in PSLZT ceramics with Zr/Ti ratio.

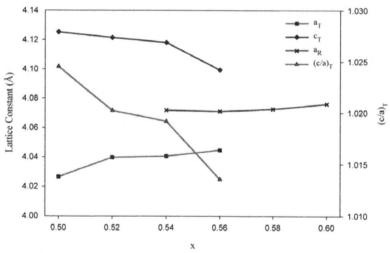

Figure 4. Variations in lattice parameters of the T and R modifications in the PSLZT ceramics with Zr/Ti ratio.

In contrast to the situation observed on grain size, the piezoelectric and dielectric properties were highly sensitive to the Zr/Ti ratio. The dielectric and piezoelectric properties of the $Pb_{0.94}Sr_{0.05}La_{0.01}(Zr_xTi_{1-x})O_3$ group ceramics, displayed in Figure 5, exhibited changes typical of MPB compositions. The piezoelectric strain coefficient $d_{33}$, the electromechanical coupling coefficient $k_p$, and the dielectric constant $K_{33}^T$ exhibited visible maxima in the interval of the Zr/Ti ratio corresponding to the MPB region of the PSLZT system examined. The maxima in these parameters were due to the coexistence of the tetragonal (T) and the rhombohedral (R) phases which lead to high levels of electromechanical response. The presence of a large number of energetically equivalent states within the transition region allows a high degree of alignment of ferroelectric dipoles and improved polarizability of the ceramic which in turn provide the enhanced piezoelectric activity in MPB[1].

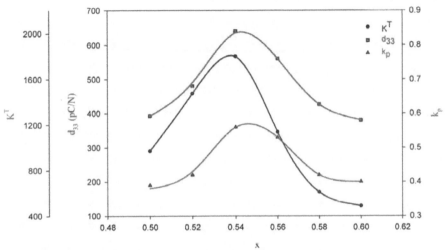

Figure 5. Variations in the dielectric and piezoelectric properties of $Pb_{0.94}Sr_{0.05}La_{0.01}(Zr_xTi_{1-x})O_3$ ceramics.

Figure 6 shows the polarization curves displaying the hysteresis behavior of the PSLZT ceramics. The coercive field values, $E_c$, in all samples were equal to or smaller than 1 kV/mm, indicating that the PSLZT ceramics could be polarized rather easily. These values are comparable to the magnitude of the $E_c$ reported earlier for PLZT ceramics[17] The ceramics with lower Zr/Ti ratio exhibited loops of ferroelectrically harder PZT ceramics owing to the dominance of the T phase. The increase in the Zr/Ti ratio resulted in higher remanance and lower coercive field, as shown in Figure 7. The area occupied by the hysteresis loop, which is a measure of polarization energy, became increasingly smaller as the Zr content was raised. These features are correlated well with the increased proportion of the rhombohedral phase in PZT compositions having relatively higher Zr level[18].

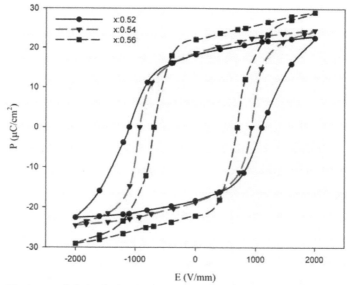

Figure 6. The hysteresis behavior in $Pb_{0.94}Sr_{0.05}La_{0.01}(Zr_xTi_{1-x})O_3$ ceramics.

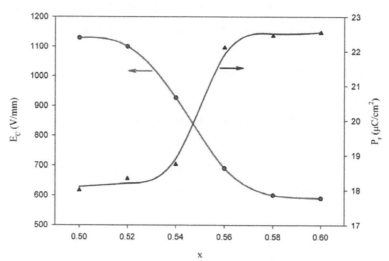

Figure 7. Variations of coercive field and remnant polarization in PSLZT ceramics with Zr/Ti ratio.

CONCLUSIONS

Sr and La co-doped PZT ceramics, represented as PSLZT, were produced with pure perovskite structure through conventional mixed oxide technique. In the co-doped ceramics, densification values higher than 97% TD could be attained due to effective development of the microstructure consisting of a mixture of large and small grains. The overall weight loss in the PSLZT discs during sintering was typically less than 0.4% indicating that the small excess of PbO due to B-site formulation was nullified during sintering, and also no PbO was gained from the bedding. Consequently, in the SEM micrographs, there was no evidence of PbO as a second phase.

The Zr/Ti ratio had great implications on the dielectric and piezoelectric properties of the PZT ceramics. From the present results, it can be revealed that the MPB coexisting the T and R phases is a broad composition region of $0.52<x<0.58$. PSLZT ceramic containing 1 at% La and 5 at% Sr exhibited high piezoelectric and dielectric properties in the vicinity of the MPB and a remarkably high strain coefficient $d_{33}$ (640 pC/N) and favorable properties of $Q_m$ (70), $K_{33}^T$ (1800), and $k_p$ (0.56) were attained at $x=0.54$. The Curie temperature of this particular ceramic was 272 °C.

ACKNOWLEDGEMENT

This work was supported by the State Planning Organization of Turkey (DPT – Project No: 2002K120510) and the Scientific Research Projects (BAP) of Selcuk University.

REFERENCES

[1]B. Jaffe, W.R. Cook, H. Jaffe, Piezoelectric Ceramics, Academic Press, London, 1971.

[2]L. Wu, C.C. Wei, T.S. Wu, H.C. Liu, Piezoelectric propertics of modified PZT ceramics, J. Phys. C: Solid State Phys. 16 (1983) 2813-21.

[3]C. Moure, M. Villegas, J. F. Fernandez, P. Duran, Microstructural and Piezoelectric Properties of Fine Grained PZT Ceramics Doped with Donor and/or Acceptor Cations, Ferroelectrics 127 (1992) 113-18.

[4]T. Yamamoto, Optimum Preparation Methods for Piezoelectric Ceramics and Their Evaluation, Am. Ceram. Soc. Bull. 71 (1992) 978-85.

[5]J.F. Fernandez, C. Moure, M. Villegas, P. Duran, M. Kosec, G. Drazic, Compositional Fluctuations and Properties of Fine-Grained Acceptor-Doped PZT Ceramics, J. Eur. Ceram. Soc. 18 (1998) 1695-1705.

[6]F. Kulcsar, Electromechanical Properties of Lead Titanate Zirconate Ceramics with Lead Partially Replaced by Calcium or Strontium, J. Am. Ceram. Soc. 42 (1959) 49-51.

[7]P.R. Chowdhury, S. B. Deshpande, Effect of dopants on the microstructure and lattice parameters of lead zirconate-titanate ceramics, J. Mater. Sci. 22 (1987) 2209-15.

[8]G.H. Haertling, C.E. Land, Hot-pressed (Pb,La)(Zr,Ti)O3 Ferroelectric Ceramics for Electrooptic Applications, J. Am. Ceram. Soc. 54 (1971) 1-11.

[9]A. Garg, D.C. Agrawal, Effect of rare earth (Er, Gd, Eu, Nd and La) and bismuth additives on the mechanical and piezoelectric properties of lead zirconate titanate ceramics, Mater. Sci. Eng. B86 (2001) 134-43.

[10]D. Kuscer, J. Korzekwa, M. Kosec, R. Skulski, A- and B-compensated PLZT x/90/10: Sintering and microstructural analysis, J. Eur. Ceram. Soc. 27 (2007) 4499-4507.

[11]M. Hammer, M.J. Hoffman, Detailed X-ray Diffraction Analyses and Correlation of Microstructural and Electromechanical Properties of La-doped PZT Ceramics, J. Electroceramics 2:2 (1998) 75-84.

[12]B. Praveenkumar, H.H. Kumar, D.K. Kharat, B.S. Murty, Investigation and characterization of La-doped PZT nanocrystalline ceramic prepared by mechanical activation route, Mater. Chem. Phys. 112 (2008) 31-34.

[13]F. Kulcsar, Electromechanical Properties of Lead Titanate Zirconate Ceramics Modified with Certain Three- or Five-Valent Additions, J. Am. Ceram. Soc. 42 (1959) 343-49.

[14]P.K. Mehta, B.D. Padalia, M.V.R. Murty, A.M. Varaprasad, K. Uchino, X-ray Structural Determinations on Sr and La Doped PZT, Ferroelectrics Letters 7 (1987) 121-29.

[15]K. Ramam, K. Chandramouli, Dielectric and Piezoelectric Properties of Combinatory Effect of A-Site Isovalent and B-Site Acceptor Doped PLZT Ceramics, Ceramics-Silikaty 53 (2009) 189-94.

[16]X. Dai, A. DiGiovanni, D. Viehland, Dielectric properties of tetragonal lanthanum modified lead zirconate titanate ceramics, J. Appl. Phys. 74 (1993) 3399-405.

[17]M.J. Hoffmann, M. Hammer, A. Endriss, D.C. Lupascu, Correlation Between Microstructure, Strain Behavior, and Acoustic Emission of Soft PZT Ceramics, Acta Materialia 49 (2001) 1301-10.

[18]S.J. Yoon, A.Joshi, K. Uchino, Effect of Additives on the Electromechanical Properties of $Pb(Zr,Ti)O_3$-$Pb(Y_{2/3}W_{1/3})O_3$ Ceramics, J. Am. Ceram. Soc. 80 (1997) 1035-39.

# Author Index